John Hancock Klippart

The Principles and Practice of Land Drainage

John Hancock Klippart

The Principles and Practice of Land Drainage

ISBN/EAN: 9783744678568

Printed in Europe, USA, Canada, Australia, Japan

Cover: Foto ©berggeist007 / pixelio.de

More available books at **www.hansebooks.com**

THE PRINCIPLES AND PRACTICE OF LAND DRAINAGE:

EMBRACING

A BRIEF HISTORY OF UNDERDRAINING; A DETAILED EXAMINATION OF ITS OPERATION AND ADVANTAGES: A DESCRIPTION OF VARIOUS KINDS OF DRAINS, WITH PRACTICAL DIRECTIONS FOR THEIR CONSTRUCTION: THE MANUFACTURE OF DRAIN-TILE, ETC.

Illustrated by nearly 100 Engravings.

BY JOHN H. KLIPPART,

Author of the "Wheat Plant," Corresponding Secretary of the Ohio State Board of Agriculture, Etc.

CINCINNATI:

ROBERT CLARKE & CO.,

PUBLISHERS AND BOOKSELLERS.

1861.

CINCINNATI:
E. MORGAN & SONS,
Stereotypers and Printers, 111 *Main St.*

PREFACE.

THIS treatise is presented to the public as a brief discussion of the most important considerations involved in Land Drainage. The writer has not aimed to produce an original work, but has endeavored to collate well ascertained facts, and present them in as brief a space as the nature of the subject would permit. The productions of the best writers on the subject in Great Britain, France and Germany, as well as the current agricultural publications of this country, whether serial or otherwise, have been consulted in the preparation of this volume; while the numerous opportunities which presented themselves to the author, in fulfilling his duties in connection with the State Board of Agriculture, for observing the practical details of the work by visiting places where draining operations were conducted, tile manufactured, etc., have been cheerfully embraced, and the results embodied in the work.

It may not be improper to state that the work was undertaken at the earnest solicitation of the Committee on Agriculture of the Ohio Legislature, during the session of 1858–9. The engravings were ordered and the greater portion of them finished before the appearance of Judge French's excellent work on the same subject; it was then too late to abandon the project.

The work is respectfully submitted to the judgment of the farmers of the Great Northwest, in the hope that it may be found not altogether unuseful to them in their endeavors to improve their property.

CONTENTS.

INTRODUCTORY.

PART I.—THEORY OF DRAINAGE.

PART II.—PRACTICE OF DRAINAGE.

LAND DRAINAGE.

INTRODUCTORY.

DEFINITION.

Drainage of land, or farm drainage, may be defined as being a process by which wet and unhealthy soils may be rendered arable and healthy, as well as to remove excessive moisture in lands not generally considered too wet.

The word *drainage*, when used in an isolated sense, means drying up, running off of stagnant water. It is also applied to a series of works which are undertaken in order to improve the sanitary condition of whole sections of country or a large city, to change the course of a river, and to protect its cultivated banks against floods. General drainage is that which constitutes a whole system of great works stretching over entire valleys, and regulate all its running water; *agricultural drainage* refers to fields only.

Drainage, as practiced at the present time, is an improvement, or a transformation of the old system for drying up moist soils by means of trenches or ditches, for the discharge of water, which was known and resorted to everywhere in former times.

It is, nevertheless, true, that the transition or change from trenches or uncovered ditches filled with stones, to the new mode of draining, was a slow process, which may explain why many persons, when they learn of the "new" drainage, will exclaim: "But this is not new, that was practiced by our fathers." These persons are right. But, as far as arts and science are concerned, all is becoming singularly perfect in our days, so that we do not always recognize the starting point. This was the case with the new system of drainage which conducts the excess of water from tillable lands through earthen pipes of moderate length, placed under ground at a depth of from three to four feet.

But how can the water be led off through such earthen pipes? we often have been asked. In 1852, Mr. Daniol, a skillful agriculturist, at Clermont Ferrant, in France, wrote to Mons. Barras, editor of an agricultural journal:

"New words used by scientific writers often cause embarrassment to their readers. When the *Journal of Practical Agriculture* indicates *drainage* as a potent system of rendering wet lands arable; when you praised the advantages which are obtained in England and Belgium, you excited my curiosity and highly engrossed my attention. But having discovered that drainage is the very thing we resort to from father to son, and which we name like OLIVER DE SERRES, *subterranean passages*, I experienced a sense of satisfaction and said to myself, is that all? what next?

"Next, you indicate, for the purpose, earthen pipes as the most efficient and economic implement! I vainly endeavored to perceive how water in excess at the surface could penetrate into these pipes. Doubtless, thought I, if this water springs up at a single place and for want of exit spreads over the whole field, it is an easy matter to collect it through waterworks introduced into the pipes and let it run off at a lower spot. But otherwise, if you have several springs to contend with; if the excess of water after the normal saturation of the soil is produced by superabundant rains during several years, how, once more, will the numerous little sources penetrate and find their way through the surface of, and into the pipes? I will suppose that the ends of the pipes are only contiguous, that at first cracks will admit water, but earth will, in course of time, fill up the joints; in both cases the work will be expensive and useless."

Such doubts expressed by a very learned agriculturist, prove that much has yet to be said in order to convince our agriculturists of the utility of drainage, and to render its effects conspicuous to them. On the other hand, numberless questions are addressed to us about the manner of making pipes, the quality of clay, the kind of machines, baking or burning, cost, results to be expected, and so forth. The work for laying the drains does not appear, in general, difficult to comprehend and execute; the numerous writings on the subject scattered through the agricultural and other journals, have enlightened the majority of cultivators. But many a point relating to the execution, remains unexplained; we, therefore, have deemed it proper to go over the whole subject without neglecting that which practical men and publications have heretofore elucidated.

Notwithstanding the origin of drainage may often have been related, a sketch of its history and progress will not be out of place here. The importance of drainage is the sole point of the subject, which really should *not* require any further discussion; it was prominently brought forward by Mr. Martinelli, of Nerac (France), at a county fair, in a few words, which we translate from the *Journal of Practical Agriculture*, viz:

> "Look at this flower pot. What is the hole at the bottom for? I ask you, because there is a complete agricultural revolution in that hole. It affords a renovation of water by a timely flow. And why must water be renovated? Because it gives either life or death: life, when it merely traverses a layer of earth which retains the fecundity with which water is pregnant, and beside dissolves nutriments conveyed to the plant; death, when it remains in the pot, because it will soon be corrupted, will cause the roots to become diseased and prevent admittance to new water."

The drainage of tillable land is a small hole at the bottom, just like that of the flower pot.

HISTORY OF DRAINAGE AMONG THE ANCIENTS.

THE new system of drainage emphatically consists in the use of covered causeways; we, therefore, will devote no space nor time to the discussion of open trenches as a means of removing superfluous water and rendering lands arable, subject to this condition. The idea of redeeming from waste and of making available for cultivation the surface occupied by gaping ditches, may be traced back to the earliest ages. The Romans were acquainted with a process of draining lands derived, doubtless, from antecedent civilization. Nevertheless, among agricultural writers, Columella is the first who speaks of underground causeways; he lived in the reigns of Augustus and Tiberius. Cato, Varro, and Virgil, advocate open trenches only. Here is Columella's text:[1]

"When the soil is moist, ditches are to be dug out in order to dry it up and let the water run off. We know of two kinds of ditches: those which are hidden and those which are wide and open; as to the hidden ditches, one will dig out trenches of three feet in depth, which shall be half filled with small pebbles or pure gravel, and then the whole will be covered with the earth which was taken out from the trench. Should there be neither stones nor gravel, then fascines formed of branches tied together, of the same shape and capacity of the trench, may be placed into it so as to fill up the cavity. When the fascines have been sunk into the bottom of the canal, they must be covered with leaves of cypress, pine, or any other tree, then shall be superadded the earth extracted from the trench, and the whole will strongly be compressed. At both ends must be placed in the form of a buttress (as it is done for small bridges), two large stones surmounted with a third one, in order to consolidate the sides of the ditch and favor the fall and exit of the water."

Palladius, who came long after Columella, thus describes the underground causeways:[2]

"When the lands are wet, they will be dried up by digging trenches

---

1 Lib. II, cap. 2. 2 Lib. VI, cap. 3.

everywhere. Every one knows how to make open trenches, but here is the way to make hidden trenches: One digs out across the field ditches of three feet in depth, which are to be half filled with small stones or gravel; after which they are filled up with the earth from the digging and leveled. But the ends of those causeways must lead in declivity unto an open ditch whither the water will run without carrying away the earth of the field. Should there be no stones, one will lay at the bottom of the ditch fascines, straw, or briars of any kind whatever."

Thus, drainage, by means of underground causeways through which the flow of water is secured by means of permeable materials, like stones or branches, is an invention that no modern has any right to claim; though drainage, such as described by Columella and Palladius, has long been used at numerous places in France. England endeavored to ascribe to Captain Walter Bligh the invention of deep trenches. Walter Bligh's only merit is the reproduction of the precepts applied before him, and perfectly elucidated by the elder French writer on agriculture, Oliver de Serres.

Even in Columella's time, the importance was fully appreciated of making the drains with sloping sides and narrow bottom. From this forward there was a slight step only to actual and thorough drainage. There is abundant evidence to prove that the ancient Romans used clay pipe as conduits for water everywhere where they established themselves; that even in lower Austria, Saxony, and other countries, a similar system of conduit by clay pipes obtained, evidence of which is yet to be found in some of the cultivated fields; there appears to be presumptive evidence, at least, that drainage by clay pipes or tiles was a Roman invention.

UNDERGROUND CAUSEWAYS AMONG THE GREEKS.

We said that the Romans were acquainted with a mode of draining lands through trenches covered and filled with stone. We did not derive the origin of it from more ancient civilization, because we do not consider the subterranean canals built by the Greeks to remove enormous reservoirs of water, which might have caused extensive floods, as mere agricultural drains. M. Jaubert de Passa, in his *Researches on Irrigation among Ancient Nations*,[1] speaks thus:

"Was the mysterious outlet of the lake Stymphalide toward the coast of Argos[2] the work of man or caprice of nature? It is known that the water of the lake did run into two abysses situated at the extremity of the valley; when these openings were obstructed, the water covered a space of over 400 stadii, or about thirty miles. The river Stymphale, which the inhabitants of Argolida named *Erasinus*, was not the only one of which the course was partly under ground. The Alpheus, after having several times disappeared from the earth's surface, plunged into the sea, according to traditions,[3] in order to go into Sicily, where it mingled its water with the spring of Arethusa. The plain of Orchomenes became marshy as soon as the subterranean ducts, regular outlet of the water from Mount Trachys, failed to be cleansed. The plain of Caphyes was sometimes overflown by the water of the Orchomenes. As a permanent protection for the country and the city, the magistrates of Caphyes caused the establishment of a causeway along the flowing canal, behind which water from various sources formed the river below.[4]

"The plain of Phenea, next to the others, remained for a long time overflown. At a remote, but unknown epoch, an earthquake, according to some, a beneficent prince, according to others, opened two abysses or *zerethra*, which let out the water and made the land healthy;[5] finally, the valley of Artemisium, situate near Mantinea, and named *Argos*, on account of its sterility, became marshy as

---

1 Vol. IV, p. 36.

2 Strabo VI, cap. 3 § 9, and VIII, cap. 9 § 4.

3 Pausanias VIII, 44, 54.

4 Pausanias VIII, p. 23.

5 Pausanias, VIII, p. 14, 19.

often as water obstructed the gulf which was its outlet. This subterranean duct extended as far as Genethlium, a city built at the head of the lake Dine."[1]

Certainly, these immense underground works of the Greeks must have had the drainage of extensive districts for their object, but they were undertaken as a public hygiene, and not to enhance the fertility of arable lands. Agricultural drainage has this last as its special object; but this can not always be attained without some general work for the drainage of a whole valley.

We read, in Walter Bligh's book, third edition, printed in 1652:[2]

"As to the drain trench, thou wilt make it deep enough so that it may reach at the bottom cold, oozing, stagnant water. Say one yard, or four feet, if thou wishest for satisfactory drain. And furthermore, having come to the layer whereat rests the oozing spring, sink further down about the depth of an iron shovel, no matter how deep thou art already, if thou wilt drain thy land throughout. ... But as to ordinary trenches, which are often dug out one or two feet, I say that it is madness and lost work, and I will spare the reader wherewithal."

These injunctions certainly are pertinent, and may serve as a guide even at this day; but one ought not to conclude from them, as some modern writers did, that, as no French agricultural author treated the subject as a specialty, and with sufficient details, all the merit of the introduction of open trenches belongs to England. Oliver de Serres, who lived before Walter Bligh, and whose *Theatre of Agriculture* was printed in 1600, gives a very complete description of the underground causeways, strongly recommending the use of them. Not only does he consider the single trench, as did Columella, but

---

1 Pausanias VIII, p. 7, 20, 21, 25.

2 "The English Improver Improved, or the Survey of Husbandry Surveyed."

he goes further; he treats of many together, he is careful to describe the main ditch as also covered, and every precaution to be taken in order to secure effective drains. As, Oliver de Serres was altogether neglected in the history of drainage, and his well-defined ideas having been attributed to divers authors, we will give, *in extenso*, a passage from tho book of this great French writer on agriculture :[1]

"To discharge noxious water, the usual way is to open ditches, especially through plains and low places, these ditches becoming inclosures for the land. Let the land then be dug around and give the ditches proper width and depth, to fulfill both objects. They must be cleansed every second year, some time previous to sowing lands, on which shall be cast this detritus from the trenches, to be used as so much manure. But, should it happen that the field be full of springs, or underground oozing sources, external ditches are no longer sufficient; then will be required another and more peculiar remedy as will be shown, in order to rid inner land of this incommodiousness. Inasmuch as the evil of too much water exceeds in destructiveness, both that of shadow and of stones, to mend the former will require greater labor than to correct the latter; of this, finally, the profit as a recompense, comes out greater than from any other reparation that can be given to the land, so fruitful is that which relieves it from water; because thereby not only are wet lands improved, but pools and swamps are converted into exquisite plow fields.

"The examples serve us as good masters to do good husbandry. Where is the farmer beholding the beautiful wheat raised on drained swamps, that does not desire, in emulation, to imitate such profitable husbandry? The cause of this comes from a superabundance of water, which prevented land from being worked for several years; at the end of which, finding itself reposed, and thereby to have acquired fertility, returns it admirably and with profit. And how much more hope you will have from this, which by the ancient subjection to the springs, was never able to produce, which you will find pregnant with fertility! Beside the income, there is no doubt that from noxious water spread here and there, on your land, when collected

[1] Theatre d'Agriculture, Second Lieu, t. 1, p. 97.

in one place, you could make a fountain spring according to places, so great and with such abundance of water, that it will suffice for the irrigation of meadows, which you will make on account of that, below the drained pieces, and indeed for erecting mills there, should the ground and other circumstances requisite be favorable.

"The ground you desire to drain must have a declivity, either small or great, without which the water could not run off. This being presupposed, a large ditch must be dug from one end to the other, always beginning at the lowest spot; into that trench many others, but smaller, may be joined on both sides, in order to discharge the water flowing from all parts of the ground. By this means, each supplying its portion, the large ditch collecting, the whole will be discharged. The large trench is, on that account, called *mother trench*, and the whole together, '*hen's paw*,' from the figure of that animal's foot, whose claws stretch in toward its trunk. The extent and surface of the land give form to the ditches, because it is fit to make them longer and wider in proportion as your land is extensive and flat, which you drain; and on the other hand, they are required shorter and narrower, if it be small and sloping; because, within a narrow compass, generally, not so much water is collected as in a large one, and as much, nay, more of it will pour out of a narrow ditch with great declivity than a wide, gently sloping trench. *About the depth of the ditches, it is not thus, for, in whatever part you dig, you must go about four feet deep, in order to cut off the source of the springs, which is the special aim of this business.* According to the nature of the place, must the trenches be disposed.

"Should there be a low vale, with high ground on both sides, the *mother* must be dug in the middle and lower spot, lengthwise, as already said, into which must fall the other ditches from both sides. But having to drain only one hillside, in that quarter there will be some small ditches running into the mother trench, and disposed as will seem fit for the best of the work and premises; as also the length of all trenches is subordinate to the plan which dictates the order of them, according to the surface and site. Having the plan, reasonable fall and extent, a proper width will also be required for the small ditches; the latter should be three feet, and the mother five feet deep; by means of this guide, your intention will be fulfilled. And to avoid any mistake, let there be as many ditches, so long, so wide, without fear of excess on this score, that no source of spring, or small fountain, be overlooked, in order to drain your land well, by the general gathering of its waters.

"*Those ditches, large or small, must be half filled with minute stones and the other half with the earth previously dug out and leveled at the top, so that no trace of it even will appear, for the commodiousness of tillage*, which should be executed very well, the plow finding depth enough of earth before reaching the stones, through which water will freely pass and flow out at the spot designed for it, leaving the surface land free of all noxious moisture and fit to bring forth all kinds of cereals.

"A similar work must be applied to all estates, vineyards, meadows, orchards, and others which produce no fruit on account of too much moisture. If you have on the spot none but large, flat stones for supplying your trenches, before using them you must break them to suit this kind of service, and they should be placed into the ditch straight (upright) and not flat, fixing them beside so skillfully that they will not be too tight to prevent the flow of water. To have this business well done, begin right, that is, artistically and with order; through ease and without confusion you will succeed very well. It will be easy to draw all your ditches, by cautiously observing the places through which they are to pass; then you begin to dig them out at the lowest spots, casting the earth all on one side of the trench, leaving the other side free, to bring thither easily the stones, which must be thrown in immediately, for fear that, by delaying, the trench might cave in by effect of the wind, trampling of beasts, or any other accident.

"Thus your undertaking shall be completed at one end, as soon as commenced, in prosecuting it until you reach the highest spot of the field. In the meanwhile the water will take its course as soon as the opening of its way has been performed, which could not take place, should you begin the work at the highest spot, for want of issue allowed to water; even this would disturb the digging by discharging into it. You will mind, also, that the issues of water be well managed, that they do not choke up afterward, because for want of issue water might retrograde and render your labor useless. This will be obviated by stones and mortar, put up by a master hand, so as to last long, especially at the spot where the main or mother trench lets out the water. You are finally advised that the extremities and ends of your small ditches, at their highest parts, need not to be as wide as in low places, not being compelled to collect there so much water as below; this, nevertheless, remains at your discretion, because they can not be too wide in any place in order to receive not only water springing from the bottom, but also that from the rain above, which shall not be overlooked.

"This work produces several advantages, since at the same time an excess of water and stones are removed from the ground, and that water is made serviceable for meadows, mills, even for fountains, the qualities of it being considered, for which usefulness it is rendered commendable; also, that improvement ought to be prosecuted by all husbandmen. Beside, nothing is lost in that performance; because the trenches being filled up to their superficies, all the land is exposed and fit for tillage, even to an inch; that can not be said when trenches are left open, which occupy much ground, and, in contradistinction to the others, are liable to need repairs from time to time.

"Should stone for replenishing ditches fail on the spot, do not have them brought from afar, at great expense, but instead use straw, which you may employ in this wise:

"The rye straw, on account of its strength, can be used, and this failing, replace it with wheat straw. *You will make with it a floor in the ditch, in order, being suspended, to cause an empty space below for the passage of the water, and above this floor you will put two feet of earth.* The empty space should be one foot high, the thickness of the floor another foot, and the two of earth will make four feet depth of the ditch. These ditches must be only two feet and a half wide, narrower by six inches than others, for the subjection of straw, for fear of choking up the empty space below, by caving in on account of its weight the earth put on the top. The mother trench, recipient of water, must not be wider than the others, considering the difficulties of the straw; but this may be overcome by using two mother trenches, or only one so deep that it will suffice to collect all the water directed to it. The straw ought to be arranged into bundles one foot thick and two and a half long, tied up even at three equidistant places.

"In order to lay these bundles as they ought be, you will make the ditch narrower at the bottom than at the top, not in declivity or slope, but perpendicular, contracting unto square at the place whereon the floor is to be laid, to rest firm and secure as upon walls. The contraction at each side must be six inches; thus the lowest place of the ditch will be one foot and a half, and two and six inches at the widest part, which is the top. Should you suspect your ditches and drains of being too small, the remedy is not to widen them, considering the difficulties of the straw, but it consists in their number; for, as already said, you can not have too many of them, and you never will remove too much water from a swamp or marshy ground. Thus you must be careful to dig a sufficient number of them and so well disposed, that they may discharge the ones upon the others by branches connecting them, in order to conduct all the water of the field into the mother trench and empty it at the proper place.

"Straw, thus employed, will last a long time; for it is admitted that, being inclosed within the earth and without the effects of air, straw remains sound over a hundred years. I am a witness that some sound straw was found entire in the midst of an old, ruined house, and the wall appeared to be the work of former ages. Therefore, use it without scruple, with the understanding that if it should rot at the end of a hundred years, those who will come then, may change it if they have a mind to."

On the subject of the straw to replenish trenches, as indicated by Oliver de Serres, Victor Yvart, in a remark added to the edition of the works of the illustrious writer, and published by the Agricultural Society at Paris, A. D. 1804, says:

"It might prove safe and cheap, in the above case, to use faggots made of small alder tree branches, which keep well in water, and for want of them, other branches, which, placed at the bottom allow, through their interstices, free exit to water, and afford all the advantages of straw, without its drawbacks."

From the above important quotation, it follows that the invention of underground causeways for draining tillable lands can not be claimed by an English author, even a Walter Bligh or an Elkington. The latter was a Warwickshire farmer, gifted with an observing mind and great perseverance, who, toward the end of the last century, drained wet lands and was so successful as to attract the attention of the parliament and to obtain very many recompenses. But his method is not much different from Oliver de Serres' stone system. Elkington's process does not admit of mother trenches, in order to lead the water out of the fields and appropriate it for divers uses.

There are three manners of disposing of the water:

1. Water sinks into permeable, inferior layers, through a well filled with stones.

2. Should the well require a depth of more than thirteen feet, its office is supplied by boring a hole with a rod, until it reaches a porous strata.

3. Water springs up, like in artesian wells, either by means of shafts or wells conveniently situated, and then removed through discharging pipes.

This method named Elkington's, consisting in the double contrivance of wells and underground ditches combined, requires special dispositions, according to the configuration of the ground; it combines, also, drains with shafts and artesian wells.

DRAIN PIPES USED IN FRANCE, A. D. 1620.

THE invention of drainage pipes has always been conceded to England. Should the statement contained in the following letter prove to be beyond controversy, we ought to say, the English have shown the importance of using underground pipes to drain the land, but this invention is of French origin:

"SIR: I read in the *Journal of Practical Agriculture* your essay on drainage scarcely begun, and already full of interest; it foretells deep attraction when it treats of its influence on crops, manures, etc.

"In the conclusions of your chapter on the history of drainage, a contrivance known to antiquity, you show three degrees in the periods of progress through which it came to us.

"The first, its origin, perhaps, was the practice of it among the Romans, as related by Columella and Palladius.

"The second, in which our priority over England is established, thanks to Oliver de Serres.

"The third, in which you abandon the whole conquest to the English, because they substituted tiles and pipes for other materials.

"The latter period is indeed capital; heretofore it was darkness; now it is a science. The last improvement elevated drainage from the rank of an agricultural drudgery to the sphere of industry; caused men of genius to cluster around; attracted the attention of men of the world and the powerful support of an enlightened, partial government.

"This lucky improvement, this starting point, I may say, shall not be denied by me to the English; it would be in bad taste to claim glory for an invention, when we failed to make it fruitful. I will only state that the same idea of this improvement was realized in 1620, about the time when Oliver de Serres published his works.

"Within the town of Maubeuge, in my own neighborhood, was a convent of monks; the epoch of its erection could easily be ascertained; its chapel is still a pure specimen of gothic style. The convent did not escape the republic of 1793, and the aspect and inmates have changed, but its wide and splendid garden was respected. Was it on account of its reputation? It is well known that, from immemorial time, it was renowned for its fertility, the beauty and earliness of its fruit and for the friability of its soil.

"The estate was sold, and last year the premises underwent repairs: the prolific garden was turned into pleasure ground, park with fountains, driving causeways, artificial elevations of ground, and so forth. This overturning disclosed the secret of its marvelous reputation.

"Two complete and regular pipe drains extended throughout the whole garden, at the depth of four feet.

"One of the drains had all its pipes radiating to a sinking well situate in a central position; the other was made of pipes all parallel, ending at a collecting pipe which discharged into a cellar.

"The owner had the kindness to give me two pipes as specimens of curiosity; they are about ten inches long and four inches in diameter; one end expands into a funnel-shape, the other tapers into a cone; they are made of an argilo-silicious composition like most of our earthenware, which is very hard and becomes very much glazed in burning, thereby becoming unalterable; all were found well preserved; they were evidently made by hand and lathe.

"When was the drain constructed? No particular data is given. MSS. left by the monks might solve the question; at any rate, some tombs, placed over the drain in 1620, show it to be anterior; here, then, is an ancient drainage, made with masterly hands three hundred and forty years ago, which, in its dimensions, system, and materials, is much like those of the present day.

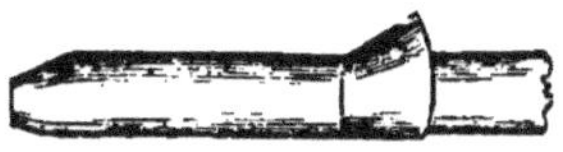

[Fig. 1. Pipe Drains of 1620, found at Maubeuge, France.]

"To vouch for the truthfulness of the facts, it remains for me to state that the particulars were given to me by Hon. Marchant, senator, owner of the estate, and by his son-in-law, my brother, a distinguished agriculturist, who was present when the excavations were made.

G. Hamoir, *Member of the Ag. Soc.*"

Having shown, at considerable length, the origin of drainage in France, it may be well to devote a few pages to the introduction of this improvement into England.

The first work published on the subject by an English author, Capt. Walter Bligh, already referred to. His work, the English Improver Improved, or, *The Survey of Husbandry Surveyed*, was published in 1650. The principles of drainage advocated by Capt. Bligh, are thus expressed by Josiah Parkes, an eminent practical drainer in England, in the 7th vol. of the *Journal of the Royal Agricultural Society:*

"In his instructions for forming the flooding and draining trenches of water-meadows, the author says of the latter: 'And for thy drayning trench, it must be made so deep, that it goe to the bottom of the cold spewing moyst water, that feeds the flagg and the rush; for the widenesse of it, use thine own liberty, but be sure to make it so wide as thou mayest goe to the bottom of it, which must be so low as any moysture lyeth, which moysture usually lyeth under the over and second swarth of the earth, in some gravel or sand, or else, where some greater stones are mixt with clay, under which thou must goe half one spade's graft deep at least. Yea, suppose this corruption that feeds and nourisheth the rush or flagg, should lie a yard or four-foot deepe; to the bottom of it thou must goe, if ever thou wilt drayn it to purpose, or make the utmost advantage of either floating or drayning, without which the water can not have its kindly operation; for though the water fatten naturally, yet still this coldnesse and moysture lies gnawing within, and not being taken clean away, it eates out what the water fattens; and so the goodnesse of the water is, as it were, riddled, screened, and strained out into the land, leaving the richnesse and the leannesse sliding away from it.' In another place, he replies to the objectors of floating, that it will breed the rush, the flagg, and mare-blab; 'only make thy drayning-trenches deep enough, and not too far off thy floating course, and I'le warrant it they drayn away that under-moysture, fylth, and venom as aforesaid, that maintains them; and then believe me, or deny Scripture, which I hope thou doust not, as Bildad said unto Job, 'Can the rush grow without mire, or the flagg without water?' Job viii, 11. That interrogation plainly showes that the rush can not grow, the water being taken from the root; for it is not the moystenesse upon the surface of the land, for then every shower should increase the rush, but it is that which lyeth at the root, which, drayned away at the bottom, leaves it naked and barren of relief.'

"The author frequently returns to this charge, explaining over and over again, the necessity of removing what we call bottom-water, and which he well designates as 'filth and venom.'

"In the course of my operations as a drainer, I have met with, or heard of so many instances of swamp-drainage, executed precisely according to the plans of this author, and sometimes in a superior manner—the conduits being formed of walling stone, at a period long antecedent to the memory of the living—that I am disposed to consider the practice of deep drainage to have originated with Capt. Bligh, and to have been preserved by imitators in various parts of the country; since a book, which passed through three editions in the time of the Commonwealth, must necessarily have had an extensive circulation, and enjoyed a high renown. Several complimentary autograph verses, written by some imitators and admirers of the ingenious Bligh, are bound up with the volume. I find, also, not unfrequently, very ancient deep drains in arable fields, and some of them still in good condition; and in a case or two, I have met with several ancient drains six feet deep, placed parallel with each other, but at so great a distance asunder, as not to have commanded a perfect drainage of the intermediate space. The author from whom I have so largely quoted, is the earliest known to me, who has had the sagacity to distinguish between the transient effect of rain, and the constant action of stagnant bottom water in maintaining land in a wet condition."

The next important step in the progress of drainage in England, was by Joseph Elkington, an illiterate Warwickshire Farmer; but a man who undoubtedly possessed more than ordinary ability, if not absolute genius. His discovery and subsequent practice created such a sensation throughout England and Scotland, but more especially in the agricultural circles, that at the solicitation of the Board of Agriculture, Parliament in 1795, voted him £1,000, as a reward for his discovery in the drainage of land.

Elkington being incapable of writing out his discovery and system of drainage, so that others might be benefited by such a work, the Board of Agriculture appointed a Mr. John Johnstone to visit Elkington's principal works, and study them carefully, and record it for the benefit of others. Mr. Johnstone accordingly studied the Elkington system of drainage, and wrote a treatise on it. Recent writers charge Mr. Johnstone with giving his own opinions in many instances, rather than those of Mr. Elkington.

He gives the following statement of Elkington's discovery:

"In the year 1763, Elkington was left by his father in the possession of a farm called Prince-Thorp, in the parish of Stretton-upon-Dunsmore, and county of Warwick. The soil of this farm was so poor, and, in many places, so extremely wet, that it was the cause of rotting several hundreds of his sheep, which first induced him, if possible, to drain it. This he begun to do, in 1764, in a field of wet clay soil, rendered almost a swamp, or *shaking* bog, by the springs which issued from an adjoining bank of gravel and sand, and overflowed the surface of the ground below. To drain this field, which was of considerable extent, he cut a trench about four or five feet deep, a little below the upper side of the bog, where the wetness began to make its appearance; and, after proceeding with it in this direction and at this depth, he found it did not reach the *principal body of subjacent water* from which the evil arose. On perceiving this, he was at a loss how to proceed, when one of his servants came to the field with an *iron crow*, or bar, for the purpose of making holes for fixing sheep hurdles in an adjoining part of the farm, as represented on the plan. Having a suspicion that his drain was not deep enough, and desirous to know what strata lay under it, he took the iron bar, and having forced it down about four feet below the bottom of the trench, on pulling it out, to his astonishment, a great quantity of water burst up through the hole he had thus made, and ran along the drain. This led him to the knowledge, that wetness may be often produced by water confined farther below the surface of the ground, than it was possible for the usual depth of drains to reach, and that an *auger* would be a useful instrument to apply in such cases. Thus, chance was the parent of this discovery, as she oftens is of other useful arts; and fortunate it is for society, when such accidents happen to those who have sense and judgment to avail themselves of hints thus fortuitously given. In this manner he soon accomplished the drainage of his whole farm, and rendered it so perfectly dry and sound, that none of his flock was ever after affected with disease.

"By the success of this experiment, Mr. Elkington's fame, as a drainer, was quickly and widely extended; and, after having successfully drained several farms in his neighborhood, he was, at last, very generally employed for that purpose, in various parts of the kingdom, till about thirty years ago, when the country had the melancholy cause to regret his loss. From his long practice and experience, he became so successful in the works he undertook, and so skillful in judging of the internal strata of the earth, and the nature of springs, that, with remarkable precision, he could ascertain where to find water, and trace the course of springs that made no appearance on the surface of the ground. During his practice of more than thirty years, he drained in various parts in England, particularly in the midland counties, many thousand acres of land, which, from being originally of little or no value, soon became as useful as any in the kingdom, by producing the most valuable kinds of grain, and feeding the best and healthiest species of stock.

"Many have erroneously entertained an idea that Elkington's skill lay solely in applying the auger for the *tapping of springs*, without attaching any merit to his method of conducting the drains. The accidental circumstance above stated, gave him the first notion of using an auger, and directed his attention to the profession and practice of draining, in the course of which he made various useful discoveries, as will be afterward explained. With regard to the use of the auger, though there is every reason to believe that he was led to employ that instrument from the circumstance already stated, and did not derive it from any other source of intelligence, yet there is no doubt that others might have hit upon the same idea without being indebted for it to him. It has happened, that, in attempts to discover mines by boring, springs have been tapped, and ground thereby drained, either by letting the water down, or by giving it vent to the surface; and that the auger has been likewise used in bringing up water in wells, to save the expense of deeper digging; but that it had been *used in draining land, before Mr. Elkington made that discovery, no one has ventured to assert.*"

Johnstone sums up this system as follows:

"Draining, according to Elkington's principles, depends chiefly upon three things:

"1. Upon discovering the main spring or source of the evil.

"2. Upon taking the subterraneous bearings; and,

"3. By making use of the auger to reach and *tap* the springs, when the depth of the drain is not sufficient for that purpose.

"The first thing, therefore, to be observed is, by examining the adjoining high grounds, to discover what strata they are composed of; and then to ascertain, as nearly as possible, the inclination of these strata, and their connection with the ground to be drained, and thereby to judge at what place the level of the spring comes nearest to where the water can be cut off, and most readily discharged. The surest way of ascertaining the lay or inclination of the different strata, is, by examining the bed of the nearest streams, and the edges of the banks that are cut through by the water, and any pits, wells, or quarries that may be in the neighborhood. After the *main spring* has been thus discovered, the next thing is to ascertain a line on the same level, to one or both sides of it, in which the drain may be conducted, which is one of the most important parts of the operation, and one on which the art of draining in a scientific manner essentially depends.

"Lastly, the use of the auger, which, in many cases, is the *sine qua non* of the business, is to reach and tap the spring when the depth of the drain does not reach it; where the level of the outlet will not admit of its being cut to a greater depth; and where the expense of such cutting would be great, and the execution of it difficult.

"According to these principles, this system of draining has been attended with extraordinary consequences, not only in laying the land dry in the vicinity of the drain, but also springs, wells, and wet ground, at a considerable distance, with which there was no apparent connection."

About the year 1810, it was deemed advisable to change the former system of draining. Flat and hollow tiles were at first adopted. Tile drainage appears to have been put in practice, for the first time, at Netherby, in Northumberland, upon the estate of Sir James Graham. "A tile and sole, with a few inches of stone, is the *ne plus ultra* of draining:" thus reads the *Journal of the Society of Agriculture of England*, vol. II, p. 293. It seems that for a period of thirty years, there appeared no possibility of improving on the method invented in 1810.

But it remained for the agriculturists of the nineteenth century to establish a system of drainage in accordance with scientific principles — to make it more general in its application; to provide apparatus and machinery for the more precise and uniform construction of the drains, as well as the tile; and the entire art has been so much improved that all previous experiments and systems vanish into almost nothingness in comparison.

The ancient mode of draining, successful as it may have been, yet inefficient as it certainly was, subjected those who practiced it to many inconveniences, while itself was subject to many liabilities from which the modern or English system of tile draining is exempt. Not only was that system attended with many inconveniences in construction, for want of proper materials, but even when constructed was liable to become deranged in a comparatively short period of time, and to repair them was attended with not only great expenditure, but great inconvenience and labor.

Beside, these drains, made of wood, stone, etc., served for the single purpose of draining the springs and stagnant surface water only, while the modern ones, made of tile, not only serve this same purpose, but also accomplish some other advantageous results.

It soon became manifest that these wooden and stone drains were sadly deficient in permanency, and great pains were taken to substitute something better. It was somewhat of an improvement when the plan was adopted of digging a canal or ditch, and covering the bottom with brick, and then placing on these brick, to the thickness of a foot or more, large-sized pebbles or brickbats. These drains proved more durable and less liable to become deranged than the previous ones, and while they drained a given quantity of water in less time, they accomplished that one object only.

This system was in turn abandoned, and a wooden pipe, or boards forming a kind of covered triangular trough, was substituted. When these troughs were fitted in the bottom of the drain, the drain itself was then closed with turf, sod or earth. But this system was found to be very expensive, and accomplished the one object only, viz : draining the surface water and the spring water, and this very imperfectly.

Another great error was committed with this system, namely, the ditches were made so shallow that even when plowing with their shallow plows, the conduit was disturbed. The result was that when the winters were severe, the drains were frozen up, and were, in consequence, not only worthless, but an actual damage, because the soil underwent all the phenomena that it does in winter—killing wheat; and it was late in the spring before they were in a serviceable condition. Hence, at the season when they should have been of the utmost importance, they were entirely useless, because at that season there is not only the most water in the soil, but it is also a period when the water produces the most injurious results to the growing plants.

Mr. Baxter, an Englishman, was perhaps the first one to indicate the disadvantages arising from shallow drains. He describes the results as follows:

"In the year 1819, I drained an eight acre field according to the old system. The pipes were laid from twenty to twenty-four inches beneath the surface, and twenty-eight feet apart. I soon became convinced that very little benefit would accrue from this system. The crops were no better than before the piece was drained—the tilth no better or easier, in spite of the best manures which I could procure. These drains were filled with stone, and covered with turf; but, being compelled to bring the stone some distance, it made the drains very expensive. In 1832, I redrained the same field according to the new system; that is, I made the drains three feet deep at parallel distances of thirty-two feet. The advantages of this system were at once apparent. The soil was sufficiently dry for purposes of cultivation much sooner than that not underdrained; the water furrows disappeared, and with them, the expense of keeping the field clean after seeding. No more manures of any kind were applied, and yet the product was fully one third more than previous to underdraining. Since I have commenced deep or thorough draining, all my crops yield fully one third more. By deep draining, the water can at once flow unhindered from the field, and the soil is consequently in a condition to be worked at a much earlier date after a rain than that which is not underdrained. A given underdrained field will sustain more cattle in a good condition than one which has not been so treated. Then, the crop, too, will ripen earlier, and the labor, in clay soil, is lightened fully one fifth for cattle or horses, while the soil itself is rendered in much better condition. The climate even is improved everywhere where deep draining is practiced. It is the most effective means of radically removing miasma that can be introduced."

The great advantages consequent upon deep draining were then made manifest by actual experiment twenty-five years ago. About the same time, another very important invention was introduced, namely, the clay pipe, or draining tiles. The credit of this most important discovery is due to Mr. Smith, of Deanston, in Scotland. A distinguished mechanic and director of a cotton factory, Mr. Smith, wondering at the infertility of a piece of ground contiguous to that factory, after a minute observation, came to the conclusion that an excess of moisture was the cause of it, and without any knowledge of what had either been written or done by former agriculturists, it occurred to him that covered drains would answer an excellent purpose to drain tillable lands. His success was great, and much talked of in the neighborhood. In 1833, he published a pamphlet under the title of *Smith's Remarks on thorough Draining;* and although he was not the first inventor of this method, he rendered to England and Scotland the service of introducing his method of drainage, which greatly increased the fertility of British lands. We must acknowledge, in honor to that country, that land owners and the government acted in this case with more promptitude than is usual with them. Even Sir Robert Peel, in 1840, had part of his estate at Drayton, Staffordshire, drained by Mr. Smith.

This new art of draining swept like a wildfire over Great Britain, and drainage soon became more and more general in England and Scotland. The great importance of this system of draining was not only acknowledged and advocated by the agriculturists of England, but the government gave the most substantial testimonials of its confidence in the system as an augmenter of products. A fund of two million pounds sterling, equal to ten millions of dollars, was appropriated as a fund to be loaned to farmers, to be expended in drainage—six and a half per cent. of the loan to be annually refunded, but at the expiration of twenty-two years, the entire fund to be extinguished.

Not only landlords drained their lands, but the tenants even, when their lease expired at the end of six years, drained the land they cultivated with advantage and profit to themselves. In consequence of the great advantages arising from drainage in England, a law was enacted regulating the amount to be allowed tenants for draining the landlord's farm, also, a law fixing the increased value of farms which are underdrained, for purposes of hypothecation, etc.

Tiles were at first made by hand. English inventive genius was not long in forwarding this industry. As drainage spread, machines came in demand to supersede manual labor. The first one, turning out flat and hollow tiles at once, was made by Irving, during the year 1842.[1] Immediately after this, the Marquis of Tweeddale, Mr. Ransome, then Mr. Etheredge, invented other machinery for the same purpose.[2] But, to manufacture subterranean pipes in two pieces was evidently useless complication. Mr. John Read was then happily inspired with the idea of substituting cylindrical pipes for tiles. Therefore, this manufacturer added to the ancient process of drainage, the last improvement which it has attained. The first machinery of this kind was exhibited in 1843, at the Derby Agricultural Fair. It won premiums of silver medals, and was the object of minute and encomiastic descriptions by Josiah Parkes, who was struck with its importance. Since that time, every year ushers in new improvements on its construction.

Elkington's attention was specially directed to springs, and the merit of his system consisted of relieving the

1 Journal of the Royal Society of Agriculture, Vol. IV, p. 370 (1843).
2 Journal of the Royal Society of Agriculture, Vol. III, p. 398.

arable soil from the effects of water issuing from below upward. Now, it is well known that, in many places as much injury is done to crops by the retention of rain water in the soil as from springs. In cases where the soil retained the rains, fields could not be drained by an auger hole and a ditch or two.

Smith, of Deanston's system consisted in cutting parallel drains at regular intervals over the entire field, without regard to springs or other sources of subterranean moisture. His characteristic views were, in brief, as follows:

"1. *Frequent* drains at intervals of from ten to twenty-four feet.

"2. *Shallow* depth—not exceeding thirty inches—designed for the single purpose of freeing that depth of soil from stagnant and injurious water.

"3. '*Parallel drains, at regular distances*, carried throughout the whole field, without reference to the wet and dry appearance of portions of the field,' in order 'to provide frequent opportunities for the water *rising from* below and falling on the surface, to pass freely and completely off.'

"4. *Direction of the minor drains* 'down the steep,' and that of the mains along the bottom of the chief hollow—tributary mains being provided for the lesser hollows. The reason assigned for the minor drains following the line of steepest descent, was, that 'the stratification generally lies in sheets, at an angle to the surface.'

"5. *As to material*—Stones preferred to tiles and pipes."

From 1833 to 1854, several systems of drainage were advocated throughout Great Britain. All, however, agreed that "*tile*" was the best material for a conduit for the water; but these systems differed from each other in the distance between as well as the depth of the drains. The merit of each of these systems will be discussed in the practical portion of this treatise.

Drainage was readily introduced into Belgium and Germany, where it produced as happy results as in England. The governments of these respective countries cheerfully extented to the system all the encouragement which could reasonably be expected. To such an extent did these governments manifest their appreciation of the system, that they actually purchased tile machines, manufactured tile, and established depots for the sale of them at low rates, so as to place them within the reach of almost all tenants and landholders. The impulse thus given by government itself soon produced the happiest results. Experiments conducted here and there, at the instance of the government, convinced even the most skeptical of the great advantages to be derived from underdraining; the praises of the system found an echo in every nook and corner of the country; and nothing ever was so universally practiced in Germany in so short a time from the period of its first introduction as thorough draining.

DRAINAGE IN FRANCE.

From returns gathered about the middle of 1856, it appears that there were then about 80,000 English acres of thorough-drained land, and 396 tile works in France. The money expended in draining, from 1850, when this improvement was begun in France, up to the summer of 1856, accordingly amounts to $8,000,000; the expense of draining being about $20 per acre.

During the draining season (autumn) of 1856, up to January 1, 1857, no less than 85,000 acres were drained.

Of these 165,000 acres, only 45,000 acres were drained by care or assistance of the government; the remainder is the work of private enterprise. When will American farmers become convinced that thorough draining is one of the *most* important aids to agriculture!

DRAINAGE IN THE UNITED STATES.

The introduction of tile drainage in the United States may be given in a very few words. The following account, prepared by a correspondent of the *New York Tribune*, and corrected and revised by the editor of the *Country Gentleman*, is perhaps the best account that has yet been written on the subject. It is true, we might state, that "Mr. John Johnston, of Geneva, N. Y., introduced tile draining on his farm in 1835; that, in 1848, John Delafield of Seneca county, N. Y., introduced the first tile machine (Scragg's patent, imported from England);" and with this passing notice proceed to the next chapter. But those for whom this treatise is written will naturally inquire, "What induced him to drain? Where did he obtain tile? How much and in what manner did he drain? *What did it cost? Did it pay?*" and a host of other questions. It will, therefore, be satisfactory, even at the expense of some space, to present a detailed statement of all the circumstances surrounding and attending his efforts.

JOHN JOHNSTON'S SYSTEM OF DRAINAGE.

Mr. John Johnson, near Geneva, N. Y., at one time esteemed a fanatic by his neighbors, has come of late years to be generally known as "the father of tile drainage in America." After thirty years of precept and twenty-two of example, he has the satisfaction of seeing his favorite theory fully accepted, and, to some extent, practically applied throughout the country. Not without labor, however, nor without much skepticism, ridicule and controversy has this end been attained; and if, now that his head is whitened and his course all but run, he finds himself respected, and appealed to by persons in every state of the Union, he does not forget that it has been by much tribulation that he has worked out this exceeding great weight of glory. Mr. Johnston is a Scotchman, who came to this country thirty-nine years ago, and purchased the farm he now occupies, on the easterly shore of Seneca lake, a short distance from Geneva. With the pertinacity of his nation, he staid where he first settled, through ill fortune and prosperity, wisely concluding that, by always bettering his farm, he would better himself, and make more money in the long run than he could by shifting uneasily from place to place in search of sudden wealth. He was poor enough at the commencement; but what did that matter to a frugal, industrious man, willing to live within his means, and work hard to increase them? And so, with unflagging zeal, he has gone on from that day to this.

HIS FARM.

His first purchase was 112 acres of land, well situated, but said to be the poorest in the county. He knew better than that, however, for although the previous tenant had all but starved upon it, and the neighbors told him such would be his own fate, he had seen poorer land forced to yield large crops in the old country, and so he concluded to try the chances for life or death. The soil was a heavy gravelly clay, with a tenacious clay subsoil, a perfectly tight reservoir for water, cold, hard-baked, and cropped down to about the last gasp. The magician commenced his work. He found in the barn-yard a great pile of manure, the accumulations of years, well rotted, black as ink, and "mellow as an ash-heap." This he put on as much land as possible, at the rate of *twenty-five loads to the acre*, plowed it in deeply, sowed his grain, cleaned out the weeds as well as he could, and the land on which he was to starve gave him about *forty bushels* of wheat per acre. The result was, as usual, attributed to luck, and anything but the real cause. To turn over such deep furrows was sheer folly, and such heavy dressings of manure would not fail to destroy the seed. But it didn't; and let our farmers remember that it never will; and if they wish to get rich, let them cut out this article, read it often, and follow the example of our fanatical Scotch friend.

This system of deep plowing and heavy manuring wrought its result in due time. Paying off his debt, putting up buildings, and purchasing stock each year, to fatten and sell, Mr. Johnston after seventeen years of hard work at last found himself ready to incur a new debt, and to commence laying tile drains. Of the benefits to be derived from drainage he had long been aware; for he recollected that when he was only ten years of age, his grandfather, a thrifty farmer in Scotland, seeing the good effects of some stone drains laid down upon his place, had said: "Varily, I believe the whole airth should be drained." This quaint saying, which needs but little qualification, made a lasting impression on the mind of the boy, that was to be tested by the man, to the permanent benefit of this country.

Without sufficient means himself, he applied for a loan to the Bank of Geneva, and the president, knowing his integrity and industry, granted his request. In 1835 tiles were not made in this country, so Mr. Johnston imported some as samples, and a quantity of the "horse-shoe" pattern were made in 1838, at Waterloo. There was no machine for producing them, so they were made by hand and molded over a stick. This slow and laborious process brought their cost to $24 per thousand, but even at this enormous price, Mr. Johnston determined to use them. His ditches were opened and his tile laid, and then what sport for the neighbors! They poked fun at the deluded man; they came and counseled with him, all the while watching his bright eye and intelligent face for signs of lunacy; they went by wagging their heads and saying, "Aha!" and one and all said he was a consummate ass to put crockery under ground and bury his money so fruitlessly. Poor Mr. Johnston! he says he really felt ashamed of himself for trying the new plan, and when people riding past the house would shout at him, and make contemptuous signs, he was sore-hearted and almost ready to conceal his crime. But what was the result? Why this: that land which was previously sodden with water, and utterly unfruitful, in one season was covered with luxuriant crops, and the jeering skeptics were utterly confounded; that in two crops all his outlay for tiles and labor was repaid, and he could start afresh and drain more land; that the profit was so manifest as to induce him to extend his operations each succeeding year, and so go on until 1856, when his labor was finished, after having laid 210,000 tiles, or more than fifty miles in length! And the fame of this individual success going forth, one and another duplicated his experiment, and were rewarded according to their deserts.

It was not long after the manufacture of the first lot of tiles that a machine was contrived which would make quite as well, and faster; and by its aid they were afforded at quite as low a price as after an English machine was imported. The horse-shoe tile has been used by Mr. Johnston almost exclusively, for the reason that they were the only kind to be procured at first, and on his hard subsoil, finding them to do as well as he could wish, he has not cared to make new experiments. He has drains that have been in function for more than twenty years without needing repair, and are apparently as efficient now as they were when first laid. In soft land, pipe or sole tiles would be preferable, or if horse-shoe were used they should be placed on strips of rough board, to prevent their sinking into the trench bottom, or being thrown out of the regular fall by being undermined by the running water. He has not used the plow for opening his trenches, for the reason that all his work has been let out by contract, and the men have opened them by the spade; charging from twelve and a half to fifteen cents per rod for opening and making the bottom ready for the tile. The laying and filling was done by the owner.

HIS PRACTICE.

His ditches are dug only two and a half feet deep, and thirteen inches wide at the top, sloping inward to the bottom, where they are just wide enough to take the tile. One main drain, in which are placed two four-inch tiles set eight inches apart, with an arch piece of tile having a nine-inch span set on top of them, was dug three and a half and four feet deep, and this serves as a conduit for the water from a large system of laterals. Drains should never be left open in winter, for the dirt dislodged by frequent frosts so fills the bottom that it will cost five or six cents per rod to clear them; and, moreover, the banks often become so crumbled away that the ditch can not be straddled by a team of horses, and thus most of the filling must be done by hand. Mr. Johnston in draining a field commences at the foot of each ditch and works up to the head. He opens his mains first, and then the lateral or small drains, but he lays the tiles in the laterals and fills them completely before laying the pipe in the mains. The object of this is to prevent the accumulation of sediment in the mains, which would naturally be washed from the laterals on their first being laid. By commencing at the foot of each ditch and working upward, he can always get and preserve the regular fall, which may be dictated by the features of his field, more easily than by working toward the outlet. A little practice teaches the ditchers how to preserve the grade almost as well as if gauges were employed; but before laying the tiles, the instrument is applied to test the bottom thoroughly.[1] The necessity of this precaution will be apparent to any one who reflects that if a tile or two in the course of a ditch be set much too high or too low at either end, the water quickly forms a basin beneath and around, sediment is washed into the adjoining pipe, and ultimately even the whole bore is filled and the drain stopped. When this happens it will be indicated after a time by the water appearing at the surface of the ground above the spot—drawn upward by capillary attraction. In such a case the ditch must be reopened and the tile relaid.

ILLUSTRATIONS.

Mr. Johnston says tile-draining pays for itself in two seasons, sometimes in one. Thus, in 1847, he bought a piece of ten acres to get an outlet for his drains. It was a perfect quagmire, covered with coarse aquatic grasses, and so unfruitful that it would not give back the seed sown upon it. In 1848 a crop of corn was taken from it, which was measured and found to be *eighty bushels* per acre, and as, because of the Irish famine, corn was worth $1 per bushel that year, this crop paid not only all the expense of drainage, but the first cost of the land as well.

Another piece of twenty acres, adjoining the farm of the late John Delafield, was wet and would never bring more than ten bushels of corn per acre. This was drained at a great cost, nearly $30 per acre. The first crop after this was 83 bushels and some odd pounds per acre. It was weighed and measured by Mr. Delafield, and the county society awarded a premium to Mr. Johnston.

---

[1] I never used a leveling instrument. I always had water, which is the best instrument.—J. J.

Eight acres and some rods of this land, at one side, averaged 94 bushels, or the trifling increase of 84 bushels per acre over what it would bear before those insignificant clay tiles were buried in the ground. But this increase of crop is not the only profit of drainage; for Mr. Johnston says that on drained land one half the usual quantity of manure suffices to give maximum crops. It is not difficult to find a reason for this. When the soil is sodden with water, air can not enter to any extent, and hence oxygen can not eat off the surfaces of soil-particles and prepared food for plants; thus the plant must in great measure depend on the manure for sustenance, and of course the more this is the case, the more manure must be applied to get good crops. This is one reason, but there are others which we might adduce if one good one were not sufficient.

Mr. Johnston says he never made money until he drained, and so convinced is he of the benefits accruing from the practice, that he would not hesitate—as he did not when the result was much more uncertain than at present—to borrow money to drain. Drains well laid, endure, but unless a farmer intends doing the job well he had best leave it alone and grow poor, and move out West, and all that sort of thing. Occupiers of apparently dry land are not safe in concluding that they need not go to the expense of draining, for if they will but dig a three-foot ditch in even the driest soil, water will be found in the bottom at the end of eight hours, and if it does come, then draining will pay for itself speedily. For instance: Mr. Johnston had a lot of thirteen acres on the shore of the lake, where the bank at the foot of the lot was perpendicular to the depth of thirty or forty feet. He supposed from this fact, and because the surface seemed very dry, that he had no need to drain it. But somehow he lost his crops continually, and as he had put them in as well as he knew how, he naturally concluded that he must lay some tile. So he engaged an Irishman to open a ditch, with a proviso that if water should come into it in eight hours, he would drain the entire piece. The top soil was so hard and dry as to need an application of the pick, but at the depth of a foot it was found to be so wet and soft that a spade could easily be sunk to the entire depth of ten inches with little force. The ditches were made, and in less than the specified time a brave lot of water flowed in. The piece was thoroughly drained, and the result was an immense crop of corn. The field has regularly borne 60 or 70 bushels since. Corn was planted for a first crop in this and the preceding instances, because a paying crop is obtained in one year, whereas if wheat were sown it would be necessary to wait two seasons. He always drains when the field is in grass, if possible, for the ditches can be made easily; and Spring is chosen that the labor may not be interfered with by frosts.

To show how necessary it is to avoid planting trees over drains, we quote a case in point. In a lot adjoining his house are four large elms which are marked to be felled, and for the reason that the lot was formerly so wet that a pond of water stood upon it in winter, and throughout the season the children skated and slid upon it. It was drained, and all went well for a time; but after seven years Mr. Johnston found his drains did not discharge properly, and that in certain places the water came to the surface, so as to destroy or greatly lessen the crop above them. He could not account for the circumstances until he dug down to the drain at each of these spots, when, to his surprise, he found the tile [two four-inch tile with a semi-circle of nine inch set on top of them,] completely choked with fibrous roots of the elms.

Mr. Johnston says he never saw one hundred acres in any one farm, but a portion of it would pay for draining. Mr. Johnston is no rich man who has carried a favorite hobby without regard to cost or profit. He is a hard-working Scotch farmer, who commenced a poor man, borrowed money to drain his land, has gradually extended his operations, and is now reaping the benefits, in having crops of forty bushels of wheat to the acre. He is a gray-haired Nestor, who, after accumulating the experience of a long life, is now at sixty-eight years of age, written to by strangers in every state of the Union for information, not only in drainage matters, but all cognate branches of farming. He sits in his homestead a veritable Humboldt in his way, dispensing information cheerfully through our agricultural papers and to private correspondents, of whom he has recorded 164 who applied to him last year. His opinions are, therefore, worth more than those of a host of theoretical men, who write without practice. He says that the retrogression of our agriculture in the older states, is to be accounted for in our lack of drainage, poor feeding of stock, which results in giving a small quantity of poor manure, and in not keeping enough to make manure. He applies twenty-five loads of manure to the acre at the beginning of a rotation, and this lasts throughout the course. He learned from his grandfather that no farmer could afford to keep any animal that did not improve on his hands, and that as soon as it was in good marketable condition it should be sold and replaced by another. This theory he has always carried out, and as a natural consequence, has always got higher prices for his beef stock, and a ready market in the dullest of times.

Although his farm is mainly devoted to wheat, yet a considerable area of meadow and some pasture has been retained. He now owns about 300 acres of land. The yield of wheat has been 40 bushels this year, and in former seasons, when his neighbors were reaping 8, 10, or 15 bushels, he has had 30 and 40. We are informed by him that there has been no such crop as the present since 1845, either in yield or quality; and the absence of weevil is remarkable. A variety of white wheat from Missouri, sown more thinly than usual, has yielded 31 bushels to something less than one bushel of seed sown. It headed out a fortnight earlier than the Soule's, but ripened later —probably because thinly sown. Mr. Johnston thinks we have been sowing too thickly for fifteen years past upon rich land, and there can be no question but that he is right. Still, it is better to take a medium course between thick and thin sowing, and thus avoid, on the one hand, rust, overcrowding, and waste of seed, and on the other, placing an entire crop at the mercy of insects which may attack it.

SALT FOR RUST.

As a sure preventive to rust, to give stiffness to the straw, and to expedite the ripening of wheat, by four or five days, Mr. Johnston sows five bushels of salt to the acre, broadcast, after seeding. He thinks, moreover, that for each of the five bushels of salt almost an extra bushel of wheat may be expected.

SIZE OF TILES FOR MAINS AND LATERALS.

A too common error with improving farmers is that of using too small tile for main drains, and too large for laterals. Those accustomed to the roomy conduits of ordinary stone drains, suppose that nothing less than a three inch bore will conduct the drainage from the surface into the mains; and curiously enough the same persons, unmindful of the large area drained by each system of laterals, err in using mains but little larger in bore than the latter. If any are willing to look into the results of the drainage on our Central Park, the most stupendous work of the kind in the country, and one of the best conducted, they will find that the one and a half inch and two inch tiles there used for laterals do not run full even after the most violent and protracted rains, and yet from a single "system" of twelve acres, the discharge after a recent rain was at the rate of 3,000 gallons per hour. This error of using too large tile Mr. Johnston fell into, and now that he has learned better after a twenty years' experience, he cautions his brother farmers against using larger than two inch tile for laterals. For mains each farmer must provide as the quantity of water to be conducted is greater or less. In many cases Mr. Johnston has used two rows of four inch, in others six inch, and in one, semicircles of eleven inches, one as top and one as bottom, making a pipe nine inches bore to discharge water. At first he had many to take up and replace with large pipe to secure a complete discharge. Main drains he makes six to eight inches deeper than those emptying into them —not with an abrupt shoulder, but leveled up, so that the descent may take place gradually in the length of two tiles—29 inches—and always giving the laterals a slight sidewise direction at the end, so that their water will be discharged down stream into the mains.

Another error he at first fell into was, in having too many drains on lowlands, and not enough on the uplands; thus seeking to carry off the effect, while the cause—the outcropping springs on the hillside—remained untouched. Where the source of the water is most abundant, the means for removing it should most abundantly be furnished. Rain water falls on hills, sinks to an impervious stratum, along which it runs until it either finds a porous section through which it can fall to a lower level, or not finding such, continues on the hard bottom of the side of the hill, where it crops out in the form of a spring. If this spring water is suffered to run down hill, it washes the hillside more or less, and coming to the lowland, sinks as far as it may into the soil, makes it sodden, and produces bad effects. To drain effectually, then, we must cut off the supply above, and fewer drains will be necessary below. Here is the whole secret of the thing, and here we see why so much money is spent to so little purpose by those who think that they should only drain wet lowland. Appearances are deceitful, and we should not suppose that a seemingly dry upland is really dry.

Tile works have been established at many places in New York state, in several places in Massachusetts, in twelve or fifteen counties in Ohio. Some five or six different tile machines are in active operation at Cleveland, and are unable to supply the demand; in fact so far as demand is concerned, the same may be said of every place at which tile are made in Ohio. Michigan, Indiana, Maryland and several other states have tile works.

Considerable draining has been done in the north-west part of Ohio, in that region more familiarly known as the *Black Swamp*—a peculiar formation extending over several counties—by means of open ditches. Brush, wood and stone drains are not unknown in Ohio; and within a few years past upward of four hundred miles of underdraining have been done in Union, Clark, Madison, Fayette, Highland and Clinton counties, by means of the so-called *mole plow*—a detailed description of this machine will be found in an appropriate portion of this work.

PART I.

THEORY OF DRAINAGE.

INTRODUCTION

The chief object of drainage is to liberate the superfluous moisture in springy land, or such lands as have an impervious strata near the surface of the soil—the carrying away of the water which accumulates on the surface, from rains, snows, or freshets, is a secondary object only of thorough drainage. Where there are springs, there is a continued tendency of the water to force through the superincumbent strata, so as to rise and spread over the surface—such land must, even in times of drought, contain more than a proper amount of moisture.

Where there is an impervious subsoil, it is there where subterranean waters accumulate and remain a given period, and then, perhaps, disappear. As a general thing, this ground water sinks the deepest, late in the summer; in autumn it begins to rise, and in winter and spring it attains its maximum hight. Now, when the winter and spring waters, from rains and snows, from the surface, find their ways down to the waters retained, and resting on the subsoil, then the entire soil becomes too thoroughly saturated with moisture to admit of tillage operations. Winter grains will not succeed at all in such a soil, and summer crops are at best very precarious. The roots of winter plants, in quest of nourishment, penetrate to the subsoil, and finding a superabundance of water there, become dropsical, and, consequently, perish; but if the roots can possibly find their way into a drier portion of the soil, even by returning toward the surface (they not unfrequently do so), yet even then they become diseased, and the plant becomes unthrifty and yields but a small product; for, according to the natural tendency, every plant pushes its roots downward, and if it does not succeed, it is prevented only by the stony or watery condition of the subsoil. But even when the water is withdrawn from the surface of the field, there is still but little to be hoped for in regard to cultivated plants; for the soil, previously softened, now hardened by the influence of the sun's rays and air, does not permit the requisite circulation of air, and prevents the extension of the roots. A very natural consequence is, that the plants become diseased and yield but little.

The same condition of things exists also with respect to summer crops upon wet grounds. Late sowing, alone, can succeed; the water-hardened soil is very difficult to work, and, therefore, affords a very incompetent nidus or bed for the growth of plants. Consequently, plants succeed badly, under all these circumstances; an entire failure of the crop, indeed, may occur, if a sudden violent rain unites its influence with the rising ground water.

Now a rational agriculture requires that the spring and ground water be removed; for, however necessary moisture in the soil may be for the successful growth of the plants, yet, as we have experienced, an excessive moisture produces the opposite effect. An excess of moisture in the soil, is recognized by certain water plants, such as bent-grass, reeds, shave-grass,moss,ranunculacæ, etc., growing, and gradually crowding out useful plants. The color and condition of the plants themselves, also indicate the superabundance of moisture in the soil. They are generally coarse and reddish, when their plants vegetate in excessive moisture. The standing of rain or snow water upon the surface, is also evidence of a superabundance of moisture, or if, afterward, rents or cracks appear, or a crust of ice form in the furrows at the slightest frost. Finally, the appearance of the soil at certain seasons, shows that it suffers on account of too much water. If, for example, the spring winds have dried the surface of the ground, so that one would think all the moisture gone, and dark spots present themselves upon the surface, this shows that much water stands there.

We will now proceed to give a chapter on soils generally, and their properties, then to state how drainage operates, and also discuss the advantages of underdraining by demonstrating—so far as *theory* (*not hypothesis*), in its proper sense, is susceptible of demonstration—that drainage,

I. Removes stagnant waters from the surface.

II. Removes surplus water from under the surface.

III. Lengthens the seasons.

IV. Deepens the soil.

V. Warms the under soil.

VI. Equalizes the temperature of the soil during the season of growth.

VII. Carries down soluble substances to the roots of plants.

VIII. Prevents "freezing out," or "heaving out."

IX. Prevents injury from drought.

X. Improves the quality and quantity of the crops.

XI. Increases the effect of manures.

XII. Prevents rust in wheat and rot in potatoes.

5

CHAPTER I.

PROPERTIES OF SOILS.

In a work of this character, it may not be necessary to describe the chemical composition of soils, although very proper to state what properties are desirable for remunerative cultivation. It not unfrequently happens, that the properties or qualities of soil are inherent: that is, the cause of productiveness is to be ascribed to the peculiar combination of substances composing the soil, which no chemical analyses have yet been able to discover, and which has not been produced by any artificial combination or process. Scientific investigations of the soil have accomplished little else than a determination of the elementary substances or constituents, as well as some inherent properties, such as color, weight, and facility of combination with other ingredients. A practical examination of the adaptation of soil for cultivation, renders a consideration of some of the other properties necessary.

The physical properties of the soil are of very great importance, so far as the culture of plants is concerned. It may, perhaps, not be asserting too much to say, that the physical properties of the soil exert a more direct influence upon the plant, upon the atmosphere in contact with it, and upon water, than do the chemical combinations of its elements. The degree of fineness of the mineral particles of the soil; its power of cohesion, moisture; its adaptation to the percolation of water, and permeation of atmosphere; its power to absorb moisture by capillary attraction, to absorb gases, to retain heat or warmth, exert, perhaps, a greater influence than is generally believed.

Therefore, it is, why soils frequently are nearly identical in their chemical analyses, yet differ so materially in their productiveness. Underdraining proposes simply to affect the physical condition of the soil without disturbing its chemical composition.

Clay—Pure clay forms a very heavy and compact soil; but if it is burned and then ground, it forms a very porous soil, and is much better adapted to the growth of crops. A soil in which silicious (flinty sand), and calcareous (limey) earths predominate, becomes so hot and parched, that the plants wither and die; on the other hand, if these same substances are finely comminuted or reduced to powder, they form a soil which absorbs entirely too much moisture, and plants suffer in consequence.

One hundred pounds of calcareous earths in an ordinary state, will absorb twenty-nine pounds of water, but when finely comminuted, will absorb eighty-five pounds.[1] Silicious earths, which usually retain no more than twenty-five per cent. of moisture, when properly prepared in a chemical laboratory, may be made to retain two hundred and eighty per cent. of moisture.

The variety of colors in soil, is not very considerable, generally brown or gray, changing into yellow; but sometimes it is found very red or black; sometimes it is strongly inclined to white, blue or green, and sometimes almost endless shades present themselves. The soils all appear much darker in the field than in the laboratory, because in the former place they are always moist, and in the latter, dry. The predominating mineral constituent, generally, imparts the color to the soil—thus, a soil in which *iron* predominates, is of a reddish hue, an alumin-

---

1 Gerardin's Views of Agriculture.

ous one, yellow, a calcareous one, bluish or whitish. When humus (decayed vegetable or organic matter) is mingled with a soil, it assumes a dark brown or blackish appearance, so that, in course of time, the original color of the soil will entirely disappear. Porphyry, mica schist, the clay slates, and the various sandstone formations produce a reddish soil. Basalt produces a brown or black; serpentine, green; phonolite or clinkstone (a feldspathic rock), white; sandstone, plaster and white lime produce a whitish gray soil. Humus (when derived from turf alone) produces at first a grayish brown, but eventually a black soil. Luster occurs in connection with color, only in instances where a moist clay has been overturned by a polished plow or other smooth metallic substance. When such polished surfaces occur on the soil, as it is being plowed, they are unmistakable evidence of comparative nonproductiveness, because they indicate a want of humus and porosity. Soils in which *mica*, or small shining particles, abound, is generally not of good quality.

The color is of great importance in practical agriculture, from the well-known fact, that dark colors always retain the heat from the sun much longer than light colored ones.

Dark soils are generally acknowledged to be more productive than light ones—but this fertility is due to other causes, perhaps, in as great degree, as to the color—they generally contain humus, or at least some organic matter. If, then, we assume the importance of the color of the soil as a fixed fact, and as a condition having an influence on temperature, we then have some data from which the amelioration or improvement in a physical aspect is to be determined.

Experience has taught that coarse dark particles of soil retain warmth longer than fine particles; hence, intelligent gardeners often mix muck, fine coal, bonedust, etc., with some calcareous soil, and distribute it among the soil in the hotbeds, and between the grapevines, when they wish to force fruits and fit them early for market. Sometimes they strew bits of slate around the plants—"this is mainly practiced on the banks of the Moselle, Nahe, Maas and the Rhine."[1] On a dark soil the vine always becomes more juicy, and contains more saccharine matter than on light soils in the same situation in all other respects.

Numerous experiments might be cited to prove that the color of the soil varies the temperature nearly fifty per cent. For example: if a calcareous clay soil is placed in a white flower pot, and exposed to the rays of the sun, it will increase sixteen degrees only in temperature, while the same soil in a *black* pot by the side of it, will have increased twenty-four degrees. Gerardin asserts that the period of ripening potatoes is varied from eight to fourteen days, by the color of the soil. In proof of this, he planted, at the same time, an equal number of varieties in different soils, and found that in white clay sixteen varieties; in yellow clay, nineteen; in whitish sandy soil, twenty, but in dark humus soil, twenty-six varieties, had fully ripened at the same time.

As the density or compactness of soil is differently understood by different parties, we shall endeavor to be as explicit as possible on this point. Generally, by density or heaviness, is understood the amount of pressure which one body exerts on another, or in other words, density and specific gravity are regarded as synonymous. But in agricultural literature, heaviness is rather synonymous with compactness or cohesiveness, than with weight. By a *heavy* soil, is meant one that is difficult to work, on

[1] Yager's Bodenkunde.

account of its adhesiveness; the term is really applicable to clay soils only, because they are capable of retaining a large amount of water, and because clay, of itself, is comparatively *heavy;* but, at the same time, a sandy soil is termed a *light* soil, although its specific gravity is greater than that of the clay. Practically, the specific gravity of a soil is of little importance. During a rise of waters, the sandy particles always settle at the bottom, while the really fertile portions are deposited on the top of the ground as the water recedes.

As a general thing, a soil of great specific gravity is porous, while those really lighter, by weight, when dried, are the heaviest to work. From the foregoing, it will be observed that cohesion is of much more importance than specific gravity. As the soil is composed of many particles of different substances, it will either be tenacious or mellow, compact or loose, in proportion as one or several of the component parts predominate; therefore, as soils are composed of almost all possible proportions of these several elements, soils will be found of all corresponding degrees of tenacity or porosity—hence, having a knowledge of the combining elements to produce a soil, we term a tenacious soil, a *heavy* one, and a mellow or porous one, a *light* one without any regard to its actual specific gravity.

The greatest degree of cohesion is termed *tenacious, strong*, or impervious. Thus, we say a tenacious clay, a strong clay, or an impervious clay. A compact soil is one in which the particles adhere so strongly to each other as to be difficult of separation, and that can not be crumbled by the fingers. A *tough* soil is always a compact one when dry; a tough soil is difficult to till when wet or moist, and no less difficult when dry. A mellow soil is one that will crumble upon slight pressure in the hand. Clay soils may be made mellow by the application of sand, humus, and by frosts in the course of cultivation, but they are never mellow without the aid of man.

The cohesion of the soil depends entirely upon the amount of clay incorporated with it—the larger the proportion of clay is, the more cohesive will the soil be, and the more sand it contains, the mellower it will be. According to Schuebler's experiments, the following named soils exhibited the degree of tenacity or cohesion placed in the corresponding columns:

Description of soil.	Degrees of cohesion.	Cohesion according to weight.
Pure clay, - - - -	100.	24.
Pipe clay, - - - - -	83.3	19.
Brickmakers' clay, - -	68.8	15.3
Common clay, - - -	57.3	13.2
Loamy clay, - - - -	33.	7.5
Slaty marl, - - - -	23.	5.2
Carbonate of magnesia, - -	11.5	2.3
Humus, - - - -	8.7	1.5
Plaster of Paris, - - -	7.3	1.2
Fine calcareous soil, - -	5.	1.
Sand, - - - - -	0.	0.

The more cohesive a soil is, the greater is its liability to be adhesive in a moist state. This adhesion often renders an otherwise productive soil very undesirable, on account of the resistance it offers in tillage. Some German writer, whose name I can not ascertain, instituted a series of experiments to determine the positive as well as comparative amount of "*adhesive resistance*" to implements generally employed in agriculture, the results of which are embodied in the following table:

Moist soils.	Degree of adhesive resistance to agricultural implements exerted by a superficial square foot, on	
	Iron.	Wood.
Pure white clay, - - -	27.	29.2
Pipe clay, - - - -	17.2	18.9
Fine calcareous earth, - -	14.3	15.6
Gypseous earth, - - -	10.7	11.8
Brickmakers' clay, - -	10.6	11.4
Humus, - - - -	8.8	9.4
Common clay, - - -	7.9	8.9
Loamy clay, - - - -	5.8	6.4
Magnesian earth, - -	5.8	7.1
Slaty marl, - - - -	4.9	5.5
Lime sand, - - -	4.1	4.4
Quartz sand, - - - -	3.8	4.3

Many other properties are connected with the cohesiveness of the soil, such as the permeability of water, capillary attraction and retention of moisture, penetrability of the atmosphere, retention of warmth, etc.

A cohesive or compact soil is, in consequence of its tenacity and retention of moisture, always cool or cold, because, in the first place, it is impermeable to the air, and does not absorb and retain the warmth of the sun, but loses its moisture through evaporation only; and it is a well-known fact, that evaporation is a cooling process. On the other hand, a mellow soil is warm, because it does not retain moisture, and is not cooled by evaporation.

A cohesive soil contracts or shrinks when dried. This contraction causes wide and sometimes very deep fissures or cracks, while a mellow soil does not perceptibly either contract or expand, but settles down and becomes more compact. A pure humus soil contracts as much if not more than clay, during a season of drought, but is held together in masses by vegetable fibers, with which it is interspersed; but whenever sand is mixed with humus, it ceases to contract. The best and cheapest method of ameliorating a clay soil is to underdrain, and expose it as thoroughly as possible to the action of frost. For this reason it should be plowed into ridges, even if very cloddy, in the fall, so that the frost may have the largest possible amount to operate on, and by spring it will be found to be much ameliorated.

Retentiveness of Moisture.—The capacity to retain moisture and exclude the permeation of the atmosphere depends entirely upon the cohesiveness of the soil. A cohesive soil is almost impervious, while a mellow soil is always porous. Pure clay will retain water until it is exhausted by evaporation, while pure sand is so porous that it may be said to swallow the water. Neither of these extremes is a desirable quality, but a medium or mean between the two is really what is requisite. It is always better to have a soil too porous than to have one too compact. A well-tilled soil is seldom so compact as to retain moisture in quantities to act injuriously upon vegetation. Every effort, therefore, which will remove surplus moisture, or such moisture as is in actual excess of the absolute amount required for vegetation, is an effort to assist nature, and consequently is in the right direction.

By porosity of the soil, must be understood not merely its adaptation to permit water to filter through it, but also the capacity to draw moisture from the subsoil by or through capillary attraction. It is a well-known fact, that mellow soils are, even in times of drought, more moist than tenacious or compact soils are; they absorb moisture from below, on the same principle that the sponge absorbs and elevates moisture. Even a very sandy soil, resting on an impervious or tenacious subsoil, is better adapted for crops during a drought than a heavy clay soil, solely on account of its capillary capacity.

But an essential quality of a good soil is, that while it

is porous enough to filter the surface water, and possesses a proper capillary capacity, that it at the same time possesses another important quality, namely, the retention of moisture. This latter quality appears to depend upon the decomposition and comminution of the mineral substances, and the decay of organic materials of which the soil is composed. Every kind or quality of soil will absorb or imbibe a certain amount of moisture, until it is completely saturated, and the remainder will drip or flow away. The amount imbibed is generally less than the weight of a given quantity of the soil.

Schuebler, it appears, took a pound of the various kinds of soil, after they were thoroughly dried, then saturated them with water and weighed them; the excess of weight when saturated over the weight when dry, of course, would give the capacity of retaining moisture. Thus, if a pound of soil when dry weighed a pound, but a pound and a half when saturated, it is very evident the absorbing capacity of that soil is 50 per cent. From Schuebler's experiments we have compiled the following table:

Soils.	Absorbing capacity per cent.	Soils.	Absorbing capacity per cent.
Quartz sand, - -	25	Common soil (what kind?)	52
Gypseous soil, - -	27	Pipe clay, - -	61
Lime sand, - -	29	Pure clay, - -	70
Slaty marl, - -	34	Fine calcareous earth,	85
Loamy clay, - -	40	Fullers' earth, - -	87
Calcareous soil, - -	47	Humus, - -	181
Brickmakers' clay, -	50	Fine magnesia, - -	256

This table presents several very striking facts. In the first place, it shows that coarse particles, like sand, retain less moisture than the same material when finely comminuted. For example, the lime, when reduced to particles of the size of common sand, will retain 29 per cent. only of moisture; while a limy soil (clay and lime) will retain 47 per cent., and the limy soil, reduced to powder, will retain 85 per cent. Now, as underdraining and culture reduce the particles of soil, it is very evident that the longer soils are cultivated the greater will be their retentive capacity.

Light clay soils appear the best adapted to retain moisture, while at the same time they appear to have a more desirable kind of porosity than either sand or heavy clay. Humus absorbs the largest proportion of water, but when it parts with it not unfrequently becomes so dry that it floats on the water; it is not an *active* absorber.

Another quality which is possessed by soils, and proper to be mentioned here, is the degree of rapidity with which soils part with the imbibed moisture. It is very evident that the sand will part with its 25 per cent., in the form either of a filtration or an evaporation, much sooner than brickmakers' clay will, with its 50 per cent., or humus its 181 per cent. To determine this point precisely, Schuebler exposed soils containing 100 parts of water to a heat of 66° F., during a period of four hours. He found the water in

	Evaporated. Per cent.		Evaporated. Per cent.
Quartz sand, -	88.4	Common soil (what kind?)	32.
Lime sand, - -	75.9	Pure clay, - -	31.9
Gypseous earth, -	71.7	Fine calcareous soil, -	28.0
Slaty marl, - -	68.8	Garden soil (what kind?)	24.3
Loamy clay, -	52.	Humus, - -	20.5
Brickmakers' clay, -	45.7	Magnesia, - -	10.8
Pipe clay, - -	34.9		

As a laboratory experiment, this table may be very valuable, but in practical agriculture we do not consider it very reliable, or of any absolute value. Every one knows that the exposure toward the north or south, east or west, would materially affect the retentive quality, so far as evaporation by the sun's rays is concerned; and equally as much would they be affected by the winds. A sharp north-west wind might "dry up" a clay soil as much as the sun would dry up a humus soil. Then, too, if furrows are plowed deep and narrow, more surface will be exposed to the action of the elements than if plowed wide and shallow. A sandy soil, covered with a mat of grass, would not evaporate moisture as rapidly as an exposed clay would.

The property of expansion and contraction of soils is intimately connected with the capacity of absorbing and retaining moisture. Some soils, when fully saturated, do not expand a particle, while others expand very much; those which expand the most when saturated, also contract the most when the moisture is exhausted. The comparative expansive or contractive capacity of soils may be very readily determined in the following manner: take a common brickmaker's mold, and fill it with thoroughly saturated soil, as compactly as possible, with the hand, then expose it either for days in unobstructed sunshine, or else expose it to artificial heat, not exceeding 212° Fahrenheit; when the soil is thoroughly dried, it will be found—according to the kind employed—to have shrunk more or less. Schuebler's investigations indicated that

1000 parts of	Will contract. Parts.	1000 parts of	Will contract. Parts.
Lime or quartz sand,	0.	Pipe clay, -	114.
Calcareous soil, -	50.	Carbonate of magnesia.	154.
Loamy clay, -	60.	Pure clay, -	183.
Brickmakers' clay, -	85.	Humus, - -	200.
Slaty marl, -	95.		

These experiments confirm repeated observations, that a soil in which clay predominates always contracts, and becomes full of fissures or cracks, when it is perfectly dry; but that in sandy soil no such change takes place. But every day's experience contradicts the statement relative to humus in the above table; it is a well known fact, that humus or turf never cracks, even in the hottest and driest weather. There is no doubt that cracking is in a very great degree due to the amount of moisture contained, and the rapidity with which it is evaporated. The same soil will contain many more fissues, if dried suddenly, than if dried slowly.

Another property inherent in soils must not be omitted, namely: the capability of absorbing moisture from the atmosphere. This property manifestly is dependent on the porosity of the soil; for it is very evident that a soil which readily absorbs a rain fall, will also absorb moisture when it is presented in the form of fog or dew, or even from the atmosphere direct. We must again refer to Schuebler to ascertain the degree in which this property is possessed by the various soils. He took 1,000 grains of dried soil of each kind, and spread each kind respectively on a surface of 50 inches, and found that

	Absorbed in			
	12 hours. Grains.	24 hours. Grains.	48 hours. Grains.	72 hours. Grains.
Quartz sand, - -	0	0	0	0
Gypseous earth, - -	1	1	1	1
Lime sand, - -	2	3	3	3
Common soil (what kind?)	16	22	23	23
Loamy clay, - -	21	26	28	28
Slaty marl, - -	24	29	32	33
Brickmakers' clay, -	25	30	34	35
Fine calcareous earth, -	26	31	35	35
Pipe clay, - -	30	36	40	41
Garden soil, having 7 per cent. humus, -	35	45	50	52
Pure clay, - -	37	42	43	49
Fine magnesia, -	69	76	80	82
Humus, - - -	80	97	110	120

It will be seen at a glance that the greatest proportion of the moisture is absorbed during the first *twelve* hours. Soils in the fields seldom, if ever, become so thoroughly dried as those employed in Schuebler's experiments; hence the absorption will necessarily be much less than the proportion stated in the table. The experiments simply confirm every day's observations, that the absorbing powers of clay are increased by the addition of sand; but *practice* does not confirm the statement with regard to humus. It is a well known fact, that a piece of humus, so dry that it will float, may lie in a damp cellar, or other moist place, for months, without absorbing a perceptible amount of moisture.

Porosity is, after all, of more importance than the property of absorbing a large quantity of moisture, because in a porous soil moisture can penetrate to a greater extent. Although quartz sand does not absorb any appreciable amount of moisture, it is a well ascertained fact that a moist atmosphere is productive of good results on a sandy soil; plants flourish and grow well, while under the same conditions they very soon die away in a heavy clay. What practical benefit is then to be derived from the great absorbing power of clay, if the moisure is confined to the surface only; while in sand, with no power of absorption, the particles of moisture can permeate everywhere? But it is asserted that the air absorbs more moisture from the soil than it imparts to it; however true this assertion may be, the advantages to growing crops of moisture imparted to the soil from the atmosphere, is acknowledged by every intelligent and observing agriculturist.

The capacity of soils to absorb gaseous elements from the atmosphere, is one of the most important properties. The fertilizing properties of gaseous elements are so well known, and generally acknowledged, as not to require any illustration or argument; the only object really accomplished by plowing is, a loosening of the soil, so as to permit the permeation of the atmosphere, and consequently absorption of gases and moisture from it by the soil. The most important, as well as most universal of these gases is oxygen; it combines chemically with moist (never with *dried*) soil, as well as it combines physically or mechanically with hydrogen to form water. "Sprouting," or germination, would be utterly impossible without oxygen; hence, seeds germinate much more readily in a properly-formed seed-bed—that is, where the soil has been reduced to mellowness and ordinarily well pulverized, than in a soil not so prepared. In proof of this assertion, we need only refer to the fact that, in forests, seeds of indigenous plants frequently become so completely excluded from the action of the atmosphere, that, when again exposed to it, after a lapse of many years, they at once germinate and grow.

Subsoils, or such soils as lie beneath the surface and beyond the influence of the atmosphere, are termed "*dead*" or "*wild*" soils, and are, as a matter of course, unproductive. But as soon as they are exposed to atmospheric influences, and especially the action of the frost, they become very productive. Some soils possess the property of absorbing gases in a much greater degree than others; blue clay, or hard pan, for instance, does not possess this property in any very considerable degree: hence, it must be exposed a very long time to atmospheric influences before it becomes fertile. We must again refer to Schuebler's experiments for the precise degrees in which the different soils possess the property.

	Absorbs. Per cent.		Absorbs. Per cent.
While humus, - -	20	Clay slate, -	11
Rich garden soil, -	18	Brickmakers' clay, -	11
Magnesia, - -	17	Fine calcareous soil,	10
Good arable soil, -	16	Yellow clay, - -	9
Pure clay, - -	15	Lime sand, -	5
Pipe clay, - -	13	Gypseous earth, -	2
Slaty marl, - -	13	Quartz sand, -	1

No less important than any of the qualities already enumerated, is the property of absorbing and retaining warmth. This property depends entirely upon the color, compactness, porosity, moisture, and the exposure to the rays of the sun. We have already referred to the fact that dark soils absorb and retain the sun's rays, while light colored soil reflect without absorbing them. In the course of the succeeding chapters, we shall fully discuss the effects of the absorption of warmth, and its consequences; nothing further need be remarked here, than to refer to Schuebler's experiments for the degrees or proportion in which the various soils possess the property of absorbing and retaining heat.

Soils.	Retentive capacity.	Soils.	Retentive capacity.
Lime sand, - -	100	Arable soil (what kind?)	70
Slaty marl, -	95	Pipe clay, -	68
Quartz sand, - -	95	Pure clay, - -	66
Hard Pan, or "Blue clay," - -	76	Garden soil (what kind?)	64
		Fine calcareous earth,	61
Gypseous earths, -	73	Humus, - -	49
Brickmakers' clay, -	71	Fine magnesian soil, -	38

From this it will be seen that a limy or sandy soil is much warmer soil than the clays: hence, a loamy soil, having a proper admixture of sand, is warmer than a clay soil.

Many persons suppose that the soil differs from the subsoil, in no other respect than that the *soil* has been cultivated, and has in consequence assumed a more porous character. While this in some cases may be correct, it can by no means be adopted as a rule. The subsoil, as a general thing, is a distinct geological formation from the soil itself—the soil may be a sandy loam, while the subsoil is an impervious clay—or the soil may be a loam, while the subsoil is gravel and sand. Where the subsoil is gravelly or sandy, as a general thing, drainage is necessary; yet, there are cases, which we will discuss in the proper place, where gravelly subsoils require drainage as imperatively as clayey ones do. If the subsoil were *always* as porous as the cultivated soil, there would be less occasion for thorough drainage, but as this is not the case, drainage becomes necessary if not indispensable.

The crust of the earth is composed of rocks, or of the material which once was rock, disposed in stata, one above the other, like the concentric *peels* of an onion, but the regularity of stratification has, in many places, been interrupted by earthquakes and volcanic action.[1]

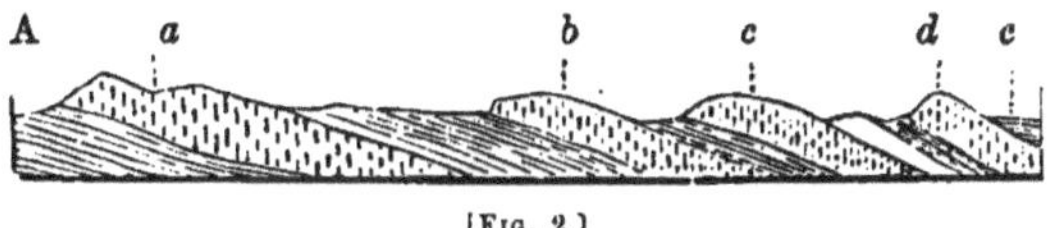

[Fig. 2.]

In passing over a region of country from A to *e*, Fig. 2, we may find at A, a deposit of shale, but it soon dis-

1 Volcanic forces have operated from beneath upon most of the older rocks, whereby they have been bent upward. The weight of the ocean, drift, etc., has bent them downward; gravity and other agencies more local, have produced a lateral pressure, especially when the strata were highly inclined; and these various agencies will account for nearly every case of flexure, not only of the laminæ, but of the beds also.—*Hitchcock's Elementary Geology, page* 18.

appears, and we find we are traveling on limestone as at *a*, then we find the limestone disappearing, and we are on a heavy clay; at *b* we find ourselves on a sandstone formation, then again on a heavy clay; then at *c* we find it gravelly, then shale, perhaps, and again a clay formation at *d*. The soil which may be represented by a line just above the upper edge of these formations, as from A to *b*, is, perhaps, a mixture of all the rocks on which it rests, and demonstrates, very clearly, why the subsoil may be different at different points, under the same kind of soil, as at A and *a*, or *b* and *c*. A farm situated at *c*, would not require any drainage; while one situated between *a* and *b*, would not be of any great value without it.

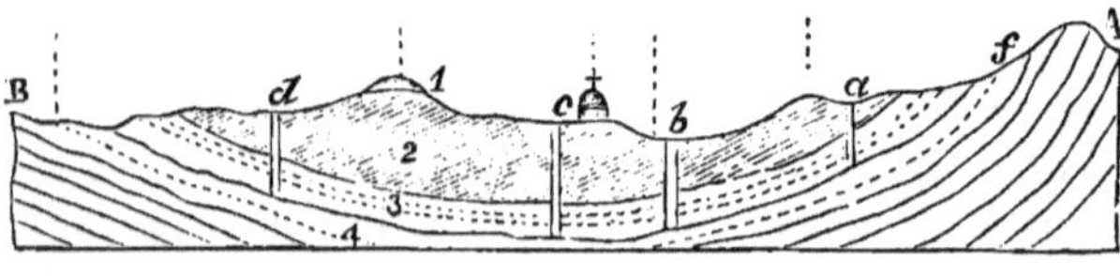

[Fig. 3.]

In the annexed Fig. 3, A and B, represent portions of strata elevated by volcanic action, forming a basin, B 3 *f*, in which the strata, 1 and 2, have subsequently been formed. Now, suppose 1 to represent a deposit of gravel, 2 a deposit or formation of blue or yellow clay, 3 a lime rock, and 4 a sandstone strata. The rains falling at B and at *f*, will readily percolate toward the center of the basin, because the strata is porous, while the rains from *a* to *d* will penetrate the earth very slowly. A field situated at B, although actually lower than one at *d*, may, nevertheless, be much drier, and in a workable condition, while the one situated at *d*, is saturated with moisture. All the rain falling between *a* and *d*, except that which flows from the surface and that evaporated, will penetrate until it reaches the limestone strata, 3, which is impervious, and of course, arrests its further progress—the result is, that at *b* a *swamp* is formed from the excess of water which can find no outlet or means of escape.

This same figure may serve to illustrate the principle of artesian wells. Strata No. 4, being porous, is constantly saturated with water, and is what is termed a *water bearing rock.* Now, if the strata 4 be penetrated at *a* or *b*, the pressure from *f* will cause the water to rise at *a* or *b*, to the same level of *f;* at *c*, the water would rise to the level of the earth only, being in the center of the basin, the water would not rise higher than the outcrop of the strata, as at B; at *d*, it would not rise to the surface, and at 1 it would remain at some distance below the surface.

CHAPTER II.

HOW DRAINAGE OPERATES—HOW IT AFFECTS THE SOIL.

It would be no difficult matter to collect a volume of experiments, made in laboratories and elsewhere, which were made to ascertain the precise workings of drainage. One of the most cheap, simple, and at the same time most satisfactory experiments, to determine the *advantage* of draining, is the following: Take two ordinary earthenware flowerpots, the one having a hole or perforation in the bottom, and the other to be without any orifice in either sides or bottom. Fill both with precisely the same quantity and quality of soil, and plant in each, either growing plants or seeds of any ordinary cultivated plants. The perforated pot will represent a drained soil, while the other represents an undrained one. Give to both the same exposure, and the same quantity of water. If *seeds* are sown in both, those in the perforated pot will germinate the soonest, and the plants become the thriftiest and hardiest; sometimes, though seldom, the plants in the other pot will not germinate at all; but generally, they do germinate, although they produce only sickly and slender plants. In this manner, the *effect* of drainage is completely demonstrated.

If both flowerpots are placed in earthen saucers or dishes, and water poured in the dishes, that pot having the perforation will absorb the water by capillary attraction—the plant will receive its due proportion, and thrive; while the unperforated pot will not absorb any water, and the plant will suffer from *drought;* thus showing the *effect or benefit* of drainage in times of drought. But the best method of demonstrating the manner in which drainage operates, is by the following apparatus:

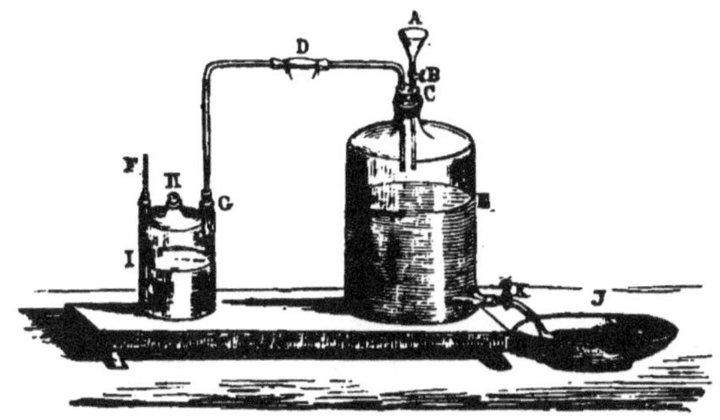

[FIG 4.]

Fill a glass vessel, E, with moistened soil, to the hight of six or more inches—the bottom of the vessel being provided with a stop-cock, K, which should penetrate several inches into the soil, so as to represent a pipe-tile as nearly as possible. The mouth of the vessel, E, should be firmly closed with a cork, C, through which is inserted a tube, whose upper portion is a *funnel*, A, provided with a stop-cock, B. This tube is for the purpose of introducing water on the soil within; and the cock, B, to prevent the introduction of air from that source, after sufficient water has been introduced. The other tube which passes through the cork, C, is luted to another tube at D. This last is inserted at G, into the vessel, I, which is partially filled with water; but the tube, G, should not be inserted so deep as to touch the water. The vessel, I, is provided with three orifices or openings; through one of these orifices a tube is inserted at F, to admit air, in such a manner, however, as to compel it to pass through the water—the air being lighter than the water, will, of course, rise through it in the form of bubbles; or rather, when bubbles are rising through the water, it is an indication that air is entering through the tube, F, from without. The orifice of the stop-cock, K, should be kept under water in the vessel, J. Having completed these arrangements, close the stop-cock, K, and open the one, B, and through the funnel, A, introduce as much water as would probably fall during an ordinary shower. It will be observed, that the water so introduced does not at once disappear, or be absorbed by the soil, but remains on the surface, or penetrates very slowly. This is the condition and action of an undrained soil.

Now, to represent the action of rain on drained soil, open the stop-cock, K, and bubbles of liberated gas will soon be seen to rise in the vessel, J; these bubbles are liberated gases from the soil—those who have analyzed these gases, state that a large amount of oxygen, in combination with gases deleterious to plants, is contained in them; and are, therefore, of opinion, that less oxygen is found in soil immediately after a shower, than before—the oxygen being restored only as soon as evaporation takes place. While the gases are being liberated through K, bubbles will be seen rising in the water in I; thus it is demonstrated that each shower furnishes drained soils with new oxygen. As soon as K is opened, the water which was on the surface of the soil, at E, sinks at once; but as soon as it is being discharged at K, into J, the bubbles cease, or nearly so, to rise in I.

A belief has obtained, that drains are of advantage or beneficial to the soil only, when they are conducting away the surplus waters from showers. It certainly is a great advantage to the plants to be relieved from surplus water, as soon as possible, but it is, at the same time, no less an advantage to be supplied with new oxygen, and to have the old removed. An undrained soil can not make these changes in its gases, for the benefit of the plant, as well as a drained soil. This aeration of the soil is absolutely necessary for the health and growth of plants. Plowing is nothing more or less than aerating the soil; and every one conversant with farming operations, is well aware, that plants grow best on a finely pulverized soil—that is, in other words, on a well aerated soil.

Oxygen is no less essential to the roots of plants, than it is to the lungs of animals; but if the oxygen is not changed, the result is very unfavorable to the plants. Every rain which falls on a porous or drained soil, brings not only new solvents of the inorganic materials which nourish the plants, that have already been oxydized, and thus prepared for the advent of another rain, but when it falls on an undrained or impermeable soil, it diminishes the amount of oxygen, and produces permanent injury to the plants, by the excessive amount of stagnant water, and by lowering the temperature for a longer period than is consistent with the health of the plant.

No fear need be entertained that any clayey or loamy soil can be *over* drained; or, in other words, that so much moisture may be drained out of the soil, as not to leave sufficient remaining for the use of any plants which may appropriately be grown in the soil.

All soils have what is termed "*capillary attraction*," that is, the power to suck up, or elevate to the surface mineral matters in solution, or moisture from the subsoil; and the finer the soil is pulverized, the stronger the capillary attraction. In proof of this position, the following, from the pen of J. H. Salisbury, an agricultural chemist of New York, is here inserted:

"From numerous observations which have been made at different times on the peculiar appearance of the surface of soils, clays, etc., during the warm summer months, and the fact that they, when covered with boards, stones, or other materials, so as to prevent them from supporting vegetation, become, in a comparatively short time, much more productive than the adjacent uncovered soil, we have been led to the belief that the soil possessed some power within itself, aside from the roots of plants, of elevating soluble materials from deep sources to the surface.[1]

"To throw some light upon the subject, in May, 1852, I sunk three boxes into the soil—one 40 inches deep; another 28 inches deep, and a third 16 inches deep. All three of the boxes were 16 inches square. I then placed in the bottom of each box, three pounds of sulphate of magnesia. The soil which was to be placed in the boxes above the sulphate of magnesia, was then thoroughly mixed, so as to be uniform throughout.

"The boxes were then filled with it. This was done on the 25th of May, 1852. After the boxes were filled, a sample of soil was taken from each box, and the percentage of magnesia which it contained accurately determined. On the 28th of June, another sample of surface soil was taken from each box, and the percentage of magnesia carefully obtained as before.

"The result in each case pointed out clearly a marked increase of magnesia. On the 17th of July, a sample of surface soil was taken a third time from each box, and carefully examined for magnesia; its percentage was found to be very perceptibly greater than on the 28th of the preceding month. On the 15th of the months of August and September following, similar examinations severally were made, with the same evident gradual increase of the magnesia in the surface soil.

"The following are the results as obtained:

	Percentage of Magnesia.		
	Box 40 in. deep.	Box 28 in. deep.	Box 16 in. deep.
May 25, - - -	0.18	0.18	0.18
June 28, - - - -	0.25	0.30	0.32
July 17, - - -	0.42	0.46	0.47
August 15, - -	0.47	0.53	0.54
September 15, -	0.51	0.58	0.61

1 Dr. Alex. H. Stephens, of New York, was, I think, the first to suggest this idea. He speaks of it in his address, delivered before the State Agricultural Society of New York, on the *Food of Plants*, in January, 1848. No accurate experiments were performed, however, to fix it with a degree of certainty, till these were made which appear in this paper.

"Before the middle of October, when it was intended to make another observation, the fall rains and frosts had commenced; on this account the observations were discontinued. The elevation of the magnesia, as shown in the above experiments, evidently depends upon a well-known and common property of matter, viz: the attraction of solids for liquids, or what is commonly denominated capillary attraction. This may be clearly illustrated by taking a series of small capillary glass tubes, and insert one extremity of them in a solution of sulphate of magnesia or chloride of ammonium, and break or cut off the upper extremities just below the hight to which the solution rises. Expose them to the sun's rays; the water of the solution evaporates, and the fixed sulphate of magnesia will be deposited just on the upper extremity of the tube. As the solution evaporates more of it rises up from below, keeping the tubes constantly full; yet no sulphate of magnesia passes off; it all, or nearly all, remains at or rises just above the evaporating surface. Just so in the soil; as the water evaporates from the surface, more water, impregnated with the soluble materials from below, rises up to supply its place. As this evaporation goes on, it leaves the fixed materials behind in the surface soil at the several points of evaporation.

"This explains why we often find, during the months of July, August and September, a crust of soluble salts covering the surface of clay deposits which are highly impregnated with the alkalies, or any of the soluble compounds of the metals, earths, or alkaline earths, also, the reason in many instances of the incrustations upon rocks that are porous and contain soluble materials. It also helps to explain the reason why manures, when applied for a short or longer time upon the surface of soil, penetrates to so slight a depth. Every agriculturist is acquainted with the fact that the soil directly under his barn-yard, two feet below the surface (that is, any soil of ordinary fineness), is quite as poor as that covered with boards or otherwise two feet below the surface in his meadow; the former having been for years directly under a manure heap, while the latter, perhaps, has never had barn-yard manure within many rods of it.

"The former has really been sending its soluble materials up to the manure and surface soil; the latter, to the surface soil and the vegetation near or upon it, if uncovered.

"The capillary attraction must vary very much in different soils; that is, some have the power of elevating soluble materials to the surface from much deeper sources than others. The pores or interstices in the soil correspond to capillary tubes; the less the diameter of the pores or tubes, the higher the materials are elevated. Hence one very important consideration to the agriculturist, when he wishes nature to aid him in keeping his soil fertile, is to secure a soil in a fine state of mechanical division, and of a highly retentive nature.

"Nothing is more common than to see soils retain their fertility with the annual addition of much less manure than certain others. In fact, a given quantity of manure on the former will serve to maintain their fertility for several years; while the latter, with a similar addition, quite lose the good effects of the manure in a single season.

"The former soils have invariably the rocks, minerals, etc., which compose them in a fine state of division; while the latter have their particles more or less coarse."

The rich, clay soil contains very many small pores, while the quartz sand, and especially the coarse, sharp sand, has larger spaces, which are not properly capillary pores. Like any other small apertures and spaces, the small pores of the soil are capable of imbibing and retaining water, contrary to the laws of its gravity. On the other hand, in the larger spaces or pores, the water moves entirely according to the laws of gravity. When we place a flowerpot, filled with earth, in a dish filled with water, the small or capillary pores will draw the water up, while the *larger* spaces will be filled with water no higher than they are under the surface of the water in the dish. And if we pour water upon the earth in a flowerpot, we may pour a certain quantity upon it without even a drop coming out of the hole in the bottom of the pot, simply because it is retained through capillarity in the fine pores. But when all the capillary pores are filled with water, then the water poured on will flow down the larger spaces, and, according to the laws of gravity, escape through the hole in the bottom of the pot. As the quantity and extension of the fine pores are very unequal in the different kinds of soil, the quantity of water which they are capable of absorbing and retaining through capillarity (or their "retentive power," as it is called), also varies greatly. The humus, or clay soil, retains most water; the coarse, sandy soil, least.

In order that we may be clearly understood, when speaking of the different kinds of soil, we have concluded to adopt the following classification of soils from the *Mark Lane Express:*

"The best classification of soils is a chemical classification, founded on their composition according to the proportion of sand separable by washing; it divides them into sands, sandy loams, loams, clay loams and clays. It subdivides these again into fine and coarse sands and sandy loams, according to the size of the particles of sand, and into gravelly sands, loams and clays, according to the proportion of pebbles or fragments of rocks. The proportion of calcareous matter indicates whether they are to be called marly or calcareous sands, loams and clays; while, if they contain a certain proportion of vegetable matter, they are called vegetable soils. Each name should express some defined proportion of sand separable by washing, and of calcareous or vegetable matter. In such a classification as we advocate we should have:

1. *Silicious soils*, containing from 90 to 95 per cent. of sand. These would be divided, on the same principle, into blowing sand, coarse sand, good agricultural sand and calcareous sand.

2. *Loamy soils*, 70 to 90 per cent. of sand separable by washing, subdivided into coarse sandy loam, fine sandy loam, loam, rich loam and calcareous loam.

3. *Clayey soils*, with 40 to 70 per cent. of sand; divided into clay loam, clay and calcareous clay.

Each of these soils termed calcareous sand, calcareous loam, etc., contain 5 per cent. of lime.

Marly soils constitute a fourth group, in which the proportion of lime ranges between 5 and 20 per cent., and are divided into sandy marls, loamy marls and clayey marls.

Calcareous soils contain more than 20 per cent. of lime. They are divided into sandy calcareous, loamy calcareous and clayey calcareous. While in calcareous sands, clays and loams, the proportion of lime does not exceed 5 per cent. The difference of composition denoted by difference of name is similar to the sulphates and sulphites of chemical nomenclature, which contain different proportions of sulphuric acid.

"According to the quantity of pebble fragments yielded by a square yard, or by a cubic foot of the soil, they might be denominated *gravels*, or *gravelly* sand, loams and clays.

"*Vegetable* soils vary from the common garden mold, which contains from 5 to 10 per cent. of vegetable matter to the peaty soil, in which the organic matter is about 60 to 70 per cent. They will be vegetable sands, loams, clays, marls, etc."

Now, a sandy or silicious soil will absorb 20 per cent. or one fifth of its own bulk of water before it is fully saturated, or the water commences to *drip* from it; a loamy soil will absorb 40 per cent., and a clayey soil will absorb from 70 to 80 per cent. The coarser non-capillary pores of the soil can not be filled with water, unless there are impediments prohibiting the water from following its gravity; thus, in the flowerpot, only when the hole below is closed; in the arable soil, only when it is resting on or inclosed by an impervious stratum; but in a properly-drained soil the water descends as regularly as in the flowerpot. Whenever the water can flow unimpeded, the larger pores are filled with air; and, as this is necessary in an arable soil, because every fertile arable soil must contain a certain quantity of air, and, to a certain extent, be in communication with the atmosphere, therefore, it follows that, in any fertile soil, the sum of the capillary pores must be in a certain proportion to the non-capillary ones, as not to exceed a certain limit, without the soil thereby assuming unfertile properties. If there is a lack of coarser pores, as, for instance, in rich clay soil; or, if the soil lacks air and communication with the atmosphere, then there will appear all the unfavorable properties characteristic of rich clay soil: wetness, coldness, a retarded decomposition of manure (inactivity), a propensity to forming acids, etc. On the contrary, if there is a lack of capillary pores in the soil, and a preponderance of the larger ones, as in a sandy soil, the soil has too little retentive power, *i. e.*, capacity of retaining water, and evaporates the little water it imbibes too soon; consequently, it is affected with drought; beside, it suffers the manuring elements to attain a state to decompose too rapidly, and allows the soluble nutriments of the plants to sink too readily with the atmospherical water into the subsoil, and the volatile nutriments to ascend, with the evaporating water, into the atmosphere. What proportion of the capillary pores to the non-capillary ones may be the most favorable in any soil, can not now be defined, but experiments for that purpose would undoubtedly result in many interesting discoveries.

Now, if we consider the distribution of atmospherical water in the soil, we might, perhaps, be led to the supposition that the uppermost strata of the soil, through their retentive power, must retain the water falling down upon them, and give nothing to those strata lying below them; thus, that the uppermost strata must be perfectly saturated with water in the capillary way, while those lying below them, being distinctly separate, must be and remain dry.

If, *e. g.*, the retentive power of a soil is equal to fifty, (or if 100 parts of the soil are capable of retaining 50 parts of water in a capillary way), and upon this soil falls a rain in such a quantity as to give one pound of water upon every square foot of the surface, then the uppermost stratum of the soil, of about one fourth of an inch in thickness (supposed to be perfectly dried), would completely retain the rain water in the capillary way, and the soil lying below it would receive none of it. If air is supposed to exist below the upper stratum of one fourth of an inch in thickness, then, of course, it would be as above stated, and no water would permeate; but if we have a subsoil, the attraction of this earth changes the condition of things; the upper soil does not remain saturated but imparts to the lower one. To what degree, and to what depth? This depends upon the quality and the kind of earth. On this subject one may readily learn much by experiments. If we, for instance, put earth in a flowerpot, and pour so much water upon it as is sufficient to saturate, in the capillary way, the uppermost layer or stratum of the soil, one inch in thickness, the examination of the quantity of water in the earth, at the different hights of the pot, will then show that the uppermost layer is fullest of water, but not saturated in the capillary way; the water has penetrated to a certain depth, and that the quantity of water is steadily decreasing from the top down to this depth. The way of diffusing the water depends on the chemical composition of the soil, as well as on its physical properties. The fine quartz-sand, for instance, when in a wet condition, parts with the water pretty quickly; but when perfectly dry, it possesses, like humus, especially when not completely decomposed, a repulsive property to water, so that the water has to act upon it a long time in order to produce saturation. If, therefore, after a drought of long duration, an extra quantity of rain falls upon a loamy soil and upon a fine sandy soil, after a time the water will be found to have penetrated pretty deep into the former, while in the latter only the uppermost layers are wet, but those lying below them remain in a state of dusty dryness. Thus, the loam soil, in spite of its far greater retentive power, diffuses the rain water more perfectly and deeper; but the fine sand, with its much inferior capillary power, retains it in the thin uppermost stratum. These relations become still stronger when vegetable remains, but little decomposed, are mixed with the sand. Also, the more or less pulverized state of the soil has an influence upon the capillary diffusion of water, but especially its being equally or unequally pulverized, so that usually the distribution is more perfect throughout strata which are equal in this respect, than throughout those which are unequal.

The distribution of the water, which is drawn up from below through capillary attraction, is as unequal as the different kinds of earth. If one puts flowerpots, filled with sand, loam and humus soil, in dishes of water, the absorption of the same will take place in a very different way; and the length of time within which the absorbed water will appear at the surface will vary very much, also.

CHAPTER III.

DRAINAGE REMOVES STAGNANT WATERS. FROM THE SURFACE.

FROM the preceding chapter it will be seen that a drained soil is necessarily more *porous* than an undrained one; consequently, when a rain falls, the water which does not immediately flow off from the surface, escapes through the pores. On an undrained soil the water becomes stagnant, because the pores are already filled with water which has no means of escape other than by evaporation. A hard impervious subsoil prevents it filtering through it, and sinking down where the roots will be uninjured by it. Furnish under currents for the water, by means of drains, and there is no longer a necessity for the water to remain above ground, until it becomes changed from a healthful to a poisonous substance, by the continued action of heat and atmospheric air upon it.

The amount of water which may be evaporated from the surface, under the various influences which cause and control this evaporation, as well as the quantity which passes downward by means of filtration through the subsoil, or into the drains, is a matter of the greatest importance to every person engaged in the cultivation of the soil.

Chemists assert that fully four times the amount of heat is required to convert water into vapor, that is required to bring it to the boiling from the freezing point. It is no uncommon occurrence that rain to the depth of *one inch* falls in the course of a shower. The amount falling on a single acre then would amount to 360 hogsheads, and to evaporate this amount of water by sunshine, would require an amount of heat that would convert upward of 1,500 hogsheads of water from the freezing to the boiling point. Every one must know that this evaporation is a very slow process, and that while it is going on the soil is kept wet, and consequently cold; that vegetation is retarded, if not absolutely checked, especially in the early spring time Now, if these 1,500 hogsheads of water were carried off by drains, this great amount of heat necessary to evaporate would be saved, and would be applied to warming the soil.

Some interesting facts, in relation to this subject, are furnished by Cuthbert W. Johnson, in a late number of the *Farmer's Magazine.* Observations were made for eight successive years, in Hertfordshire, and the mean amount of rain which fell, was found to be for each year 26½ inches, of which over 11 inches passed into the soil and was filtrated, and over 15 inches were evaporated from the surface. During the colder months, the amount filtrated was from three to six times as great as the quantity which passed off in the form of vapor. On the other hand, the quantity evaporated during the hottest months, was more than fifty times as great as the amount filtrated, the latter indeed, not amounting during a whole month to the twentieth of an inch.

The greatest quantity evaporated, in a single year, was about 1,800 tuns per acre, and the greatest quantity filtrated was over 1,400.

The rate of evaporation is influenced by the amount of moisture required by the different soils for saturation, and the degree of exposure to sun and winds. Even the direction of the prevailing winds, characterized by the moisture they contain, has a material influence. Several examples are given, by which it appears that the average amount of rain at the places of observation, was about 25 inches per year; that the evaporation from *water* exposed to both sun and wind, was about 35 inches per year; shaded from the sun, but exposed to the wind, it was about 23 inches; from *soil*, when drained, about 20 inches; and from undrained soil, saturated with water, about 33 inches, an excess of 13 inches of water to be charged against an undrained soil.

These experiments were made with bare earth, free from herbage of any kind. By means of other experiments made with plants in pots, it was found that 22 square inches of surface of bare mold, evaporates in twelve days, 1,600 grains of moisture, while a pot of the same size, containing a polyanthus, evaporated 5,250 grains; showing conclusively the great rapidity with which plants carry off moisture, and the great error of those who suppose that weeds can be of any use in shading the soil.

Many persons presume that a comparatively small amount only of the water which falls in rain, on the surface of the earth, is retained by the soil, or is evaporated, but are of opinion that nearly all finds its way into rivers or smaller streams.

Some writers assert that almost the entire mass of water, from rains, is absorbed in supplying springs, and other subterranean streams. Marriotte, a celebrated French writer, has examined the point, with direct reference to whether the quantity of rain water is sufficient to feed all the springs and rivers, and so far from finding a deficiency, he concludes upon the amount being so great as to render it difficult to conceive how it is expended. According to observations which have been made, there falls annually upon the surface of the earth, about 19 inches of water; but to render his calculation still more convincing, Marriotte supposes only 15, which makes 45 cubic feet per square toise, and 238,050,000 cubic feet per square league of 2,300 toises, in each direction. Now, the rivers and springs which feed the Seine, before it arrives at the Pont-Royal at Paris, embrace an extent of territory about sixty leagues in length, and fifty in breadth, making 3,000 leagues of superficial area; by which, if 238,050,000, be multiplied, he have for the product 714,150,000,000 for the cubic feet of water which fall, at the lowest estimate, on the above extent of territory. Let us now examine the quantity of water annually furnished by the Seine. The river above the Pont-Royal, when at its mean hight, is 400 feet broad and five deep; when the river is in this state, the velocity of the water is estimated at 100 feet per minute, taking a mean between the velocity at the surface and that at the bottom. If the product of 400 feet in breath by five in depth, or 2,000 feet square, be multiplied by 100 feet, we shall have 200,000 cubic feet for the quantity of water which passes, in a minute, through that section of the Seine above the Pont-Royal. The quantity in an hour will be 12,000,000; in a day 288,000,000; and in a year 105,120,000,000 cubic feet. This is not the *seventh* part of the water which, as previously stated, falls on the extent of country that supplies the Seine; the large remainder, not received by the river, being taken up by evaporation, beside a prodigious quantity employed for the nutrition of plants.[1]

Now, if this astounding calculation is true of France, what must be the condition of Ohio, and many other states where the annual rainfall is about 40 inches, or nearly three times the amount assumed by Marriotte. Think for a moment of the entire surface of Ohio, being annually covered more than a yard deep, with rain water! The autumn rains average about 10 inches, and generally thoroughly saturate the earth with water, so that when

[1] Gallery of Nature, page 263.

the winter precipitations take place they can not infiltrate, or penetrate the soil—neither does evaporation take place during this period of the year; so that when spring returns the task upon the heat from the sun is not only to evaporate so much of the 10 inches of spring rain as has not flowed off by surface drainage, but the 8 inches of the winter precipitations, and much of the autumn rains—is it any wonder that the soil is not in a workable condition much before the middle of May? Think of the spring sun being obliged to evaporate about 3,000 hogsheads of water from every acre of arable soil!

The following table shows the amount of rain and (melted) snow which falls at fifteen points, in different portions of the State of Ohio:

NOTE—The rainfall is stated in inches and hundredths in the columns of the respective months—thus, at Marietta the rainfall for the month of July, is 4.56 inches, or a little more than four and a half inches; for the year at the same place it is nearly 43 inches.

	Latitude.	Longitude.	January.	February.	March.	April.	May.	June.	July.	August.	September.	October.	November.	December.	The Year.	Spring.	Summer.	Autumn.	Winter.		Hight above Atlantic.
																					feet.
Oberlin, Lorain Co.	41° 20′	82° 15′	1.41	1.84	1.83	3.48	3.41	4.45	3.21	3.36	3.11	2.49	3.23	1.63	33.48	8.72	11.05	8.83	4.88	Prof. Fairchilds, mean of 8 yrs.	800
Perrysburg, Wood Co.	41° 35′	83° 30′	2.00	1.37	2.50	4.68	3.00	8.20	9.81	2.44	7.50	2.25	7.00	4.50	55.25	10.18	20.45	16.75	7.87	Mr. Hollenbach, for yr. 1855 only.	610
Cleveland, Cuyahoga Co.	41° 31′	81° 46′	1.71	2.33	1.20	1.92	3.08	5.28	2.33	11.05	2.67	2.84	3.91	3.55	34.33	6.20	10.37	9.42	7.58	Mean of 4 yrs.	650
Urbana, Champaign Co.			2.30	2.85	3.15	3.45	4.92	4.99	3.50	3.00	4.04	2.07	4.31	3.01	44.14	11.52	12.24	11.32	9.06	M. G. Williams, mean of 10 yrs.	
Steubenville, Jefferson Co.	40° 25′	80° 40′	2.71	2.53	3.33	3.10	3.95	4.03	3.77	3.94	3.02	2.96	3.13	3.30	39.76	10.37	11.74	9.11	8.54	Roswell Marsh, mean of 26 yrs.	670
Germantown, Montgomery Co.	39° 30′	84° 10′	2.45	3.81	3.28	3.44	4.03	3.12	3.90	3.00	2.44	2.28	3.01	3.19	38.88	10.76	10.02	8.63	9.46	L. Groneweg, mean of 5 yrs.	720
Cincinnati, Hamilton Co.	39° 06′	84° 30′	3.35	3.51	3.93	3.60	4.55	5.01	4.37	4.32	3.10	3.32	3.48	4.29	46.89	12.14	13.70	9.00	11.15	Ray, mean of 20 years.	543
Portsmouth, Scioto Co.	38° 45′	82° 56′	3.00	2.90	2.90	3.20	3.90	4.50	4.40	2.70	2.40	2.90	2.80	2.60	38.20	10.00	11.60	8.10	8.50	Hempstead, mean of 15 yrs.	540
Marietta, Washington Co.	39° 25′	81° 31′	2.82	3.04	2.92	3.28	4.13	4.04	4.56	3.89	3.18	3.12	3.27	3.80	42.97	10.33	13.40	9.57	9.66	S. P. Hildreth, mean of 30 yrs.	630
Columbus, Franklin Co.															48.16	10.58	16.58	11.85	9.14	Prof. Mather's MS.	
Granville, Licking Co.			2.55	3.88	3.22	3.65	3.60	5.60	4.70	3.70	2.80	3.10	4.07	4.81	45.83	10.47	14.09	10.03	11.24	S. N. Sanford, mean of 8 yrs.	
Massillon, Stark Co.			1.05	1.78	1.53	1.08	3.08	5.15	3.59	2.84	2.35	2.11	3.87	2.08	33.51	7.19	11.58	8.33	6.41	J. Watson, mean of 3 yrs.	
Springfield, Clark Co.			3.58	3.46	1.45	2.43	4.45	3.05	4.20	3.37	2.34	3.31	5.98	3.02	41.54	8.33	10.62	11.03	10.96	J. T. Warder, for 1857.	
Hudson, Summit Co.	41° 15′	81° 25′	2.62	2.42	2.92	3.06	2.81	4.48	4.34	4.37	3.14	3.20	2.41	2.64	38.45	8.80	13.09	8.76	7.70	Mooney & Emerson, mean 13 yrs.	1131
Lebanon, Warren Co.			4.15	2.08	5.87	4.64	2.67	3.07	7.25	3.42	2.58	3.10	2.70	4.70	47.25	13.18	13.74	8.44	11.89	For the years 1849 and 1850.	

How much of the amount of rain which falls can be carried off by the drains is an all important question; and upon the answer to this question depends, in a very great degree, the benefit, or disadvantage of underdrainage.

The following table, copied from observations at Tharand, in Saxony, will serve to show the influence of rains on the discharge of water from drains:

	Temperature of air.		Quantity of Rain per day, per acre, in gallons.	Discharge of Drain Water, per day, per acre, in gallons.	Increase compared with the preceding day.
	Minimum.	Maximum.			
1853.					
May 13	37.4	54.5	16091	1198	401
" 14	39.2	59.	——	2568	1370
June 15	50.	71.6	12105	101	——
" 16	53.	69.8	29967	1793	1692
" 17	55.4	66.2	13212	5222	3429
" 18	54.5	68.	29	5267	45
" 23	48.2	64.4	5167	2228	1405
" 24	49.	62.6	17429	10852	8624
1854.					
May 14	45.5	71.	——	38	——
" 15	51.	65.5	27754	103	65
" 16	47.5	59.	23324	9132	9029
" 29	49.	60.	27089	15919	13851
June 29	56.5	71.6	49528	15291	12280
" 30	50.	73.4	43844	15203	——
July 1	46.4	59.	8562	15708	505

In the *Journal of the Royal Agricultural Society*, vol. 5, page 151, Josiah Parkes, a celebrated land drainer in England, publishes a table, embracing observations during a space of eight years, in which he finds the amount of water *filtrated*, that is, passed into the earth and absorbed by drains, roots of plants, and retained in the soil, to vary from 36 to 57 per cent. Annexed is the table prepared by Mr. Parkes:

Years.	Rain. Inches.	Filtration. Per cent.	Evaporation. Per cent.	Rain, per acre. Tuns.
1836,	31.	56.9	43.1	3139
1837,	21.10	32.9	67.1	2137
1838,	23.13	37.0	63.0	2342
1839,	31.28	47.6	52.4	3168
1840,	21.44	38.2	61.8	2171
1841,	32.10	44.2	55.8	3251
1842,	26.43	44.4	55.6	2676
1843,	26.47	36.0	64.0	2680
Average,	26.61	42.4	57.6	2695

In vol. 20, page 292, of the same journal, Mr. J. Bailey Denton, an agricultural engineer, and who is, perhaps, oftener quoted than any other agricultural engineer at the present day, shows the discharge from drains from the 1st of October 1856, to the 31st of May 1857, to have varied from less than one fourth, to more than one half of the entire rainfall during that period.

"The following observations on evaporation and filtration,[1] for which we are indebted to the patient and carefully conducted experiments of Mr. Charles Charnock, of Holmfield House, near Ferry-Bridge (one of the vice-presidents of the Meteorological Society of London), present some valuable facts for consideration. (pp. 80–1.)

"In these experiments, the evaporation from *saturated* soil was determined thus: 'A leaden vessel of 13 inches deep, and a foot square, was filled to within an inch of the top with soil, and placed in the ground, in the same manner as the previous vessel, with a pipe level with the surface of the soil to carry off the excess of top-water into a receiver. The same quantity of water was then daily supplied to this soil as the evaporating dish of column 2 showed was evaporated. The soil was stirred as in the former case, and thus represented wet and undrained land.'

[1] Quoted by J. H. Charnock, Esq., Assistant Commissioner under the Drainage Acts, in a paper "On Suiting the Depth of Drainage to the Circumstances of the Soil," given in the *Journal of the Royal Agricultural Society*, vol. x, pt. ii, pp. 515 to 518.

An Account of Observations made, through a series of five years, at Holmfield House, near Ferrybridge, in the county of York, by Charles Charnock, Esq., with a view to determine the amount of Evaporation and Filtration under the several circumstances, on the Magnesian Limestone Soil.

Explanation.—Column 1, shows the Depth of Rain fallen, as registered by the ordinary Rain Gauge.

Column 2, is the Amount of Evaporation from a Surface of Water fully exposed to both Sun and Wind.

Column 3, is the Evaporation from Water shaded from the Sun, but exposed to the Wind.

Column 4, is the Evaporation from what represented drained or dry land.

Column 5, is the Evaporation from the same when saturated.

Column 6, is the Amount of Water which filtered through the soil.

Into a leaden vessel, of a foot square and three feet deep, was put two feet of gravel and calcareous sand, so as to represent the substratum of the farm, and the remainder filled up, to within an inch of the top, with an average quality of soil. At the bottom a pipe was inserted, which conveyed all the water which was filtered through into a bottle, which was regularly emptied and registered. The vessel was inserted in the ground to within an inch of the surface, keeping the level of the soil, inside and outside, alike, with an inch of the vessel above, to prevent any communication of water from without. The soil was kept free from weeds, and occasionally stirred, that it might not be more than ordinarily compact.

Months.	1842.						1843.					
	Rain.	Evaporation.				Filtration.	Rain.	Evaporation.				Filtration.
	1	2	3	4	5	6	1	2	3	4	5	6
	On the surface.	From Water		From Soil.		Through the Soil from the Drain 3 ft. deep.	On the surface.	From Water.		From Soil.		Through the Soil from the Drain 3 ft. deep.
		Exposed to both Sun and Wind.	Shaded from sun, but exposed to Wind.	When drained.	When Saturated.			Exposed to both Sun and Wind.	Shaded from Sun, but exposed to Wind.	When Drained.	When Saturated.	
January,	2.70	1.69	1.13	1.66	1.59	1.04	1.48	2.57	1.71	0.69	1.87	0.79
February,	0.76	1.23	0.81	0.68	1.04	0.08	3.25	2.65	1.10	2.29	2.78	0.96
March,	3.48	1.92	1.28	2.40	2.52	1.08	0.95	3.05	2.03	0.72	2.43	0.23
April,	1.51	2.98	1.99	1.11	2.31	0.40	2.19	3.22	2.05	1.84	2.39	0.25
May,	2.98	4.14	2.76	2.89	3.82	0.09	2.81	2.91	1.94	2.47	2.65	0.34
June,	1.94	4.18	2.79	1.94	4.27	—	2.31	5.12	3.41	2.10	4.86	0.21
July,	3.74	4.16	2.73	3.26	3.89	0.48	2.70	3.76	2.50	2.55	3.49	0.15
August,	1.49	3.36	2.24	1.37	2.39	0.12	3.99	3.71	2.57	3.77	6.74	0.22
September,	2.44	2.39	1.74	2.24	2.88	0.20	1.07	2.06	1.34	0.90	2.18	0.17
October,	1.12	2.00	1.37	0.92	1.99	0.20	1.10	1.80	1.20	0.82	1.93	0.28
November,	3.19	2.26	1.50	2.49	1.83	0.70	2.30	1.64	1.09	1.69	1.48	0.67
December,	0.76	3.00	2.14	0.60	1.49	0.16	0.28	1.68	1.78	0.27	1.39	0.01
Totals,	26.11	33.61	22.48	21.56	30.02	4.55	24.49	34.17	22.72	20.11	31.19	4.28

ACCOUNT OF OBSERVATIONS—*Continued*.

MONTHS.	1844.						1845.						1846.					
	Rain.	Evaporation.				Filtration.	Rain.	Evaporation.				Filtration.	Rain.	Evaporation.				Filtration.
	1	2	3	4	5	6	1	2	3	4	5	6	1	2	3	4	5	6
	On the surface.	From Water		rom Soil.		Through the Soil from the Drain 3 ft. deep.	On the surface.	From Water.		From Soil.		Through the Soil from the Drain 3 ft. deep.	On the surface.	From Water.		From Soil.		Through the Soil from the Drain 3 ft. deep.
		Exposed to both Sun and Wind.	Shaded from Sun, but exposed to Wind.	When drained.	When Saturated.			Exposed to both Sun and Wind.	Shaded from Sun, but exposed to Wind.	When Drained.	When Saturated.			Exposed to both Sun and Wind.	Shaded from sun, but exposed to Wind.	When Drained.	When Saturated.	
January,	1.31	1.01	1.08	0.85	1.50	0.46	1.74	1.53	1.02	1.28	1.40	0.46	2.18	2.07	1.58	1.22	1.94	0.96
February,	2.22	1.31	0.88	1.66	1.11	0.56	0.73	0.71	0.47	0.43	0.94	0.30	0.47	2.59	1.09	0.44	2.09	0.03
March,	2.27	2.13	1.42	1.58	1.50	0.09	1.88	2.01	1.94	1.33	2.89	0.55	0.93	2.56	1.55	0.86	2.16	0.07
April,	0.27	3.83	2.55	0.27	3.42	—	1.54	4.79	3.19	1.05	4.06	0.39	5.97	1.91	1.27	2.08	1.49	2.99
May,	0.42	5.77	3.85	0.42	4.86	—	2.24	2.84	1.09	1.97	3.26	0.27	0.82	4.52	3.02	0.73	3.89	0.09
June,	1.24	5.31	3.58	1.20	4.95	0.04	3.18	3.10	2.06	2.93	2.98	0.25	1.65	4.88	3.25	1.58	4.73	0.07
July,	2.76	4.17	2.78	2.43	4.28	0.33	3.49	2.86	1.89	8.30	2.79	0.19	2.90	4.44	2.96	2.74	4.39	0.16
August,	2.85	4.70	3.14	2.44	4.63	0.41	4.61	2.56	1.70	4.24	2.43	0.37	2.65	3.08	2.05	2.46	3.28	0.19
September,	1.92	4.91	3.28	1.02	3.96	0.30	1.36	2.79	1.86	0.95	2.29	0.41	1.07	2.99	1.99	1.00	3.14	0.07
October,	1.41	2.79	1.86	1.17	2.93	0.24	3.36	2.77	1.84	2.59	2.82	0.67	4.09	2.23	1.49	2.40	2.76	1.69
November,	1.98	2.84	1.80	1.47	2.78	0.51	1.01	2.13	1.41	0.73	2.20	0.28	1.15	1.63	1.09	0.88	1.74	0.17
December,	0.35	0.79	0.53	0.29	0.93	0.06	3.04	3.57	2.38	2.36	2.87	0.68	1.36	1.79	1.10	1.09	1.67	0.27
Totals,	19.00	40.10	26.75	15.40	37.85	3.60	28.18	32.50	21.75	23.26	31.09	4.92	25.24	34.69	23.04	18.38	33.28	0.70

"In the first place, it is observable how much greater is the amount of evaporation from water than from land, and how near, as shown by columns 2 and 5, the evaporation from wet land is to that from water itself: hence, the wetter the land the greater the evaporation, and, as the well-known consequence, the greater its excess of coldness. We have a familiar illustration of nature's process in this particular, in the method often adopted to cool our wine on a hot summer's day, by wrapping a wet napkin round the bottle, and exposing it to the full sun: as the moisture from the napkin is evaporated, the temperature of the wine declines to almost freezing point. The school boy's experiment of producing ice before a fire, by incasing the vessel in wet flannel, and adding a portion of salt to the water, is a similar example, with this additional lesson to the farmer, that to apply certain limes to wet land is only increasing the evil.

"You will then, in the second place, notice how much less the evaporation is in the shade than in the sun, and consequently that wet land must be the warmest when there is the least sun. From which cause, no doubt, arises that too vigorous growth of young wheat, so often observable on such land in the winter and spring months, which never fails to produce serious injury to the crop in all its subsequent stages. And, thirdly, you will remark how comparatively small a proportion of the rain which falls is shown to be carried off by filtration. Taking the average of the five years' experiments, it will be seen that only 4·82 inches out of 24·6 inches of rain passed through the land to the depth of three feet. We might, therefore, be led at the first glance to infer that land, in general, stands less in need of drainage, or may be drained by a less perfect system, than is supposed to be requisite, did not daily experience oppose such a conclusion. We must, therefore, endeavor to reconcile this seeming incongruity, and deduce at the same time, from the facts disclosed, such data, as may guide us in determining the essential requisites to ensure completeness of effect in drainage.

"Now, although there can be no reason to question the accuracy of the experiments on filtration made by Mr. Dickinson, and recorded in the *Journal of the Royal Agricultural Society*, of England, vol. v, part 1, yet there is very considerable difference in the aggregate result, as shown by them and the account before us. 'The first important fact disclosed,' says the commentator, page 148, 'is, that of the whole annual rain, about 42½ per cent., or 11 3–10 inches out of 26 6–10, have filtered through the soil:' whereas, in the Holmfield House experiments there is only shown, as we have already said, 4·82 inches out of 24·6, or about 5 1-10 per cent against 42½ per cent. This is certainly a very great and somewhat irreconcilable difference in the result of two experiments made professedly to ascertain the same fact. Now, on referring to the '*Memoirs of the Literary and Philosophical Society of Manchester*,' vol. v, part 2, you will find a paper on rain, evaporation, etc., from the pen of the celebrated Dr. John Dalton (the father of the science of meteorology), wherein he explains a series of experiments made by himself and his friend, Mr. Thomas Hoyle, jun., to ascertain the amount of evaporation and filtration, and giving the following table of results:

Months.	Water through the two Pipes.			Mean.	Mean Rain.	Mean Evaporation.
	1796.	1797.	1798.			
January, -	1.897	.680	1.744	1.450	2.458	1.008
February, - -	1.778	.918	1.122	1.273	1.801	.528
March, - -	.431	.070	.335	.279	.902	.623
April, - -	.220	.295	.180	.232	1.717	1.485
May, - -	2.027	2.443	.010	1.493	4.177	2.684
June, - -	.171	.726	——	.299	2.483	2.184
July, - -	.153	.025	——	.059	4.154	4.095
August, - -	——	——	.504	.168	3.554	3.386
September, -	——	.976	——	.325	3.279	2.954
October, - -	——	.680	——	.227	2.899	2.672
November, -	——	1.044	1.594	.879	2.934	2.055
December, -	.200	3.077	1.878	1.718	3.202	1.484
	6.877	10.934	7.379	8.402	33.560	25.158
Rain, - -	30.629	38.791	31.259			
Evaporation,	23.725	27.857	23.862			

"'Having got a cylindrical vessel of tinned iron,' says the doctor, 'ten inches in diameter, and three feet deep, there were inserted into it two pipes, turned downward, for the water to run off into bottles: the one pipe was near the bottom of the vessel, the other was an inch from the top. The vessel was filled up, for a few inches, with gravel and sand, and all the rest with good fresh soil. Things being thus circumstanced, a regular register has been kept of the quantity of rain water that ran off from the surface of the earth through the upper pipe (while that took place), and also of the quantity of that which sank down through the three feet of earth, and ran out through the lower pipe. A rain-gauge of the same diameter was kept close by, to find the quantity of rain for any corresponding time.'

"You will notice that the general result of these experiments accords, pretty nearly, with that of the Holmfield account; and yet it may be readily conceived that circumstances of situation and stratification may often occasion as wide a difference in the amount of filtration as is shown between Mr. Dickinson's and Mr. Charnock's observations.

"On an examination of the *details* registered in the account before us, it will be evident that the amount of filtration is not exclusively dependent on the fall of rain; but that a variety of other causes combine to affect its proportion. For instance, in March, April, May, June, and July, of 1842, the fall of rain was 13·65 inches, and the filtration for the same period was only 2·05 inches; while in April, 1846, there was 5·97 of rain, and 2·09 of filtration. Similar instances are also noticeable in Mr. Dickinson's details. From March to October, inclusive, of 1840, a fall of 11·52 inches of rain is recorded, without any filtration; but in November 1842, the rain was 5·77, with 5 inches of filtration. Dr. Dalton's table also shows the same variations. The lesson, therefore, derivable from these experiments, so far as regards filtration by drains, is one rather of a speculative than of a definite character: for, although we are assured filtration must be secured, we are left with a large and varying margin as to the proportion. We must not, however, overlook the fact, that all the registered details show occasionally an amount of filtration nearly equal to the rain that falls, and, therefore, in determining the size of pipe to be used, the ready exit of this *maximum* quantity must be provided for."

Perhaps, the most accurate observations to determine the amount of rain carried off by drains, were made in Prussia, at Tharand, by Dr. Hugo Schober, of the Agricultural School and Experimental Farm at that place. We subjoin the following from the "*Jahrbuch der Akadamie zu Tharand fur* 1855."

"These experiments were made on three several tracts; two were upland, and the other was partly a garden, and partly a meadow.

"The first tract was an upland experiment field, and contained about 3½ acres.

"The second tract was an upland experiment field, and contained about 5½ acres.

"The third tract was part garden and part meadow, and contained about 2¼ acres.

"The drain pipe was laid at a depth of four feet in each tract, but in the first the drains were three rods apart, while in the other two they were two rods only. The fall was well adapted to test the workings of the drains; and, therefore, the minor drains were laid with 1½ inch pipe, while the sub-main and other drains had 2½ inch pipe.

"The first tract had 89 rods of minor, and 16 rods of sub-main, making a total of 105 rods per acre. In the 2d and 3d tracts were an average of 145 of minor, and 21 rods of sub-main, making an aggregate of 166 rods per acre of drains.

"The operations of these were in the highest degree satisfactory. These tracts are situated at an elevation of 714 French feet above the Elba, or 1028 above the North Sea. It was hazardous to grow winter crops on these tracts, on account of the excessive moisture they contained—the crops being liable to winter-kill, but since they have been underdrained, are as reliable for winter crops, as any other fields in the kingdom. It was the rule that it was very late in the spring, before they were in a condition to be cultivated; but since they have been underdrained, they have become workable at as early a period in the spring as any other *terrains* in the district. The crops on these tracts are remarkable for their vigor and evenness.

"So far as the annexed tabular statements are concerned, it may be necessary to state that the quantity of water from the main drain of each tract, was daily measured, regularly at 8 o'clock A. M., and 4 o'clock P. M., and the hourly discharge per acre computed from this data. It is true that this method does not give the exact or precise amount, yet sufficiently so for all practical purposes. The rain-gauge was observed at 8 A. M., and 8 P. M.; the snow was melted, and the resultant water measured in the rain-gauge.

AGGREGATE AMOUNT OF RAIN PER ACRE; ALSO, THE AGGREGATE AMOUNT OF WATER DISCHARGED BY THE DRAINS; ALSO THE PER CENT. OF RAIN WATER DISCHARGED BY DRAINS.

	Amount of Rain. Galls.	Discharge by Drains. Galls.	Per cent. of Rain water discharged.
1853.			
February, - -	57381.2	26932.6	46
March, - -	40840.4	71025.5	173
April, - - -	153486.2	124659.7	80
May, - - -	91546.2	53297.	58
June, - - -	173896,	69922.8	40
July, - - -	123949.4	40656.3	32
August, - - -	85667.2	1014.	1
September, - -	136271.8	20588.6	15
October, - - -	73324.9	17073.9	27
November, - -	44712.8	3706.8	8
December, - -	16888.1	1224.7	7
1854.			
January, - -	31142.	10699.9	34
February, - -	78449.8	46381.6	59
March, - -	61160.6	102612.9	167
April, - - -	81828.	55379.4	67
May, - - -	215545.9	91680.9	42
June, - - -	266136.7	74928.4	28
July, - - -	225024.4	188964.7	83
August, - - -	174698.3	19925.1	11
September, - -	30159.4	1536.8	5
October, - - -	48907.8	873.2	1
November, - -	104751.5	967.3	0.9
December, - -	204106.	185729.2	90
1855.			
January, - -	46678.7	61420.9	131
Aggregate from Feb. 1, '53, to Jan. 31, '54,	1029656.6	440802.5	
Feb.1, '54, to Jan.31, '55,	1537694.3	830400.8	

There are three instances only in which the drains discharge more water during the month than the amount of rain which fell during the same period; but the excess of discharge is readily explained; by reference to the table it will be observed that in each instance, during the month previous, a greater amount of rain fell than during the month in which the discharge was excessive. During the year commencing February 1, 1853, and ending January 31, 1854, the proportion of water discharged by the drains was 42 per cent. of the amount of rain; for the year end ing January 31, 1855, the drainage amounted to 55 per cent., or an average for the two years of 48·5 per cent.

AVERAGE DISCHARGE PER HOUR, PER ACRE, OF WATER FROM THE DRAINS.

	First Tract. Galls.	Second Tract. Galls.	Third Tract. Galls.
1853.			
February,	23.4	26.6	51.3
March,	59.4	66.8	161.3
April,	191.4	171.5	156.4
May,	47.1	51.9	115.8
June,	62.7	81.5	147.1
July,	45.8	46.8	71.2
August,	.5	1.4	2.1
September,	22.4	19.6	43.6
October,	19.7	16.7	32.4
November,	—	2.9	12.5
December,	—	.5	4.4
1854.			
January,	12.3	9.7	21.
February,	58.6	47.4	100.9
March,	122.2	105.7	185.8
April,	57.6	89.9	83.4
May,	70.9	83.1	213.
June,	42.9	94.4	174.8
July,	82.1	270.2	409.
August,	2.4	22.3	55.6
September,	0.4	0.8	5.1
October,	—	—	3.5
November,	0.1	0.2	3.6
December,	265.1	199.2	283.9
1855.			
January,	68.0	91.0	88.5
Average from Feb. 1, '53, to Jan. 31, '54,	45.6	46.8	77.5
Average from Feb. 1, '54, to Jan. 31, '55,	64.2	83.8	134.1
Excess in '54,	18.6	37.0	56.6

In the second table we find that the largest quantity discharged in an hour from an acre was 409 gallons; this would amount to about 156 hogsheads in 24 hours; therefore it would require two and about one third days to drain a rainfall of one inch, or of 360 hogsheads per acre. The entire amount of the 10 inches of spring rains in Ohio would then be carried off by the drains in about 24 days, provided none of it escaped by surface drainage or evaporation; but at least one half of the amount precipitated escapes by these means; it is, therefore, very certain that the drains would remove the remainder in less than twelve days.

How long would evaporation require to remove this amount of water?

It is well known that evaporation commences whenever the thermometer is above 32° F. by means of solar heat, but the winds very often evaporate or "dry up" more moisture than the warmest summer day.

The evaporation from a reservoir surface at Baltimore, during the summer months, was assumed by Colonel Abert to be to the quantity of rain as two to one.

Dr. Holyoke assigns the annual quantity evaporated at Salem, Mass., at 56 inches; and Colonel Abert quotes several authorities at Cambridge, Mass., stating the quantity at 56 inches. These facts are given by Mr. Blodget, and also the table below:

QUANTITY OF WATER EVAPORATED, IN INCHES, VERTICAL DEPTH.

	Jan.	Feb.	Mar.	Apr.	May.	June.	J'y.	Aug.	Sept.	Oct.	Nov.	Dec.	Year.
Whitehaven, Eng., mean of 6 yrs.	0.88	1.04	1.77	2.54	4.14	4.54	4.20	3.40	3.12	1.93	1.32	1.09	30.03
Ogdensburg, N. Y., 1 yr.,	1.65	0.83	2.07	1.63	7.10	6.74	7.79	5.41	7.40	3.95	3.66	1.15	49.37
Syracuse, N. Y., 1 yr.,	0.67	1.48	2.24	3.42	7.31	7.60	9.08	6.85	5.33	3.02	1.33	1.86	50.20

The quantity for Whitehaven, England, is reported by J. F. Miller. It was very carefully observed, from 1843 to 1848—the evaporation being from a copper vessel, protected from rain. The district is one of the wettest of England—the mean quantity of rain, for the same time, having been 45·25 inches.[1]

If, then, the atmosphere of Ohio has the evaporating capacity of that of Ogdensburg, N. Y., it would require

[1] French on Farm Drainage.

the entire months of March, April and May to evaporate the amount of spring rains—that is, if none of the precipitation escaped by infiltration or surface drainage; or, in other words, underdrains will accomplish in 24 days the same removal of water for which evaporation requires 92 days.

A few of the more obvious advantages of draining over evaporation may be briefly enumerated thus:

In undrained ground the season of growth is shortened by the time occupied in evaporation, always a long and tedious process. In drained lands, on the contrary, much time is gained, not only by permitting an earlier working, but in the better adaptation of the ground to germination. In undrained ground the water, passing off in the form of vapor, carries with it a certain quantity of the latent heat of the earth, and this heat is in proportion to the amount of vapor formed. Thus, the land is left colder than it was when covered or saturated with water, and by so much germination is retarded. But in land properly drained the water passes off without being converted into vapor. The temperature of the land at the surface remains the same, and the temperature of the subsoil, through which the water passes, becomes as warm as the surface. Thus the depth of heated earth is increased, and the surface is less liable to be affected by change of temperature.

In evaporation, organic and mineral matters, in the form of gases, pass off with the vapor, thus leaving the ground poorer; while in filtration, accomplished by draining, these substances become fixed in the earth for the nourishment of the future plant.

Undrained lands suffer from hot and dry weather. For, though there may be water within a few inches of the surface, the ground becomes so compact and baked that it is not sufficiently porous to draw up moisture. Drained land, on the other hand, is open to the action of the atmosphere to a great extent; it becomes finely comminuted, the hard pans and stiff clays are broken up and rendered sufficiently porous to imbibe water from below; and also, having a greater surface exposed to the air, it receives more moisture in the form of dew.

In winter and spring, wet land heaves up, under the influence of frost and heat, thus exposing such grains as have been planted directly to the weather; for this reason wheat and other grains are liable to winter-kill, and, instead of them, spring up wild grasses and noxious weeds. Draining, to a great extent, prevents this. When land is dry, the variations it experiences under unequal temperatures are very slight, compared with the changes produced by the same variations on wet land. The result of extreme heat and extreme cold is to increase the bulk of water to a considerable extent. This expansion on the surface of the ground is seen in little hillocks, with cracks running in all directions. By the evaporation of moisture from this frozen ground, many particles of earth are left unsupported and fall, thus leaving the tender roots exposed to the weather when protection is most needed.

Messrs. Waege and Von Mollendorf, of Gorlitz, Prussia, have published, in the *Zeitschrift fur Drainirung*, No. 23, 1855, a series of observations on the discharge of drain water from different kinds of soil. They employed the Dalton apparatus for percolation. The experimental boxes were filled to the depth of four feet of soils taken from fields in which drains were placed at the same depth. These boxes contained respectively:

I. Box No. 1, a clay soil, consisting of 88 per cent. of clay and 12 per cent. of sand; box No. 2, loamy soil, 41·7 per cent. of clay, humus, etc., 58·3 per cent. sand; box

No. 3, a loamy sand soil, 19·2 per cent. of clay, humus, etc., 80·8 per cent. sand.

II. The soil in the system A, of the *Moholz estates*, is loamy soil, corresponding to that of box No. 2; in the system B, loamy sand soil, corresponding to that of box No. 3.

III. The plan of *Kuestner*, on the field of Gorlitz. The drains have, with a very cloddy (much cut) ground, a fall of 18-¾ inches to 10 perches. They are 4 feet deep and 4 perches apart. The soil consists alternately of strata of clay, of loam and of gravel, and corresponds on an average with No. 3 of the experimental boxes.

The corresponding observations of the depth of rain were made at Gorlitz.

The monthly average, derived from daily observations, for the meteorological year 1854 (Dec. 1, 1853—Dec. 1, 1854), are given in the first table on pp. 92–3.

Influence of the kind of soil on the quantity of drained water.—In confirmation of former observations the loamy soil drained the largest amount of water of the three different kinds of soil employed in the Gorlitz experimental boxes. The results at Tharand were similar. The mouth of the main drain of the third part of the estate at that place—being loamy soil—discharged, in monthly average, 309·9 cans per acre per hour; while the mouths of the first and second division—in clay soil—yielded only 182·4 and 187·3 cans, respectively. The drain water, according to the measurements in the apparatus at Gorlitz, amounts to 15 per cent. of rain water in the clay soil, and 33·4 per cent. in the loamy soil. Supposing the falls, etc., to be equal, the same capacity of pipes would suffice for about two acres of clay soil that is required for one acre of loamy soil (also loamy sand soil).

YEAR 1854.	Winter.			Spring.		
	Dec.	Jan.	Feb.	Mar.	Ap'l.	May.
Rain fall in Prussian cubic inches on 1 sq. foot Prussian, at Gorlitz, -	61.74	181.15	376.43	180.50	138.02	411.57
Drain water in Prussian cubic inches, on 1 Prussian sq. foot land:						
I. Gorlitz, experimental boxes. 1, clay soil, -	—	—	168.8	115.9	—	3.6
2, loam soil,	—	—	9.5	299.6	30.6	9.6
3, loamy sand soil,	—	—	3.4	295.2	16.7	6.8
II. Moholz, A, loam soil,	35.78	62.42	333.13	319.34	68.49	33.67
B, loamy sand soil,	29.59	51.27	292.29	297.86	76.54	43.69
III. Kuestner's plan: clay, loam, sand,	—	—	—	—	—	—
Average of the six Silesian stations,	13.07	22.74	161.42	265.58	28.47	19.47
Drain water in per cent. of rain:						
I. Gorlitz, experimental boxes. 1, clay soil,	—	—	49.36	64.21	—	0.87
2, loam soil, -	—	—	2.5	165.98	22.17	2.33
3, loamy sand soil,	—	—	0.9	163.55	12.10	1.65
II. Moholz, A, loam soil, -	57.95	34.46	88.5	176.92	49.62	8.18
B, loamy sand soil,	47.93	28.3	77.65	165.02	55.46	10.61
III. Kuestner's plan: clay, loam, sand,	—	—	—	—	—	—
Average of the six Silesian stations,	21.17	12.25	43.78	147.14	27.87	4.73

Influence of the season on the discharge of water.[1]—These observations are in harmony with the former observations at Tharand, and also agree with the *average* of the Silesian stations, inasmuch as the drains in *spring* flow strongest, compared with the quantity of rain water (44·3 per cent. of the rain water). The least discharge took place, on an *average*, in *fall*—thus deviating from former observations at the Gorlitz boxes, and agreeing

[1] The discharge is, according to *John*, changed according to the *time of day*. A comparison of the observations made, three times a day, by Gropp, at Isterbies, 1852, resulted as follows:

	Number of Observations.	Morning. Prussian quarts.	Noon. Prussian quarts.	Evening. Prussian quarts.
February, - -	29	1848	1828	1810
March, - -	31	1163	1160	1149
April, - -	30	826	821	823
May, - - -	31	1205	1206	1193
June, - -	30	592	537	532
July (first half), -	15	13 1-2	12 3-5	12 3-5
In the 5 1-2 months,	166	5647 1-2	5564 3-5	5519 3-5
Average,	- - -	5577 quarts of drain water.		

Summer.			Fall.			Winter.	Spring.	Summer	Fall.	YEAR
June.	July.	Aug.	Sept.	Oct.	Nov.					1854.
657.74	472.53	611.37	116.59	120.68	375.34	619.32	730.09	1741.64	612.61	3703.66
9.6	153.8	80.9	—	—	30.9	168.8	119.5	244.3	30.9	363.5
516.9	188.6	112.3	6.2	2.3	60.2	9.5	339.8	817.8	68.7	1235.8
360.4	266.0	70.6	—	1.3	26.1	3.4	318.7	697.0	27.4	1046.5
140.52	311.27	143.61	72.86	29.54	136.20	431.33	431.50	595.40	238.60	1686.83
134.22	337.85	216.30	107.26	42.95	189.95	373.15	418.09	688.37	340.16	1819.77
9.39	37.94	2.26	0.09	—	0.69	—	—	49.59	0.78	—
195.17	215.91	104.33	31.07	12.68	74.01	164.36	323.52	515.41	117.76	1270.5
1.46	32.55	13.23	—	—	8.23	30.16	16.37	14.0	5.0	15.0
78.59	39.91	18.37	5.33	1.91	16.04	1.53	46.55	47.0	11.2	33.4
54.79	56.29	11.55	—	1.08	6.95	0.55	43.66	40.0	4.4	28.3
21.36	65.87	23.49	62.49	24.48	36.29	69.65	58.42	34.2	38.9	45.6
20.41	71.50	35.38	92.	35.59	50.61	60.28	57.26	39.5	55.5	49.2
1.43	8.03	0.37	0.08	—	0.18	—	—	2.8	0.1	—
29.67	45.69	17.06	26.65	10.51	19.71	32.43	44.32	29.6	19.2	34.3

with the observations at Tharand. Quite considerable deviations, however, occur sometimes, which can be explained only by continued and increased observations.

Time in which drain water arrives at the pipes in various kinds of soil.—The following table, from the Gorlitz boxes, is confirmed by numerous experiments:

	No. of rain-days.	The Drains discharged.				The Drains were humid.			
		No. 1.	No. 2.	No. 3.	Average No. of days.	No. 1.	No. 2	No. 3.	Average No. of days.
December,	—	—	—	—	—	—	—	—	—
January,	—	—	—	—	—	—	—	—	—
February,	22	2	4	4	3	3	5	3	4
March,	17	6	11	11	9	2	2	3	2
April,	7	—	3	2	2	1	3	2	2
May,	13	2	3	4	3	9	6	8	8
June,	20	5	26	22	18	22	3	7	11
July,	9	15	16	16	16	9	5	8	7
August,	21	10	12	8	10	4	—	2	2
September,	9	—	3	—	1	4	2	2	—
October,	14	—	1	—	—	—	1	—	—
November,	21	9	16	9	11	7	5	8	7
In 1854,	153	49	95	76	73	61	32	41	45
May to Nov. 1853,	104	84	106	118	103	53	27	10	30

There were discharged, during 166 observations in the morning, 127 9-10 quarts more than in the evening; these observations show a discharge for the 166 days greater, by 184,176 quarts, or 6,821 cubic feet, than the evening observations.

This disproportion, according to *John*, must be explained partly by the smaller amount of evaporation during the night, and partly by the fact that the rainfall during the day (which, perhaps, exerts a greater influence on tho discharge of water from the drain pipes tho next morning than the *immediately* preceding night rain), seems to exceed the rainfall during the night (at Crefeld the ratio of night rain to the day rain was as 159·28 to 181·6 in the years 1850–'54). The influence of the time of day is to be taken in consideration for the obvervations of the quantities of drain water; if an observation for three times a day can not be made, the noon time should be preferred, as its results come nearest to the average, according to the table.

In the clay soil No. 1, as compared with the sand soil No. 2, and the loamy sand soil No. 3, the drain was flowing the least number of days, but kept humid the longest.

Condition of moisture of the soil.—With regard to the question, whether draining might not dry too much; experiments were made again in the boxes, in a depth of two feet, at a time when the drains had just ceased to carry away water. The contents of moisture amounted to:

	In clay soil. Per cent.	In loam soil. Per cent.	In loamy sand soil. Per cent.	In the average of the three kinds of soil.
On May 6, 1854, -	18.6	20.	20.9	19.8
Sept. 5, 1854, -	18.5	19.	19.5	19.
May and Aug., 1853,	20.5	19.3	15.6	18.5
Oct. and Nov., 1853,	20.5	18.5	14.	17.6

Permanent moisture has, therefore, in heavy soil, diminished since 1853, perhaps owing to an increase of drying crevices, and it has considerably increased in lighter soil, perhaps owing to its having become more compact.

The above-mentioned *V. Mollendorf* has, beside, published a *summary comparison of the quantities of rain and drain water according to German and English observations*, with the remark that the German measurements (which are not specified), that served for computation, had been made at Tharand (Saxony), Gorlitz, Moholz, Grosskrauscha, Deutsch-Paulsdorf, Ullersdorf (all of them in Silesia), and at Suisheim (Baden).—(*Wilda, Centralblatt*, 1856, I, No. 14.) *See table at top of pp.* 96–7.

These figures, on the whole, confirm the conclusions drawn from the investigations made in common with *Waege;* the difference between the discharge of drain water of loam soil and loamy sand soil, and that of clay soil, is found to be less considerable.

The atmospheric precipitations of a large aggregate of ponds, ditches, and other works, on a surface of 1.43 geographical square miles, near Leipsic, are collected and employed as spring water in the mining districts. The water is measured every week by the rotations of the water wheel. If we compare the discharge of water computed therefrom, with the rains from 1830–'51, the result shows that there has been an annual average (during these 22 years) of rain equal to 24·55 Prussian inches.

In spring,	61·8	per cent.	(64·73.)
" summer,	31·1	"	(36·82.)
" fall,	39·5	"	(27·84.)
" winter,	76·7	"	(37·4.)
In the year,	47·7	per cent.	(41·64.)

Observation.—The figures in parentheses—average per cent. of the rain water discharge by drains according to German observation.

	Spring.				Sum-	
	Mar.	Ap'l.	May.	Total.	June.	July.
A. CLAY SOIL. German observations.						
Rainfall, - - Prussian inches,	1.31	2.20	3.22	6.73	4.48	3.97
Discharged drain water, " "	1.87	1.41	0.97	4.25	0.99	1.68
Drain water, per cent. of rainfall,	142.7	62.3	30.2	63.2	22.1	42.3
B. LOAM SOIL. *a.* English observations.						
Rainfall, - - Prussian inches,	1.57	1.41	1.81	4.79	2.15	2.22
Drain water, - - " "	1.05	0.29	0.11	1.45	0.04	0.04
Drain water, per cent. of rainfall,	66.6	21.	5.9	30.3	1.8	1.9
b. German observations.						
Rainfall, - - Prussian inches,	1.35	1.81	3.05	6.21	4.25	3.73
Drain water, - - " "	2.34	1.3	1.61	5.25	1.77	2.27
Drain water, per cent. of rainfall,	173.3	71.8	52.8	84.5	41.6	63.5
C. LOAMY SAND SOIL. German observations.						
Rainfall, - - Prussian inches,	1.35	1.16	2.68	5.19	3.7	3.41
Drain water, - - " "	1.23	0.51	0.51	2.25	1.38	2.07
Drain water, per cent. of rainfall,	91.1	44.	19.	43.4	37.3	60.7
D. LIMY SOIL. English observations.						
Rainfall, - - Prussian inches,	—	—	—	—	—	—
Drain water, - " "	—	—	—	—	—	—
Drain water, per cent. of rainfall,	—	—	—	—	—	—
Average from German observations:						
Rainfall, - - Prussian inches,	1.34	1.72	2.98	6.04	4.14	3.7
Drain water, - - " "	1.81	1.07	1.03	3.91	1.71	2.01
Drain water, per cent. of rainfall,	135.	62.2	34.57	64.73	41.3	59.9

The excess of drain water over rain water prevailing in the German observations, in the first spring month, or March, is probably caused by the circumstances that this month must remove the meteorical precipitations of the winter months, collected in the form of ice and snow; a circumstance which does not occur in England with its milder winter. The difference between the discharge of the drains in England and Germany during the summer is to be accounted for by the prevalence of the summer rains in the German climate; the fact that autumn furnishes more rain in England than in Germany, is in consequence of the prevalence of the fall rains in the English climate, and of the cloudy quality of its fall atmosphere, which retards evaporation.

…mer.		Fall.				Winter.				Total Year.
Aug.	Total.	Sept.	Oct.	Nov.	Total.	Dec.	Jan.	Feb.	Total.	
3.52	11.97	2.01	1.54	1.58	5.13	2.48	1.21	1.95	5.64	29.47
0.67	3.34	0.37	0.63	0.38	1.38	1.24	0.24	1.02	2.5	11.47
19.	27.9	18.4	40.9	24.	26.9	50.	19.8	52.3	44.3	38.9
2.36	6.73	2.56	2.74	3.73	9.03	1.68	1.79	1.92	5.39	25.94
0.03	0.11	0.36	1.36	3.16	4.88	1.76	1.26	1.5	4.52	10.96
1.5	1.7	14.	49.6	84.9	54.1	104.6	70.4	78.5	83.9	42.3
3.75	11.73	2.	1.33	1.81	5.14	3.06	1.21	1.98	6.25	29.33
1.04	5.18	0.52	0.53	0.39	1.44	2.	0.43	0.9	3.34	15.21
27.7	44.2	26.	39.9	21.5	28.	65.7	35.5	45.4	53.4	51.9
3.83	10.94	2.12	1.31	1.59	5.02	2.42	1.12	1.98	5.52	26.67
0.86	4.31	0.63	0.65	0.16	1.44	0.55	0.	0.43	0.98	8.98
22.5	39.4	29.7	49.6	10.1	28.7	22.7	0.	21.7	17.7	33.7
—	—	—	—	—	—	—	—	—	—	23.88
—	—	—	—	—	—	—	—	—	—	4.68
—	—	—	—	—	—	—	—	—	—	19.6
3.7	11.54	2.04	1.4	1.66	5.1	2.65	1.18	1.97	5.8	28.48
0.85	4.26	0.51	0.6	0.31	1.42	1.27	0.22	0.78	2.27	11.86
23.	36.82	25.	42.57	18.7	27.84	47.54	18.64	39.6	37.4	41.64

CHAPTER IV.

DRAINAGE REMOVES SURPLUS WATER FROM UNDER THE SURFACE.

THERE is a body of stagnant water below the surface of the ground, as those who work clayey soils will not have failed to observe. This water sometimes settles in the bottom of the furrows, even when the surface of the land was sufficiently dry to work. In porous soils this body lies at a great depth, but in clayey soils usually within a foot or two of the surface, and is known among drainers as the "water line." This body of water not only saturates the soil, and consequently excludes the *air*, whose presence, on a previous page, we have shown to be very necessary, but it keeps the soil cold and retards vegetation. In the words of Dr. Hobbs, quoted by Judge French, in his admirable treatise on "Farm Drainage," "A knowledge of the depth to which this water table should be removed, and of the means of removing it, constitutes the science of draining."

In the annexed engraving (Fig. 5), suppose the right-hand portion, 1, to represent a loamy soil, and the left-hand portion, 2, a heavy clayey soil. If in the clayey soil a drain be sunk to 7,5, the water table will ordinarily assume the direction 3,8,7, when the drain commences to act, leaving that portion of the soil indicated by the lighter shade, 2,3,8,7, in good workable condition, and ready to supply nutriment to the roots of plants; while in a loamy soil the direction will be 5.4.f. In other words, the table at 3 will not sink to a level with 7 as rapidly as from 1 to 5. Hence, in clayey soils the drains should either be deeper or closer together, to effect the same object as in loamy soils.

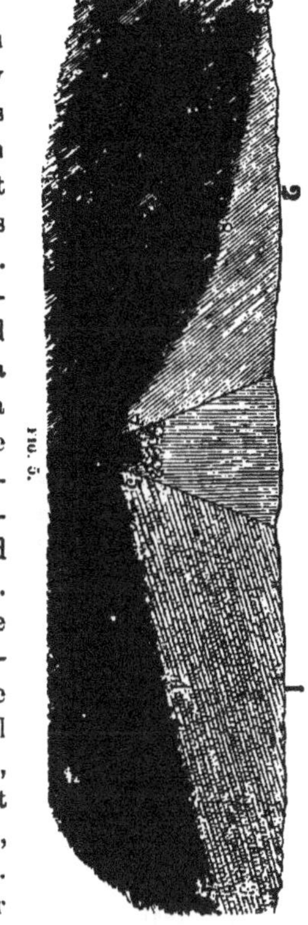

This ground water arises from several causes, viz: There may be an impermeable or impervious strata at a comparative short depth from the surface, which will not permit the waters from the rains to pass through it, as at 5, Fig. 6. Suppose the contour of the surface of a field to be represented by A, B, C, D, Fig. 6, and 5 is a stratum of impervious clay; 4, a stratum of "*hard pan*," or blue clay; and 3 a stratum of compact white clay, resting on a stratum, 2, of sand or gravel; and this last on an impervious bed, 1. It is very evident that all the rains falling from A to C will collect from the surface at B, while that which penetrates the soil will flow along the top of the stratum, 5, until it reaches the lowest point under B; consequently, at B, there will be a swamp or morass. Even should the surface water

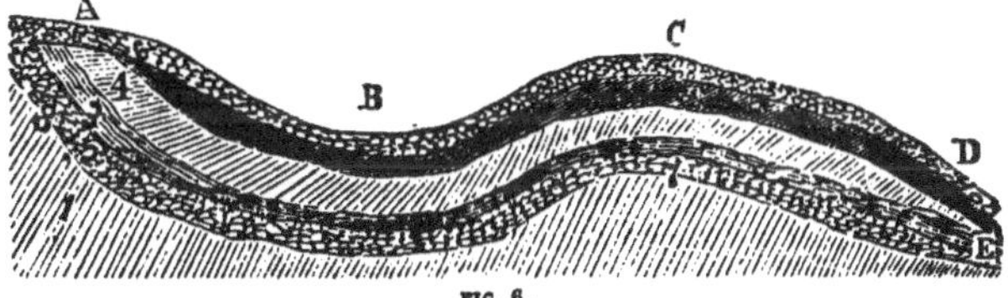

FIG. 6.

from the rain be evaporated, the swamp would still be supplied with water inherent in the strata from A to C. A well sunk at C into the gravel at 7 would be well supplied with water, because 2, being a water bearing stratum, would receive its supply from below A, and 5, being an impervious stratum, would not permit water to escape at B; but a well sunk at B, into the stratum 2, would seriously affect the well at C, and perhaps "dry it up;" at E there would be springs, swamp or morass, according to the nature of the ground. In this case, the ground water, or *water of pressure*, as Judge French terms it. is that which is found in strata 4 and 5.

There may be a nice distinction in law, and, perhaps, in *very* scientific treatises on drainage, between ground water and springs; yet, for all practical purposes, so far as drainage is concerned, they amount to about the same thing—for the reason that the great object of drainage is to cut off this supply of superabundant subterranean water—derived originally, no doubt, from the same source. And the only difference which really can exist is this, viz: a spring is a body of subterranean water, *collected* in a reservoir, and flowing from thence in a larger or smaller stream; while the ground water is a body of subterranean water *diffused* throughout the strata, and, when collected and discharged by drains, is as much spring water as if nature herself discharged it in the form of a regular spring.

The strata beneath the soil are not always conformable to the surface, as those in Fig. 6, but frequently lie nearly level with the horizon, or making but a small angle with it, while the surface itself may be full of undulations. In illustration of the formation of springs and the action of rain on such a district or region of country as represented by Fig. 7, we copy from *Morton's Cyclopedia of Agriculture:*

"When rain falls on a tract of country, part of it flows over the surface, and makes its escape by the numerous natural and artificial courses which may exist, while another portion is absorbed by the soil and the porous strata which lie under it.

"Let the following diagram represent such a tract of country, and

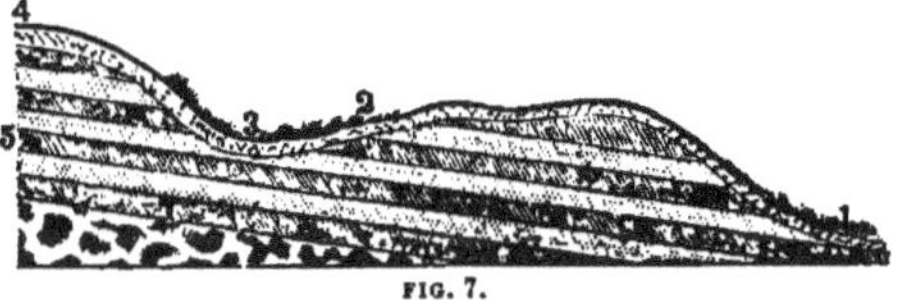

FIG. 7.

let the dark portions represent clay or other impervious strata, while the lighter portions represent layers of gravel, sand or chalk, permitting a free passage to water.

"When rain falls in such a district, after sinking through the surface layer (represented in the diagram by a narrow band), it reaches the stratified layers beneath. Through these it still further sinks, if they are porous, until it reaches some impervious stratum, which arrests its directly downward course, and compels it to find its way along its upper surface. Thus, the rain which falls on the space represented between 2 and 4, is compelled, by the impervious strata, to flow toward 3. Here it is at once absorbed, but is again immediately arrested by the impervious layer 5; it is, therefore, compelled to pass through the porous stratum 3, along the surface of 5 to 1, where it pours forth in a fountain, or forms a morass or swamp, proportionate in size or extent to the tract of country between 2 and 4, or the quantity of rain which falls upon it. In such a case as is here represented it will be obvious that the spring may often be at a great distance from the district from which it derives its supplies; and this accounts for the fact, that drainage works on a large scale sometimes materially lessen the supply of water at places remote from the scene of operations.

"In the instance given above, the water forming the spring is represented as gaining access to the porous stratum, at a point where it crops out from beneath an impervious one, and as passing along to its point of discharge at a considerable depth, and under several layers of various characters. Sometimes, in an undulating country, large tracts may rest immediately upon some highly porous stratum, rendering the necessity for draining less apparent; while the adjoining parts of the country may be full of springs and marshes, arising partly from the rain itself, which falls in these latter districts, being unable to find a way of escape, and partly from the natural drainage of the more porous soils adjoining being discharged upon it.

"Again: the higher parts of hilly ground are sometimes composed of very porous and absorbent strata (1, 2, 3, Fig. 8), while the lower portions, 4, 5, are more impervious, the soil and subsoil being of a very stiff and retentive description. In this case, the water collected by the porous layers is prevented from finding a ready exit, when it reaches the impervious layers, by the stiff surface soil. The water is by this means dammed up, in some measure, and acquires a considerable degree of pressure, and, forcing itself to the day at various places, it forms those extensive "weeping" banks, as at 6, 6, 7, which have such an injurious effect upon many of our mountain pastures. This was the form of spring or swamp, to the removal of which Elkington principally turned his attention; and the following diagram, taken from a description of his system of draining, will explain the stratification and springs referred to more clearly:

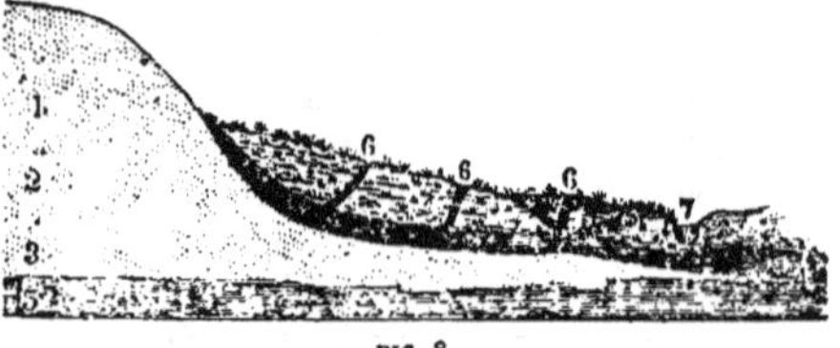

FIG. 8.

Fig. 8, although copied from *Morton's Cyclopedia of Agriculture*, was, in all probability, an ideal section, yet it is a correct representation of a portion of the country between Canton and Massillon, in Stark county. About three miles west of Canton the railroad passes through what is familiarly known in the locality as Buck's Hill, composed of drift (sand and gravel), as represented at 1, 2, 3; this drift rests upon another drift formation, "hard pan," or blue clay, 5. Wells in the immediate vicinity of the hill have been sunk 66 feet, in pure sand and gravel, before reaching the stratum, 5. Originally (in 1800, up to 1818), most of the land lying along the line of railroad, from the bed of the Nimishillen, 7, for a mile or more toward the hill, was a perfect morass, as represented in the figure; but occasional open drains and wells have now rendered it good arable land, although, in the immediate vicinity of the creek, evidences of the former marshy condition yet remain.

The precise *quantity* of water *required* for the agricultural purposes of any district depends upon the nature of the soil and the crops, and the position of the district in relation to the surrounding country. Thus, if a permeable soil occupy an elevated site, the water deposited upon it will pass rapidly, and, perhaps, before serving for the germination or nutriment of the plant. If, on the other hand, as is the far more common case in this country, the soil be of a retentive character, and the site low in relation to other districts, the water will be kept while the soil becomes saturated to so great an extent, that the processes of vegetable germination and growth are greatly impeded. The soil exists in one of three conditions: 1. In the form of clay, being a dense mass of finely comminuted particles, but all of a highly tenacious kind; in a state of slight moisture, it becomes a clammy paste, and is never found so utterly devoid of moisture that its constituent particles are separable; it affords no passages for water, receiving it with difficulty, and retaining it in the same way. 2. In the form of sand or gravel, the particles of which are seldom or never united, and the soil is, therefore, full of passages or canals for water. Soil of this kind has no power either to oppose the admission or effect the retention of water poured upon it. And, 3. Existing in the form of a mixture of the aluminous, silicious, and calcareous elements, in endless variety of proportions, found as *clods*, and in this state affording two classes of passages for the ingress and permeation of water, viz.: those remaining between the particles which are congelated in each clod, and those formed by the spaces between the clods. The former are sometimes called *pores*, the latter *canals*. The power of admitting and retaining or discharging water, exerted by these mixed soils, will exist in an endless variety of degrees, according to the mechanical formation of the constituent particles and clods. The state of soil which is most favorable for the germination and development of the plant, is that of *moistures*, capable of being readily crumbled by the hand, and equally removed from the adhesive extreme of *mud*, and the volatile one of *dust*. In this condition it will be found that the *pores* are filled with water, but the *canals* are not—these latter serving as passages for the air, which is one of the feeders of vegetable life; and we can, therefore, readily understand that, when water exists in such quantity that the soil is saturated, and all the pores or canals filled, its condition is unhealthy for the growth and development of plants.

The following extract from an admirable lecture on agricultural science, by Dr. Madden, quoted by the General Board of Health in their "*Minutes of Information*," although of considerable length, claims a space here, for the valuable information it conveys on the fitness of soil for promoting vegetable germination.

"The first thing which occurs after the sowing of the seed is, of course, *germination*; and before we examine how this process may be influenced by the condition of the soil, we must necessarily obtain some correct idea of the process itself. The most careful examination has proved that the process of germination consists essentially of various chemical changes, which require, for their development, the presence of air, moisture, and a certain degree of warmth. Now it is obviously unnecessary for our present purpose, that we should have the least idea of the nature of these processes; all we require to do, is to ascertain the conditions under which they take place; having detected these, we know at once what is required to make a seed grow. These, we have seen, are air, moisture, and a certain degree of warmth; and it consequently results, that whenever a seed is placed in these circumstances, germination will take place. Viewing matters in this light, it appears that soil does not act *chemically* in the process of germination; that its sole action is confined to its being the vehicle by means of which a supply of air and moisture, and warmth, can be continually kept up. With this simple statement in view, we are quite prepared to consider the various conditions of soil, for the purpose of determining how far these will influence the future prospects of the crop, and we shall accordingly at once proceed to examine, carefully, into the *mechanical relations of soil.*

"Soil, examined mechanically, is found to consist entirely of particles of all shapes and sizes, from stones and pebbles, down to the finest powder; and on account of their extreme irregularity of shape, they can not be so close to one another, as to prevent there being passages between them, owing to which circumstance soil in the mass is always more or less *porous.* If, however, we proceed to examine one of the smallest particles of which soil is made up, we shall find that even this is not always solid, but is much more frequently *porous,* like soil in the mass. A considerable portion of this finely divided part of soil, the *impalpable matter* as it is generally called, is found, by the aid of the microscope, to consist of *broken down vegetable tissue,* so that when a small portion of the finest dust from a garden or field, is placed under the microscope, we have exhibited to us particles of every variety of shape and structure, of which a certain part is evidently of vegetable origin.

"On examining a *perfectly dry* soil, we perceive that there are two distinct classes of pores: 1. The large ones, which exist between the particles of soil; and, 2. The very minute ones, which occur in the particles themselves; and, whereas, all the larger pores, those between the particles of soil, communicate most freely with each other, so that they form canals, the small pores, however freely they may communicate with one another in the interior of the particle in which they occur, have no direct connection with the pores of the surrounding particles. Let us now, therefore, trace the effect of this arrangement. If the soil is *perfectly dry,* the canals communicating freely at the surface with the surrounding atmosphere, the whole of these canals and pores will, of course, be filled with air If, in this condition, a seed be placed in the soil, you at once per ceive that it is freely supplied with air, *but there is no moisture;* therefore, when soil is *perfectly dry*, a seed can not grow.

"Let us turn our attention now to that state of the soil in which water has taken the place of air, or, in other words, the soil is *very wet.* If we observe our seed now, we find it abundantly supplied with water, but *no air.* Here again, therefore, germination can not take place. It may be well to state here, that this can never occur exactly in nature, because water has the power of dissolving air to a certain extent; the seed is, in fact, supplied with a certain amount of this necessary substance; and, owing to this, germination does take place, although by no means under such advantageous circumstances as it would, were the soil in a better condition.

"We pass on to a different state of matters. Let us suppose the canals are open, and freely supplied with air, while the pores are filled with water. While the seed now has quite enough of air from the canals, it can never be without moisture, as every particle of soil which touches it is well supplied with this necessary ingredient. This, then, is the proper condition of the plant for germination, and, in fact, for every period of the plant's development; and this condition occurs when the soil is *moist*, but not *wet*—that is to say, when it has the color and appearance of being well watered, but when it is still capable of being crumbled to pieces by the hands, without any of its particles adhering together in the familiar form of mud.

"Let us observe still another condition of soil: in this instance, as far as water is concerned, the soil is in its healthy condition—it is moist, but not wet, the pores alone being filled with water. But where are the canals? We see them in a few places, but in, by far, the greater part of the soil none are to be perceived; this is owing to the particles of soil having adhered together, and thus, so far, obliterated interstitial canals, that they appear only like pores. This is the state of matters in every *clod of earth;* and you will at once perceive, on comparing it with a stone, that it differs from it, only in possessing a few pores; which latter, while they may form a reservoir for moisture, can never act as vehicles for the *food* of plants, as the roots are not capable of extending their fibers into the interior of a clod, but are at all times confined to the interstitial canals.

"With these four conditions before us, let us endeavor to apply them *practically*, to ascertain when they occur in our fields, and how those which are injurious may be obviated.

"The first of them is a state of too great dryness, *a very rare* condition, in this climate at least; in fact, the only case in which it is likely to occur is in very coarse sands, where the soil, being chiefly made up of pure sand and particles of flinty matter, contains comparitively much fewer pores, and, from the large size of the individual particles, assisted by their irregularity, the canals are wider, the circulation of air freer, and, consequently, the whole is much more easily dried. When this state of matters exists, the best treatment is to leave all the stones which occur on the surface of the field, as they cast shades, and thus retard the evaporation of water.

"We will not, however, make any further observations on this very rare case, but will rather proceed to much more frequent, and, in every respect, more important condition of soil—an *excess of water*.

"When water is added to perfectly dry soil, it, of course, in the first instance, fills the intestitial canals, and from these enters the pores of each particle; and if the supply of water be not too great the canals speedily become empty, so that the whole of the fluid is taken up by the pores; this, as we have already seen, is the *healthy* condition of soil. If, however, the supply of water be too great, as is the case when a spring gains admission into the soil, or when the sinking of the fluid through the canals to a sufficient depth below the surface is prevented, it is clear that these also must get filled with water so soon as the pores have become saturated. This, then, is the condition of *undrained soil*.

"Not only are the pores filled, but the interstitial canals are likewise full; and the consequence is, that the whole process of the germination and growth of vegetables is materially interfered with. We shall here, therefore, briefly state the injurious effects of an excess of water, for the purpose of impressing more strongly on your minds the necessity of thorough draining, as the first and most essential step toward the improvement of your soil.

The *first* great effect of an excess of water is, that it produces a corresponding diminution of the amount of air beneath the surface, which air is of the greatest possible consequence in the nutrition of plants; in fact, if entirely excluded, germination could not take place, and the seed sown would, of course, either decay or lie dormant.

"*Secondly*, an excess of water is most hurtful, by reducing considerably the *temperature* of the soil; this I find, by careful experiment, to be to the extent of 6½ degrees Fahrenheit, in summer, which amount is equivalent to an elevation above the level of the sea of 1,950 feet. So that, supposing two fields lying side by side, the one drained, the other undrained, and supposing them both equally well cultivated, there will be nearly as much difference in the amount and value of their respective crops, as if the drained one was situated at the level of the sea, and the other had an elevation as high as the most lofty of the Pentland Hills. But, beside this, and what is nearly equally bad, the temperature is rendered unnaturally high during winter; whereas, it has been proved that one great source of health and vigor in vegetation is the great difference which exists between the temperature of summer and winter, which difference amounts, in dry soil, to between thirty and forty degrees; while in soil, very much injured by an excess of water, the whole range of the thermometer throughout the year will probably not exceed from six to ten degrees.

"These are the chief injuries of an excess of water in soil which affect the soil itself. There are very many others affecting the climate, etc.; but these are not so connected with the subject in hand as to call for an explanation here.

"Of course all these injurious effects are at once overcome by thorough draining, the result of which is to establish a direct communication between the interstitial canals and the drains, by which means it follows that no water can remain any length of time in these canals, without, by its gravitation, finding its way to the drains.

"Too much can not be said in favor of pulverizing the soil; even thorough draining itself will not supersede the necessity of performing this most necessary operation. The whole valuable effects of plowing, harrowing, grubbing, etc., may be reduced to this; and almost the whole superiority of *garden* over *field* produce, is referable to the greater perfection to which this pulverizing of the soil can be carried.

"The celebrated Jethro Tull has the honor of having first directed the farmer's attention forcibly to the subject; and so deeply impressed was he with its infinite importance, that he believed the use of manure could be entirely superseded were this pulverizing carried to a sufficient extent.

"The whole success of the drill husbandry is owing, in a great measure, to its enabling you to stir up the soil well during the progress of your crop; which stirring up is of no value beyond its effect in more minutely pulverizing the soil, increasing, as far as possible, the size and number of the interstitial canals.

"Lest any one should suppose that the contents of these interstitial canals must be so minute that their whole amount can be of but little consequence, I may here notice the fact, that in moderately well pulverized soil they amount to no less than one fourth of the whole bulk of the soil itself; for example, 100 cubic inches of *moist* soil (that is, of soil in which the pores are filled with water, while the canals are filled with air), contain no less than 25 cubic inches of air. According to this calculation, in a field pulverized to the depth of eight inches, a depth perfectly attainable on most soils by careful tillage, every imperial acre will retain beneath its surface no less than 12,545,280 inches of air. A familiar illustration of the space occupied by the spaces between the particles of loosened soil is afforded by the fact that when soil is disturbed it more than fills the space it previously occupied.

"Taking into calculation the weight of soil, we find that with every additional inch which you reduce to powder (by plowing, for example, nine inches in place of eight), you call into activity 235 tuns of soil, and render it capable of retaining beneath the surface 1,568,160 additional cubic inches of air. And to take one more element into the calculation, supposing the soil were not properly drained, the sufficient pulverizing would increase the escape of water from the surface by upward of 100 gallons a day.

So far as the legal distinction between *water of pressure* and springs is concerned, Judge French says:

"As we find it in our field, it is neither rain water, which has there fallen, nor spring water, in any sense. It has been appropriately termed the *water of pressure*, to distinguish it from both rain and spring water; and the recognition of this term will certainly be found convenient to all who are engaged in the discussion of drainage.

"The distinction is important in a legal point of view, as relating to the right of the land owner to divert the sources of supply to mill streams, or to adjacent lower lands. It often happens that an owner of land on a slope may desire to drain his field, while the adjacent owner below, may not only refuse to join in the drainage, but may believe that he derives an advantage from the surface washing or the percolation from his higher neighbor. He may believe that, by deep drainage above, his land will be dried up and rendered worthless; or, he may desire to collect the water which thus percolates, into his land, and use it for irrigation, or for a water ram, or for the supply of his barn-yard. May the upper owner legally proceed with the drainage of his own land, if he thus interfere with the interests of the man below?

"Again: wherever drains have been opened, we already hear complaints of their effects upon wells. In our good town of Exeter, there seems to be a general impression on one street, that the drainage of a swamp, formerly owned by the author, has drawn down the wells on that street, situated many rods distant from the drains. Those wells are upon a sandy plain, with underlying clay, and the drains are cut down upon the clay, and into it, and may possibly draw off the water a foot or two lower through the whole village—if we can regard the water line running through it as the surface of a pond, and the swamp as a dam across its outlet.

"The rights of land owners, as to running water over their premises, have been fruitful of litigation, but are now well defined. In general, in the language of Judge Story—

"'Every proprietor upon each bank of a river is entitled to the land covered with water in front of his bank to the middle thread of the stream, etc. In virtue of this ownership, he has a right to the use of the water flowing over it in its natural current, without diminution or obstruction. The consequence of this principle is, that no proprietor has a right to use the water to the prejudice of another. It is wholly immaterial whether the party be a proprietor above or below, in the course of the river, the right being common to all the proprietors on the river. No one has a right to diminish the quantity which will, according to the natural current, flow to the proprietor below, or to throw it back upon a proprietor above.'

"Chief Justice Richardson, of New Hampshire, thus briefly states the same position:

"'In general, every man has a right to the use of the water flowing in a stream through his land, and if any one divert the water from its natural channel, or throw it back, so as to deprive him of the use of it, the law will give him redress. But one man may acquire, by grant, a right to throw the water back upon the land of another, and long usage may be evidence of such a grant. It is, however, well settled that a man acquires no such right by merely being the first to make use of the water.'

"We are not aware that it has ever been held by any court of law, or even asserted, that a land owner may not intercept the percolating water in his soil for any purpose and at his pleasure; nor have we in mind any case in which the draining out of water from a well, by drainage for agricultural purposes, has subjected the owner of the land to compensation.

"It is believed that a land owner has the right to follow the rules of good husbandry in the drainage of his land, so far as the water of pressure is concerned, without responsibility for remote consequences to adjacent owners, to the owners of distant wells or springs, that may be affected, or to mill owners.

"In considering the effect of drainage on streams and rivers, it appears that the results of such operations, so far as they can be appreciated, are to lessen the value of water powers, by increasing the flow of water in times of freshets, and lessening it in times of drought. It is supposed in this country, that clearing the land of timber has sensibly affected the value of 'mill privileges,' by increasing evaporation, and diminishing the streams. No mill owner has been hardy enough to contend that a land owner may not legally cut down his own timber, whatever the effect on the streams. So, we trust, no court will ever be found, which will restrict the land owner in the highest culture of his soil, because his drainage may affect the capacity of a mill stream to turn the water wheels."

CHAPTER V.

DRAINAGE LENGTHENS THE SEASONS.

We have already shown that drainage removes stagnant surface water, and surplus ground water. The removal of these waters, as a necessary consequence, prepares the soil at an earlier period, for the labors of the farmer, than if left to natural causes alone.

The time required for the "settling of the soil," after the winter frost passes from it, depends, to a great extent, upon its porous or its retentive character, is everywhere known and conceded. The deep gravelly loam is seen to be very soon free from water, while the heavy clay requires a long time to become fit for cultivation. In the one case the soil is fully drained—in the other the water mostly passes off by the slow process of evaporation. The water being removed, prepares the ground to receive to its fullest extent, the beneficial influence of the sun's warming rays, to impart to the soil the proper temperature for the germination and growth of seeds and plants. Thoroughly drained soil is not unfrequently ready for the plow from ten to fifteen days earlier, than a similar undrained one. Ten days of advanced growth of corn, barley, oats, wheat or potatoes, have often protected these crops from the effects of drought, early frost, or insects. Ten to fifteen days of advanced maturity, would fully protect the earlier varieties of wheat grown in Ohio, from the ravages of the midge (*cecidomyia tritici*), or the blighting effects of rust. The same advanced growth may secure the entire corn crop against early frosts, because in less time than that assumed, corn passes from the milky stage, when it would not require a severe frost to ruin it, to the glazed stage, when it is perfectly secure from the action of ordinary fall frosts. Potatoes, in the same time, would be so far advanced as to be enabled, much better, to withstand the drought—such as we had in Ohio in 1854 and 1859—when the *very* early potatoes did well enough, but the late ones were almost an entire failure. Fifteen days of advanced maturity of the oat crop in Ohio, in 1858, would have saved the entire crop from the blight or rust; and thorough drainage will place these fifteen days at the farmer's disposal.

In what manner will drainage prepare the soil fifteen days sooner than an undrained soil? In the first place, the autumn rains will not so thoroughly stagnate in the soil, as they necessarily must, in an undrained one, because they will drain off all the superfluous moisture between the drain and the frozen surface during the fall and winter; then, in the spring time, drain off all the moisture from the surface, as it finds its way into the soil.

In order to demonstrate why drained soil is in order to be cultivated in the spring time, so much sooner than undrained, we introduce the following table copied from the Saxony experiments at *Tharand*:

1853. Feb.	Temperature of air.	Temperature of water.	1853. Mar.	Temperature of air.	Temperature of drain.	1853. July.	Temperature of air.	Temperature of drain.	1853. Aug.	Temperature of air.	Temperature of drain.
1	36.5	36.5	1	24.	33.8	1	67.	54.5	1	56.5	56.7
2	31.	37	2	23.	33.	2	54.5	54.5	2	56.5	56.7
3	29.	37.5	3	26.6	33.	3	55.4	54.5	3	71.	56.7
4	31.	37.4	4	20.7	33.	4	53.6	54.5	4	64.4	56.5
5	31.	38.	5	24.	33.	5	60.	54.5	5	64.4	56.7
6	31.	38.	6	33.	33.5	6	63.5	54.5	6	55.4	56.5
7	30.	38.	7	35.	33.8	7	74.5	54.5	7	55.4	56.
8	32.	38.	8	38.	35.	8	78.8	54.5	8	54.5	56.
9	32.	38.	9	31.	35.	9	80.	55.	9	48.2	55.4
10	25.	38.	10	32.	35.	10	78.8	55.	10	55.4	56.
11	27.5	38.	11	27.	36.	11	62.	55.	11	53.6	56.
12	28.5	37.4	12	29.	36.	12	62.	55.	12	55.4	56.
13	22.	37.5	13	33.	36.5	13	73.4	54.	13	54.5	56.
14	23.	37.	14	33.	36.	14	71.6	53.6	14	56.	56.
15	21.	36.5	15	36.5	36.5	15	64.4	54.5	15	57.	55.4
16	21.	35.6	16	24.	36.	16	62.6	54.	16	56.5	55.
17	21.	34.2	17	17.6	36.	17	69.	54.	17	57.2	55.
18	15.8	34.2	18	17.6	36.	18	64.4	54.	18	56.5	55.
19	19.5	35.	19	19.4	36.	19	65.5	54.5	19	57.2	55.
20	21.	35.	20	17.6	36.	20	55.4	54.	20	66.2	54.5
21	23.	34.	21	21.2	35.	21	56.	54.	21	71.	54.5
22	24.8	33.8	22	24.	36.	22	69.	54.5	22	78.8	54.5
23	26.6	33.8	23	23.	36.	23	60.8	53.6	23	82.4	54.
24	26.6	33.8	24	22.	36.	24	62.	54.	24	78.	54.5
25	24.	33.8	25	23.	36.	25	77.	54.5	25	68.	54.5
26	16.5	33.8	26	21.2	36.	26	69.8	54.5	26	67.	54.5
27	28.5	33.8	27	17.6	36.	27	69.8	55.5	27	57.2	55.
28	27.5	34.	28	17.6	36.	28	77.	56.	28	60.	55.
			29	20.	36.	29	73.4	56.5	29	62.	55.
			30	26.6	36.	30	68.	56.7	30	52.	56.5
			31	27.5	36.	31	64.4	57.	31	67.	56.5

This table shows, that during the months of February and March, the water issuing from the drains had a much higher temperature than the atmosphere, and that during the months of July and August, the drain water was much cooler than the atmosphere. On the 18th of February, the temperature of the air was 15·8° Fahr., or 16·2° *below* freezing point, while the drain water stood at 34·2°, or 18·0° above the air.

The following table gives the mean monthly temperature of the atmosphere at 8 A. M., and 4 P. M., also of the drainage water at the same periods at Tharand, Saxony:

	Mean temperature of the air at 8 A. M.	Mean temperature of the drainage water at 8 A. M.	Mean temperature of air at 4 P. M.	Mean temperature of drainage water at 4 P. M.
1853.				
February,	26.	35.6	27.	35.6
March,	25.	35.	31.	35.
April,	38.7	38.	42.	38.
May,	51.	44.6	61.	44.6
June,	62.	50.	66.	50.
July,	67.	53.6	72.5	53.6
August,	61.	54.5	67.	54.
September,	56.7	53.6	60.5	53.6
October,	49.	49.	55.	49.
November,	37.4	46.4	38.5	47.
December,	24.	37.4	24.	37.4
1854.				
January,	29.	34.2	31.	34.5
Mean for the year,	43.	44.	46.5	44.4

From this we learn that the temperature of the air for the months December, January, February and March, was below *freezing* point (32°), while the temperature of the drain water, during this entire period, was from 2·2° to 3·6°, above the freezing point; consequently, the drains were discharging water during the entire winter, and when spring came, the soil was not saturated with autumn and winter rains, as undrained soils necessarily are.

There is no doubt that temperature has considerable influence on the discharge of water from the drains. The observations recorded in the following table, were made at Tharand, in Saxony, to determine the influence of temperature upon the discharge of water from drains.

The following statements and observations are given as *theory* only—that is, theory in its true sense—inferences based upon facts or actual observations, in contradistinction to the usual definition of theory, viz: speculations, or hypothesis to explain facts or phenomenon.

TABULAR STATEMENT SHOWING THE INFLUENCE OF TEMPERATURE UPON THE DISCHARGE OF WATER FROM THE DRAINS; ALSO, THE INFLUENCE OF RAINS UPON THE DISCHARGE FROM DRAINS.

	Temp. of atmosphere.		Quantity of Rain per day, per acre, in gallons.	Discharge of Drainage Water, daily, per acre.	Increase compared with the preceding day.
	Minimum.	Maximum.			
1853.					
March 7	32.	35.	——	280	38
" 8	35.	42.8	1624	2061	1781
" 9	33.5	42.	1919	3515	1454
" 10	32.	41.	295	4137	622
" 11	26.6	42.	——	5614	1477
April 1	24.8	46.4	——	2158	1264
" 2	37.5	45.5	1255	8605	6447
" 3	33.8	47.5	295	9410	805
1854.					
Feb. 6	31.	44.	14541	5653	5018
" 7	44.6	47.	14614	11462	5809
March 2	28.4	40.	——	5245	4831
" 3	28.4	41.	——	7619	2374
" 9	39.2	44.6	5905	12154	4535
Dec. 15	32.	49.	32772	7084	5837
" 16	35.	50.	54178	21463	14379

We will now, in addition to this theory, present some facts, or rather the testimony of various practical farmers on this question, and none more to the point than the following: At an agricultural meeting in Boston, Mr. B. F. Nourse, of Orington, Me., who was present, said that drainage on his farm "had put his springy, cold lands, in good working condition, earlier in the season, than any other in the neighborhood. One lot drained in 1852, was in good working condition as soon as the frost was out. Before drainage, cattle could not cross it early in June without miring. It enabled the *later* as well as the *earlier* cultivation of the land. He had plowed as late as the 20th of November."

Mr. French, in his essay on drainage, refers also to

Mr. Nourse's experience, making mention of a piece of corn he planted in this land on a drizzling rain, after a storm of two days. The corn came up and grew well; although on a clayey loam, formerly as wet as the adjoining grass field, over which oxen and carts could not pass on the day of this planting, without cutting through the turf and miring deeply.

Messrs. Maxwell Brothers, of Geneva, N. Y., in a statement of draining done on their farm in 1855, and which received the first premium of the State Agricultural Society, say they underdrained one clayey lot, which previously "it was quite impracticable to plow or cultivate in a wet time, and consequently it was very difficult to get in a spring crop in season." After underdraining "they could cultivate immediately after rains with advantage," and, of course, get in their crops much more seasonably than before. Mr. Yeomans, another central New York farmer and nurseryman, states that on his drained lands "the ground becomes almost as dry, in two or three days after the frost comes out in spring, or after a heavy rain, as it would do in as many weeks before draining," and the frost leaves drained land at least a week sooner than that which remains undrained.

It is a weighty argument for draining, that it relieves the ground of surplus water early in the spring, and so enables the work of the farmer and gardener to commence earlier than it otherwise could. It also makes that work easier and pleasanter. When the ground is undrained, it can not become dry, except by evaporation, or by the oozing away of the water, particle by particle, through a long reach of stiff soil, into some natural outlet. Meanwhile, the farmer must sit with folded hands in comparative idleness, knowing that by the time his land had become dry, his work will accumulate and press upon him ith a burden he can hardly bear. It would not be strange if some of that work should be left undone, or be slighted. Let but suitable drains be cut through that land, and the melting snows and drenching rains would speedily find their way in these channels, and leave the ground dry and warm, and ready for tillage several weeks earlier than fields not so treated. It would tend to relieve farm life of a great objection to it, in many minds, viz: that it imposes such hurrying and exhausting labors at particular seasons, and especially in spring. It would enable the farmer to get certain crops into the ground earlier, and so make sure of a vigorous growth before the droughts of midsummer, and of maturity before the frosts of autumn. The farmer at the extreme north, who sometimes repines at the shortness of the growing season, and the coldness of his soil, would thus practically gain almost a degree of south latitude, without the necessity of selling his farm and moving his household gods.

CHAPTER VI.

DRAINAGE DEEPENS THE SOIL.

It is a most important fact that drainage "deepens the soil," but in what manner this desideratum is accomplished none of the writers on the subject (whose works we have had the privilege of perusing) appear to explain. One writer[1] says:

"The effects of thorough draining in deepening the soil, are readily seen in a comparison of the characteristics of those wet and retentive, with those either naturally or artificially of a porous nature.

"All heavy soils must be shallow from the influence of stagnant water—of water which saturates the surface, not being able to pass away by filtration. Every fall of water gives a mortar-like consistency to such a soil, and as the moisture passes off by the slow process of evaporation, it becomes baked and brick-like, instead of light and friable. If plowed when wet, it is entirely unfit for the growth of crops; if stirred when dry, it turns up in clods and lumps; in either case, it is only after much labor that any finely pulverized earth is obtained to support and nourish vegetable growth, and an inferior crop is ever the result. *Saturation, without filtration, kills the productive power of any soil*—makes it hard, shallow and sterile, however rich in every element of fertility it may be, when differently situated in the single circumstance of drainage.

"Porous, or well drained soils, on the contrary, never retain, even if they become saturated with water. The surplus moisture filtrates at once into the drains, leaving the surface loose and friable. Such a soil can be plowed at any seasonable time, and turns up mellow earth, readily fitted as a seed bed for any crop. Such a soil invites the roots of plants down, offering them food instead of a stone-like earth, and every year deepens the area of vegetable growth, until the full depth is reached to which it has been drained."

---

1 Editor, *Country Gentleman*.

But then, after all, we have no explanation of how the soil is deepened, other than it becomes so by thorough drainage. The editor then produces the following capital statement of the fact, if any were needed, to prove that drainage deepens the soil:

"That draining deepens the soil, we will bring a single instance to show—one which confirms every point stated above. It is condensed from a letter from that pioneer drainer and pioneer of good farmers, John Johnston, near Geneva, N. Y., and was published in the *Country Gentleman*, of January 19, 1854. He says:

"'Last spring I concluded to plow a clayey field, containing forty acres, only once for wheat, and that after harvest. Previous to draining, it was one of my wettest fields, and in dry weather, even in April and May, was very hard to plow, often having to get the coulters and shares sharped every day, when we used wrought iron shares. Owing to the great drought before, during and after harvest, I got a large plow made, so that I could put two or more yokes of cattle, and a pair of horses to it if necessary. Immediately after harvest we started for the field, oxen and drivers, plowmen and horses; and beside new shares on the plows, took other new shares along, expecting to be obliged to change every day.

"'When we got to the field, I had one man put a pair of horses before the large plow, and try to open the land with a shallow furrow. He went seventy rods away and back without even a stop, expect when the clover choked the plow. I then put the plow down to eight inches, and after one round, to nearly ten, and we went around without any trouble. His furrow was over nine inches deep, and laid as perfect as could be. I then had one yoke of oxen put behind my smallest horses, and a pair of horses before each of my other plows, and they plowed the field with perfect ease, only changing shares twice.

"'Although the field was undoubtedly plowed at the rate of nine inches deep, yet the clover roots went deeper, and the land plowed up as mellow as any loam; whereas, had it not been drained, it would have broke up in lumps as large as the heads of horses or oxen. A few years ago, a neighbor broke up a field about the same season, and similar land, but not drained, and after cultivating, rolling and harrowing, he had to employ men and mallets to break the lumps, before he could get mold to cover the seed; and after all did not get the third of a crop of either wheat or straw. My wheat looks as well as any I ever saw, and I doubt not it will be a good crop.'

"Those farmers, and they are not few, who have had experience in the cultivation of clayey soils when dry, or in any state, will not wonder that Mr. Johnston exclaimed, on finding this great change in the depth and friability of this clay bed: 'I never was more agreeably surprised in my life—in fact, had my men been plowing in gold dust, as they do in California, I should have been no more pleased.' This great change was the simple effect of thorough drainage—the soil, no longer compelled to remain saturated with water, lost its brick and morter character, and became a *live*, or at least an active and productive soil, ready to reward the labor of the farmer."

A correspondent, reviewing Judge French's work on drainage, says:

"In his chapter on the effects of drainage upon the condition of the soil, the author of '*Farm Drainage*,' announces the proposition that 'drainage deepens the soil.' How this effect is produced we are not told, though a hint is given that it 'lowers the line of standing water,' and then we are treated to a page or so of reasoning and authorities to show that plants send their roots to a great depth if the soil admits. We are tóld also that the roots of few plants, except those of an aquatic character, will grow or live in stagnant water. Hence we are left to conclude that 'drainage deepens the soil.'"

And thus it is with all authorities; they state the fact, and this is, perhaps, all that the present locomotive, telegraph, lightning press, steamship age requires; but the inducement for those to underdrain who are deprived of the privilege of seeing and examining the practical workings of underdraining, will be much stronger if the *chain* of the argument is complete—not having a single defective link in it.

Alderman Isaac J. Mechi, speaking of draining heavy lands, says:[1] "I consider it an axiom that the friability of the soil will be in proportion to the rapidity of percola-

[1] *How to Farm Profitably*, page 49.

tion. Filtration may be too sudden, as is well enough shown by our hot sands and gravels; but I apprehend no one will ever fear rendering strong clays too porous and manageable. The object of drainage is to impart to such soils the mellowness and dark color of self-drained, rich and friable soil. That perfect drainage and cultivation will ultimately do this is a well-known fact. I know it in the case of my own garden. How it does so I am not chemist enough to explain in detail; but it is evident the effect is produced by the fibers of the growing crops intersecting every particle of the soil, which they never could do before draining; these, with their excretions, decompose on the removal of the crop, and are acted on by the alternating air and water, which also decompose and change, in degree, the inorganic substances of the soil. Thereby drained land, which was before impervious to air and water, and consequently unavailable to air and roots, to worms, or to vegetable or animal life, becomes, by drainage, populated by both, and is a great chemical laboratory, as our own atmosphere, subject to all the changes produced by animated nature."

The explanation given above is not much more satisfactory than the preceding ones, with this exception; he hints at a *chemical* mean, while the others are rather inclined to attribute it entirely to a *mechanical* one; but both leave us to infer that the roots of the crops have much to do with producing the mellow condition of the soil. A correspondent of the *Country Gentleman*, over the signature of "A READING FARMER," concludes that the roots of a crop are not essential to produce this condition, for he says:

"A simple experiment will convince any farmer that the best mean of permanently deepening and mellowing his soil, is by thorough drainage, to afford a ready exit for all surplus moisture. Let him take, in spring, while wet, a quantity of his hardest soil—such as it is almost impossible to plow in summer—such as presents a baked and brick-like character, under the influence of drought—and place it in a box or barrel open at the bottom, and frequently during the season let him saturate it with water. He will find it gradually becoming more and more porous and friable—holding water less and less perfectly as the experiment proceeds, and in the end it will attain a state best suited to the growth of plants from its deep and mellow character. Here we have the result of drainage as it acts upon the soil. It may require time to act—probably on heavy clays more than a single season, but it will act in conjunction with other natural influences to give depth and friability to the soil."

Now, so far as ourself is concerned, we agree with all the authorities we have quoted, that drainage "deepens the soil," by "lowering the line of standing water," "by roots of crops," by making the subsoil accessible to "vegetable and animal life;" but, at the same time, we are of opinion that there is another agent, which we find no where mentioned in this connection, whose influence in mellowing and deepening the soil is just as powerful as the causes just cited.

Some years since, in conversation with a very intelligent gentleman, a farmer, but an Englishman by birth, we expressed our surprise that while he spoke so very highly of everything *English*, that he did not use a Crosskill's clod crusher, that being eminently English. "For the very reason," replied he, "that in this country (U. S.) we have an infinitely cheaper, and much better '*clod crusher*' than Crosskill's, or any the English can invent, unless the gulf stream should change its course." What relation possibly could exist between the gulf stream and a clod crusher we could certainly not divine, and as we looked inquiringly into our friend's face to assure ourself of his sanity, he laughingly replied: "I mean JACK FROST." True they have frosts in England, but no comparison to the frosts we have here—their frosts are rather mild and good natured, like John Bull himself, after he has just risen from the enjoyment of his favorite *roast beef*, while the frosts here act with the force of a locomotive running eighty miles an hour, breaking up, tearing up, and smashing things generally. Plow up a stiff clay in the fall, and let Jack Frost have fair play on it until next April, and Crosskill's clod crusher might be very thankful if the next harvest awarded it a *second* premium, while Jack Frost would take the first over all competitors.

Jack Frost, in our opinion, is the certain other agent above alluded to. In a drained clay, the late fall or early winter rains always leave moisture enough for the frost to congeal, and every congelation separates particles of the soil; then, when a thaw takes place, the water is borne off into the drain. In fact, in three-feet drains the process of thawing is going on from below upward all the while; hence, in springtime, drained soil is dry, or ready to work in a few days. Jack Frost, in this manner, institutes beneath the soil, in the words of Alderman Mechi, "as great a chemical laboratory as our own atmosphere." Our own observation is, that mild winters are never succeeded by as bountiful harvests as more severe ones are; but then winters may be *too* severe, as well as too mild; yet drained soils, in either event, fare much better than undrained ones.

We do not think that a heavy clay soil, drained in the spring and plowed in the fall of the same year, would plow so much easier, as Johnston, just above quoted, states that his did, or as that did in the following extract from a correspondent to the *Country Gentleman:*

"A Western New York farmer had a wet soil, thoroughly underdrained with tile—a field of forty acres. It had always been very hard and difficult to plow in the summer, taking a strong force of teams, and wearing out the plows very rapidly, and still the work was done in a very imperfect manner. After draining, he concluded to plow it once after harvest for wheat, as it had lain for some time in clover. He went on with it with his triple teams and large plows, but found that a single team could turn a furrow ten inches deep with perfect ease. The land plowed up as mellow as any loam, where, previous to draining, at that season it would have broken up in lumps as large as the heads of his horses. To drainage he attributed the change, and we have no doubt that the deep mellow state of the soil resulted entirely from 'lowering the line of standing water;' from affording it opportunity for *filtering* rapidly through the soil, instead of rising slowly as *evaporated* by the heat of summer."

Deepening the soil is certainly a great desideratum with every intelligent agriculturist, since every inch of depth of soil gives 100 tuns of active soil per acre for the nutrition of the growing crop. Those who argue that the cultivated plants obtain all their nourishment in a soil 8 or 10 inches deep, certainly have never investigated the subject. The writer remembers distinctly that many years ago he measured the roots of the oat plant, found in the course of digging a cellar, which had penetrated the soil to a depth of four feet. At another time, under similar circumstances, the roots of the wheat plant were found to have penetrated to a depth of five feet and several inches. Roots of the corn plant were found at a depth of three feet. Mr. Denton, an English writer, quoted by Judge French, says:

"I have evidence now before me that the roots of the wheat plant, the mangold wurzel, the cabbage and the white turnip frequently descend into the soil to the depth of three feet. I have myself traced the roots of wheat nine feet deep. I have discovered the roots of perennial grasses in drains four feet deep; and I may refer to Mr. Mercer, of Newton, in Lancashire, who has traced the roots of rye grass running for many feet along a small pipe drain, after descending four feet through the soil. Mr. Hetley, of Orton, assures me that he discovered the roots of the mangolds, in a recently made drain, five feet deep; and the late Sir John Conroy had many newly-made drains, four feet deep, stopped by the roots of the same plants."

It certainly requires no argument to prove that the roots of plants can not penetrate to these depths in a "water-logged," compact clay soil.

The following cuts (Figs. 9 and 10) *prove* that drainage deepens the soil. In Fig. 9, *a* represents the surface soil;

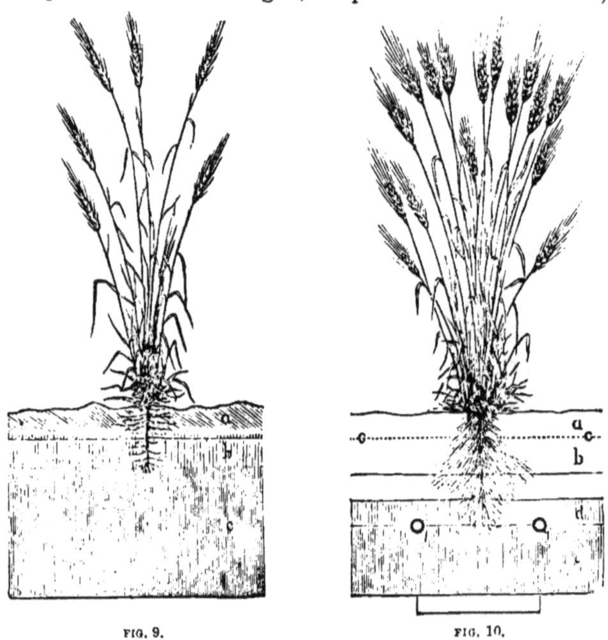

FIG. 9. FIG. 10.

the line between *a* and *b* the line of capillary attraction; *c*, ground water, and the line between *b* and *c* is the ground water line. The plant is weak, has few heads and very few roots. In Fig. 10, *a* is the surface soil; *c*, the ground water; *f f*, drains, 3 feet deep; *d*, water of capillary attraction. It is a well-known fact, that the roots go down to the water table, whether it is at *b*, Fig. 9, at 10 inches depth, or at *f*, Fig. 10, at 3 feet depth. The plant in

Fig. 10 is thrifty, has large and many roots and numerous heads. We have known a single grain of wheat, on such soils in Ohio produce 60 perfect heads of wheat. In the springtime the roots and foliage of wheat on such soil are represented by Fig. 11.

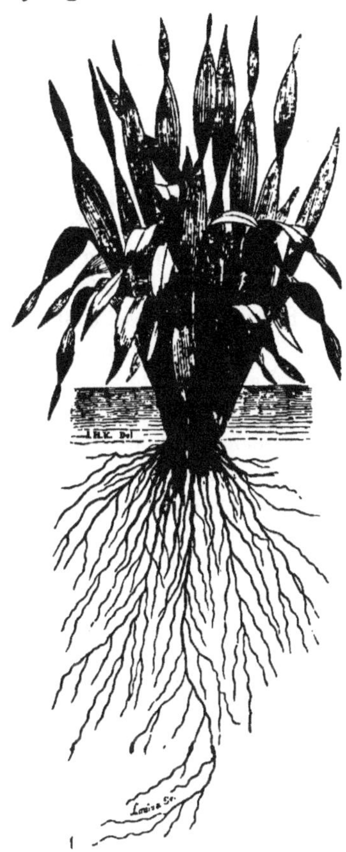

FIG. 11.—Appearance of foliage and roots of the wheat plant in the spring, on an underdrained soil.

CHAPTER VII.

DRAINAGE WARMS THE UNDER SOIL.

PROFESSOR JOHNSTON says:

"As the rain falls through the air it acquires the temperature of the atmosphere. If this be higher than that of the surface soil, the latter is warmed by it; and if the rains be copious, and sink easily into the subsoil, they will carry this warmth with them to the depth of the drains. Thus the under soil, in well-drained land, is not only warmer, because the evaporation is less, but because the rains in the summer season actually bring down warmth from the heavens to add to their natural heat."

There are two reasons why wet lands are always cold, and especially so in the spring: one is, the slow conduction of heat downward through a body of water; the other is, the heat lost in the evaporation of water. When heat is applied to the bottom of a vessel containing water, the portions that are first heated expand, become specifically lighter, ascend, and give place to heavier and colder portions. These, in turn, are heated and ascend, and thus a constant circuit is maintained until the whole is equally heated. When heat is applied to the top of a vessel of water, the upper stratum is heated; but this remains on the top, and no movement in the liquid brings unheated portions in contact with the fire, consequently heat passes downward through water slowly, and with great difficulty. Hence, a wet piece of land, which receives its heat from the sun's rays acting on the surface, will be a long time in being warmed to any depth.

The removal of surplus water by evaporation interferes still more with the warming of the soil in the spring. The vapor of water may be described as a compound of water and heat. Not more certain is it that water is taken from the soil by evaporation, than that a definite proportion of heat passes off with it. And hence, lands from which a large amount of water has to be removed in the spring by evaporation, are kept cold until the process is finished. The cooling effects of evaporation are highly beneficial, under some circumstances, but on wet lands, in the spring of the year, they are anything but desirable.

It follows, therefore, that the way to make cold, wet lands warm, is to resort to thorough drainage; and this, experience has shown, has the effect to raise the temperature of the soil many degrees through the whole spring. Not only is the heat communicated by the sun's rays retained, instead of being carried off in converting water into vapor, but dry soil, or any other solid bodies, will transmit heat downward better than liquids; in addition to this, a dry soil has its interstices filled with an air which diffuses heat, and helps to elevate the temperature.

A warm soil in the spring has a great advantage over a cold one. The seeds, planted or sown germinate at once, and don't permit the hardier seeds of weeds to get the start of the crop. Or, if the farmer chooses to stir the soil two or three times before putting in the crop, the seeds of weeds have the opportunity to germinate early, so that the young weeds may be killed by subsequent workings before the intended crop is sown. It is the experience of all, that drained lands are much easiest kept clean. But the temperature of the soil has a controlling influence on the growth of corn and some other spring crops. Corn will germinate when the temperature of the soil is about 55° Fahrenheit; a few degrees lower it will rot in the ground. In fact, fault is often found with seed corn, when the only difficulty is in the want of sufficient warmth in the soil. Barley, oats, and spring wheat may be sown on drained lands almost as soon as the frost is out, and there is little or no risk of the seed perishing. If the temperature of the atmosphere is not sufficient to justify growth above ground, if the soil be only warm enough, the plant will make the better growth of roots below.

The following illustration of the manner in which drainage warms the under soil, we find in the Patent Office Report for 1856, over the signature "D. J. B." (D. J. Browne), but the same illustration, engraving and text, "*verbatim, spellatim et punctuatim,*" are credited to the *Horticulturist*, of November, 1856, by Judge French. Not knowing, therefore, which one of the authorities is entitled to the credit, we accordingly "*divide the honors*" between them.

"The reason why drained land gains heat, and water-logged land is always cold, consists in the well-known fact that heat can not be transmitted *downward* through water. This may readily be seen by the following experiments:

" *Experiment No* 1.—A square box was made, of the form represented by the annexed diagram, eighteen inches deep, eleven inches wide at top, and six inches wide at bottom. It was filled with peat, saturated with water to e, forming to that depth (twelve and a half inches) a sort of artificial bog. The box was then filled with water to f. A thermometer, c, was plunged, so that its bulb was within one inch and a half of the bottom. The temperature of the whole mass of peat and water was found to be 39½° Fahr. A gallon of boiling water was then added; it raised the surface of the water to g. In five minutes the thermometer, c, rose to 44° owing to the conduction of heat by the thermometer and its guard tube; at ten minutes from the introduction of the hot water, the thermometer, c,

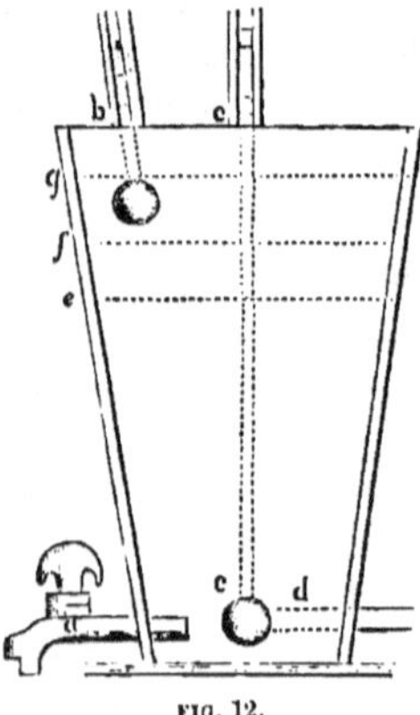

FIG. 12.

rose to 46°, and it subsequently rose no higher. Another thermometer, b, dipping under the surface of the water at g, was then introduced, and the following are the indications of the two thermometers at the respective intervals, reckoning from the time the hot water was supplied:

	Thermometer, b.	Thermometer, c.
20 minutes, - -	150 deg.	46 deg.
1 hour 30 " - - -	101 "	45 "
2 hours 30 " - - -	80½ "	42 "
12 " 40 " - - -	45 "	40 "

"The mean temperature of the external air to which the box was exposed, during the above period, was 42°, the maximum being 47°, and the minimum 37°.

"*Experiment No.* 2.—With the same arrangement as in the preceding case, a gallon of boiling water was introduced above the peat and water, when the thermometer, c, was at 36°; in ten minutes it rose to 40°. The cock was then turned for the purpose of drainage, which was but slowly effected; and, at the end of twenty minutes, the thermometer, c, indicated 40°; at twenty-five minutes, 42°, while the thermometer, b, was 142°. At thirty minutes, the cock was withdrawn from the box, and more free egress of water being thus afforded, at thirty-five minutes the flow was no longer continuous, and the thermometer, b, indicated 48°. The mass was drained, and permeable to a fresh supply of water. Accordingly, another gallon of boiling water was poured over it; and, in

3 minutes, the thermometer, c,	rose to - -	77 deg.
5 " "	fell to -	76½ "
15 " "	" - -	70½ "
20 " "	remained at	71 "
1 hour 50 " "	" " -	70½ "

"In these two experiments, the thermometer at the bottom of the box suddenly rose a few degrees immediately after the hot water was added; and it might be inferred that the heat was carried downward by the water. But, in reality, the rise was owing to the action of the hot water on the thermometer, and not to its action upon the cold water. To prove this, the perpendicular thermometers were removed; the box was filled with peat and water to within three inches of the top, a horizontal thermometer, c d, having been previously secured through a hole made in the side of the box, by means of a tight-fitting cork, in which tho naked stem of the thermometer was grooved. A gallon of boiling water was then added. The thermometer, a very delicate one, was *not in the least affected* by the boiling water in the top of the box.

"In this experiment the wooden box may be supposed to be a field; the peat and cold water represent the water-logged portion; rain falls on the surface, and becomes warmed by contact with the soil, and, thus heated, descends; but it is stopped by the cold water, and the heat will go no further. But, if the soil is drained, and not water-logged, the warm rain trickles through the crevices of the earth, carrying to the drain level the high temperature it had gained on the surface, parts with it to the soil as it passes down, and thus produces that bottom heat which is so essential to plants, although so few suspect its existence."

CHAPTER VIII.

DRAINAGE EQUALIZES THE TEMPERATURE OF THE SOIL DURING THE SEASON OF GROWTH.

WE have already shown that the discharges of water from drains are always several degrees above freezing point; and as heat naturally tends *upward*, that the soil is measurably warmed from below during the germinating season of the seeds, so that not unfrequently the soil is, during the entire month of April, much warmer than the air at night, although, perhaps, colder than the air during the day. The soil of a drained field is, therefore, free from the extremes of temperature of day and night air, during the same time. Again we have shown, by the tables copied from observations at Tharand, in Saxony, that the drains have a lower temperature during July and August, than the air. Thus, while drains equalize the temperature of the soil, and exempt it from the extremes of day and night temperature; it also equalizes it, and exempts it from the extremes of winter and summer temperatures. Prof. Johnston says:

"The sun beats upon the surface of the soil, and gradually warms it. Yet, even in summer, this direct heat descends only a few inches beneath the surface. But when the rain falls upon the warm surface, and finds an easy descent as it does in open soils, it becomes itself warmer, and carries its heat down to the under soil. Then the roots of plants are warmed, and general growth is stimulated."

CHAPTER IX.

DRAINAGE CARRIES DOWN SOLUBLE SUBSTANCES TO THE ROOTS OF PLANTS.

PROF. Johnston says: "When rain falls upon heavy undrained land, or upon any land into which it does not readily sink, it runs over the surface, dissolves any soluble matter it may meet with, and carries it to the nearest ditch or brook. Rain thus robs and impoverishes such land."

It must be self-evident to every observing person, that the advantage of rain to growing crops, is to furnish them with new supplies of nutrition. In undrained lands the "water line" being near the surface, most of the benefits to growing crops, which might be derived from rains are lost, either by evaporation of volatile substances, or, being in a soluble condition, they are washed to lower levels of surface, or into streams. The ground being already saturated from below, prevents the entrance from the surface of the desired elements; but in a drained soil on account of its greater porosity or mellowness and lower level of the water line, these substances can readily penetrate from the surface. The subject of the "absorbing qualities of arable soil" having been extensively discussed, within the last few years, by the best agricultural chemists of England and the continent, it may not be inappropriate to present a summary of the experiments and the conclusions deduced from them.

In 1848, Messrs. Huxtable and Thompson discovered, in arable lands, the property of fixing some of the elements of manures. Mr. Huxtable, having filtered some, barn-yard liquid through some earth, obtained it deprived of color and bad odor.

At the same time H. S. Thompson found the earth to possess the faculty of retaining, in an indissoluble state, the alkali of an ammoniacal solution, and even of solutions in which the bases were no longer in a free state, but were in combination with hydrochloric acid, and sulphate and nitrate of ammonia.

Mr. J. T. Way having been made acquainted with these remarkable results, undertook a long series of researches with the view of determining the cause and conditions of this absorption; he found that the absorbing property of the earth is not confined to ammonia only, but that it is extended to all alkaline and earthy bases, which are necessary for the growth of vegetation; such as soda, potash, magnesia, and lime, either free or involved in combinations.

After numerous experiments, which were repeated with the view of ascertaining whether the property of retaining the elements of manures, really existed in the arable soil, Mr. Way endeavored to express by numbers, the value of this absorption. For that purpose he caused a quantity of earth to be digested in a solution of the compound he wished to operate upon, the difference in degrees which the composition of the liquid indicated after its contact with the earth, showed what had been absorbed.

By this operation, Mr. Way found that 1000 grains of either clay or earth could absorb from a solution which contained 3173 per cent. of ammonia (or $3\frac{173}{1000}$ grains of ammonia in 1000 grains of solution) quantities of alkali ranging from $1\frac{570}{1000}$ grains to $3\frac{021}{1000}$ grains, differing only with the difference in the soils or clays, the per cent. being uniform in repeated experiments with the same earth.

There is nothing absolute in these figures; they are modified by the degree of concentration of the liquid and the quantities of either earth or liquid which are used.

The absorption takes place with great rapidity, and is as complete in half an hour, as it is after remaining in contact during fifteen hours; when the experiment is made with a salt of ammonia, it is decomposed, the bases only are fixed, while the acid is totally eliminated in a state of calcareous salt, in the decanted liquid.

What is the cause of this phenomenon? Is it caused by the presence of the lime, or of the organic matter? Is it owing to the free aluminium which may be in the soil? Mr. Way does not think so, because no greater absorption of ammonia is obtained by adding carbonate of lime to clay, free from this basis in a carbonated state, but containing a small proportion of the calcareous element; on the other hand, the incineration of the clay could not entirely destroy the decomposing and absorbing power of this earth, either on salts or bases; finally, the treatment by hydrochloric acid causes the solution of the aluminum to diminish, but does not destroy the decomposing and absorbing power of clay.

The rapidity with which the absorption of the bases by earth is effected, led Mr. Way to suppose that a chemical combination was formed; in such a case, the absorption of an alkali other than ammonia ought to take place, in the proportion of the two elements.

When acting with a salt of potash, 1000 grains of earth retained $4\frac{366}{1000}$ grains of potash; the absorption was then more considerable, without being, as it ought to have been, in the provision of the hypothesis; and, nevertheless, the liquid contained potash (substituted for ammonia), in true proportion indicated by the equivalents, that is: $8\frac{255}{1000}$ of potash to 1000 grains of solution.— In this case, as with ammonia, the more concentrated liquids produce a greater absorption; and when the salt was replaced by a solution of caustic potash, 1000 grains of earth retained $11\frac{716}{1000}$ grains of alkali; this absorption was still increased, when the earth had previously been boiled in an acid.

As it was easy to foresee, the salts of lime in solution underwent no modification when filtered through the earth; but lime water, according to the quantities which are employed, imparts, in similar circumstances, a quantity of alkali, which, in 1000 grains of earth, varies from $2\frac{31}{100}$ grains, to $14\frac{68}{100}$ grains.

When the lime is in the form of a carbonate, dissolved in water containing carbonic acid, the absorption is $7\frac{31}{100}$ grains of carbonate of lime for 1000 grains of earth.

The salts of soda and magnesia undergo a transformation in the soil, similar to that of the other alkaline compounds; but the action is less conspicuous, and gives no occasion for numerical determinations.

The bases being retained by the soil, the acids which, like phosphoric acid, form with lime insoluble compounds, ought to be insoluble also; this idea was confirmed by two operations of Mr. Way: in the one, water in which flax had been steeped; and in the other, sink water was employed; both were filtered through earth; the soil retained precisely all the substances which are the most useful to vegetation, such as organic matters in general, or nitrogenous substances only, the phosphoric acid, potash and magnesia; while the others were absorbed only in part, or were found in greater proportion in the filtered water.

Mr. Way was led, by these experiments, to the following conclusions: 1. The plants do not absorb the manure in a state of solution. Next, the form under which mineral matters and ammoniacal salts are applied, is different; because the soil possesses the power of bringing them back to a special state, in which these substances are presented to the plants, an important circumstance for the agriculturist who endeavors to introduce an alkali as manure into the soil; the salt that will supply that alkali at the lowest price, will, of course, obtain the preference.

It was also found, by Mr. Way, that clay possesses antiseptic qualities, because urine filtered through clay did not undergo putrid fermentation. From this fact, we have reason to infer that plants do assimilate nitrogenous substances other than ammonia and nitric acid.

Finally, Mr. Way thinks that manures, and by all means artificial ones, ought to be spread with great evenness, in order to secure everywhere uniform vegetation; because capillarity would be unable to disseminate the fertilizing substances, by means of diffusion, on account of their insolubility. The same cause would permit the application of large quantities of manure, without fear of any loss through drain water, because a good soil may retain, without waste, sixty times as much of the fertilizing elements as are applied with the manures.

Messrs. W. Henneberg and F. Stokmann, after repeating Way's experiments, confirmed in every point his conclusions; they found the results so perfectly regular, that Mr. Bœdecker was enabled, from their data, to establish algebraic formulæ, showing the value of absorption on any given quantity of earth, of liquid, and degree of solution.

Liebig resumed Way's experiments, and confining himself to the study of arable lands, ascertained that all soils possessed very nearly the same absorbing power; he found, as did Mr. Way, that soils, either rich or poor in carbonate of lime, did not manifest that power with the same intensity with all bases; thus, in filtering barn-yard liquid, pòtash was retained with more readiness than soda. $\frac{125}{1000}$ grammes of barn-yard liquid which contained, before being filtered:

Potash, - -	grains.	0.0867
Soda, - -	"	0.0168

Contained, after being filtered:

Potash, - -	grains.	0.0056
Soda, - -	"	0.0118

Whereas, the whole of the ammonia was retained.

The alkaline silicate of potash is acted upon by the earth like other salts of potash; the basis is absorbed, and in the same time a quantity of the silica is retained. While the absorption of the basis by various soils offers but small differences, the absorption of the silica appears to be in an inverted ratio, as the organic substances present in the soil, which are mostly of an acid reaction; and, strongly saturating the earthy bases, such as lime and magnesia, they present an obstacle to the fixation of silica.

The solution of silicate of potash, which was acted upon, contained per quart:

Potash, - -	grains.	1.166
Silica, - -	"	2.780

The absorption for 1,000 cubic centimetres of various soils, was as follows:

	Potash.	Silica.
Earth from a forest, - - -	gr. 0.951	gr. 0.115
" from a garden, - - -	" 1.055	" 1.081
" from Bogenhausen, - -	" 1.148	" 2.007
" from Hungary, - - -	" 1.151	" 2.644

The soil from the forest was full of organic detritus,

being mixed with milk of lime up to the alkaline reaction, and then dried, absorbed:

Potash,	- -	grains,	0.987
Silica,	- -	"	3.169

Liebig thinks that this absorption is owing partly to the chemical action of silicates and hydrate of aluminum upon the silicate of potassium, and beside that it must be connected with the physical state of the soil.

Phosphate of lime, dissolved by means of carbonic acid, is entirely retained by the soil; the same happens with ammoniaco-magnesian phosphate; the only soil of Tchernosem made an exception with phosphates; but Liebig found that it was saturated with them.

In view of the above facts, and considering the small proportion of mineral substances which are dissolved by drain water, from the analysis of Messrs. Way and Krocker, Liebig comes to the same conclusions as Mr. Way, that manures are presented to plants under a special form, and that on account of the insolubility of the new compounds which are formed, there must be in the roots a peculiar force, allowing them to select and assimilate substances which they are unable to obtain from a solution.

Liebig thinks, however, that aquatic plants, such as *lemna trisulea*, the roots of which swim in water without direct communication with the soil, are submitted to other laws, and absorb their nutriment in a state of solution. It was proved by analysis, that it was so, because the water in which these plants grow, contains in solution all the mineral matter which is found in the ashes.

The alimentation of plants would not then be as simple as physiologists and agriculturists did suppose; nor would it be the same in process for all species. The importance of the result in the agricultural point of view, and also the deep interest involved in the study of the absorbing power of soils, are a sufficient stimulus for renewing and enlarging investigations.

Mr. Boussingault instrusted F. Brustlein, of the Conservatory of Arts of Paris, with the work of continuing the investigation. He reports that "it was performed in his laboratory, at the Conservatoire des Arts et Metiers." We present a translation of this report:

"The rapid and perfect exactness with which ammonia is measured with Boussingault's apparatus, the importance of this alkali as an agent of fertilization, and the identity of its reactions with those of the other bases in the soil, suggested the exclusive use of ammonia for the experiments.

"Three specimens of soil were tested, and each of them possessed a physical character widely different. The first, taken from Bechelbronn, is a tenaceous compact clay, rich in carbonate of lime, retaining water, and when dried, very hard. The second, from Mittelhausborgen, is the *lehm* (loam) of the fertile neighborhood of Strasbourg, very rich in carbonate of lime, with little plasticity but very homogeneous. The third is the soil of Liebfrauenberg's orchard, sandy, quartzose, very rich in organic detritus, remains of ancient and very powerful manuring."

* * * * * *

From the above experiments we may derive the following conclusions: The property of absorbing ammonia by arable lands is exclusively dependent on the physical constitution of the mineral substances, and even on the organic matters with which they are formed. This was made evident by the action of humus, peat and animal-black upon an ammoniacal solution; the former two decompose at the same time a noticeable proportion of alkali. The existence of a carbonate in the soil is necessary, so that the earth may decompose an ammoniacal salt in retaining the basis of it; animal-black is possessed of that faculty when carbonate of lime has been incorporated. The decomposition being stopped strictly at the quantity of salt, the ammonia of which has been fixed, the force which impels the absorption is powerful enough to provoke this double decomposition.

We know with what readiness ammoniacal salts are decomposed in presence of carbonate of lime. Mr. Boussingault demonstrated that moist carbonate of lime, in presence of a fixed salt of ammonia, sets free, in time, and dries up all the ammonia in a state of volatile carbonate; the same happens, when a very weak solution of hydrochloric acid boils in presence of carbonate of calcium.

The absorption of ammonia by the soil, in an atmosphere which is loaded with it, is considerable, as stated by Mr. Way. When the air, very limited in ammonia, is passed through a long column of earth, the latter absorbs almost the whole quantity of the ammonia, and loses it again, in great part, by the action of currents of moist air. These experiments do not permit us to draw conclusions as to the absorption of ammonia by the earth from the atmosphere; because, in the experiment, when that alkali was most minutely diffused through the air, it was found that the air which traversed the apparatus contained 225 times as much ammonia as the air which circulates at the surface of the globe.

With the earth loaded with ammonia, and exposed to the air when moistened, there was a production of azotic acid; but this production was not large enough, if we compare it with that which took place in the experiments on nitrification of arable land, made at Liebfrauenberg, in 1857, by Mr. Boussingault, so that we can not affirm that it was owing to the transformation of the volatile alkali.

The ammonia absorbed by the earth is of great stability as long as the earth is dry; but as soon as water intervenes, it causes, by evaporation, the dissipation of the ammonia. This phenomenon is well known to agriculturists who park their sheep; because urine, impregnating the earth's surface, putrefies within 24 hours, with a temperature of 15° centigrade; it then emits ammoniacal vapors, which are wastefully rising, unless promptly checked by timely plowing. This volatilization of ammonia in arable land proper, is a fact which was constantly observed by Mr. Boussingault, in his researches on the atmosphere confined in the soil.

The earth, according to its richness in ammonia and its force of retention, imparts to water greater or lesser quantities of alkali, independent, in some measure, of the proportion of the liquid. Water, containing very minute quantity of ammonia, seems to be endowed with the property of circulating through the soil; because, in the above experiments, water was never entirely deprived of its alkali by the earth, even when contained in extremely limited proportions. Taking into account the feeble dose of ammonia which exists in the soil, its solubility, how small soever, and then its diffusion—knowing that the reactions of the other alkalies, except the volatility, are identical with those of ammonia—it seems very probable that plants select the largest part of their nutriment from very weak solutions, in which is found the azotic element, so necessary to them, in a state of ammonia or nitric acid. It can not be doubted that such is the fact; aquatic vegetables stand as a proof of it; and Boussingault, by beautiful experiments, did establish that a plant acquires a complete growth in a soil formed with a sand of pure quartz previously calcined, provided it be supplied with nitrate of potash, phosphates and alkaline ashes. In this condition the vegetable is then compelled to derive its nutriment from a solution.

CHAPTER X.

DRAINAGE PREVENTS "HEAVING OUT," "FREEZING OUT," OR "WINTER-KILLING."

We have frequently heard farmers complain of portions of fields, and sometimes of entire fields, that winter crops would "winter-kill" or "freeze out" on them. In point of fact, we have known of several farms which, in other respects, were good farms, sold at a reduced price, because certain portions were "*spouty.*" This "spoutiness" is easily remedied by underdraining. The cause and process of freezing out are explained as follows: In places where it occurs there is a stratum of clay, or hard pan, at the depth of a few inches, or a foot from the surface. This stratum is nearly impervious to water. The soil has been pulverized by plowing, and absorbs like a sponge the autumn rains, and the melting snows in the springtime; it not only absorbs but retains these waters, because there is no way for them to escape except by evaporation, and this takes place to a very limited extent only in cold weather. The ground is frozen during the night and the water converted into ice; this process of freezing not only severs the smaller roots and sometimes the larger ones, but throws up the soil in *scales* or spicules, thus drawing the clover or wheat roots from their beds. When a thaw comes, the saturated soil settles down and exposes the roots, and a few repetitions of this process, which occurs every winter, leave the plants dead upon the field in the spring.

Underdraining affords a certain outlet or escape for the waters. Winter grain crops seldom suffer from freezing when the soil is dry. We know of several instances where "freezing out" was effectually remedied by underdraining; but, as we prefer the evidence of others to our own observation, we again quote from the *Country Gentleman:*

"A case coming under our observation the past winter will well illustrate the subject. A field of five acres, seeded to clover two years ago upon rye, owing in part to the presence of snow upon the ground the greater part of the first winter and spring, escaped with slight injury from this cause, and gave a very good growth of clover. But the past winter, the weather being of a different character, the grass on about three acres of the field was entirely destroyed, every root of clover, being pulled up or thrown out, laid loose upon the surface of the ground the present spring. This was an example of heaving out' of unmistakable character.

"The evil lies in *a saturated soil.* It matters little whether the surface be clay or sandy—it did not in the case above mentioned—if the subsoil is of an impervious character. We were much surprised to find in a slight depression, some three or four rods across, where the surface soil was a light sand, that the clover was as badly winter-killed as on the clayey part of the field; and the clayey part, it is well to mention, had good surface drainage from the descent or slope of the ground—at least an inch in a foot. This sandy corner was underlaid by an impervious hard pan, holding water equally as well as the clay; and we believe this will generally be found to be the case in all loams which suffer from heaving or freezing out.

"We hâve shown, in a previous article, that 'draining deepens the soil,' and hence it is *the* remedy for freezing out in all cases. Water no longer saturates the surface soil in such quantity as to form honeycomb ice every time it freezes; the plants are no longer confined to short roots, but have a better hold upon the soil, and it has been found that no loss whatever results from this cause, however unfavorable the season, on a thoroughly drained soil.

"A little testimony on this point may not be out of place here. Maxwell Brothers, of Geneva, tell us, in the *Transactions of the N. Y. State Agricultural Society*, for 1855, about draining a clay field, which previously could not be worked for spring crops in season for sowing, and *heaved so badly as to ruin winter crops*, which draining has rendered as mellow and productive as can be desired, so that they can cultivate immediately after heavy rains, and grow wheat and clover without loss from frost. John Johnston, of Seneca

14

county, has given pointed evidence on the subject. By draining he has so improved his clayey farm that no loss is suffered from this cause, though formerly it was a source of great injury to the crops in the low lands, entirely ruining wheat, and destroying it in many places upon the higher parts of the farm. Many like cases of the beneficial results of draining in this respect could be given, were it needful."

CHAPTER XI.

DRAINAGE PREVENTS INJURY FROM DROUGHT.

THE year 1854, was a year of severe drought in the state of Ohio. But notwithstanding the drought there were some excellent crops grown in the state. Mr. Crosby, of Ahstabula county, states (under oath) that in that year he grew three acres in wheat, which produced thirty bushels per acre. Messrs. F. and W. Donaldson, of Clerment county, grew thirty-two bushels per acre. Mr. J. A. Webster, of Meigs county, produced thirty-six and one third bushels per acre, on 4½ acres, and Mr. A. Edgel, of Washington county, produced thirty-eight and one third bushels per acre, on three acres, while G. Dana and son, of the same county, produced thirty-six bushels per acre, on nine acres. These gentlemen have all made their statements under oath. In their account of culture they unanimously state that they *plowed deep.*

In the same year, Mr. Standiford, of Allen county, states that he produced ninety-four bushels of corn per acre. Mr. Chaffee, of Ashtabula county, ninety-two bushels. Mr. Hewitt, of Hancock, ninety-five and one half bushels per acre, and Mr. C. Shepard, of Washington, one hundred and eight bushels per acre. The statements of these gentlemen are subscribed under oath, and they all agree that they that season *plowed deep.*

"We recollect," says the *Genesee Farmer*, "walking through a magnificent field of corn on the thoroughly underdrained farm of our friend John Johnston. One of the underdrains was choked up, and *there the crop was a failure.* Corn delights in a loose, dry, warm soil. If it is surcharged with water, all the sunshine of our hottest summers can not make it warm, and all the manure that can be put on it will not make the corn yield a maximum crop. In passing along the various railroads, we have often been saddened to see thousands of acres of land planted to corn, which, by a little underdraining, would have produced magnificent crops of this grandest of cereals, but which presented a miserable spectacle of yellow, sickly, stunted, half-starved plants, struggling for very life. We have ever been willing to apologize for the shortcomings of American farmers. We know the difficulties under which many of them labor. We *do* believe them to be, as a whole, 'intelligent and enterprising.' But these sickly cornfields are well calculated to create a very different impression. We have frequently to repeat the German proverb—'to know is not to be able.' These farmers know how to raise good corn, but they are not always able to put in practice improved methods of cultivation. There is scarcely a plant which does not thrive much better in a loose, deep soil, than in a shallow, compact one; but in no case is this fact susceptible of more ready verification than in the corn plant."

One instance only may be cited to illustrate the effects of deep culture. There is in the immediate vicinity of Columbus a tract of "*Scioto bottom land*," which has for upward of forty years been cultivated in corn annually. In 1851, Mr. John L. Gill of Columbus, anxious to test the effect of deep culture on corn, plowed eleven acres and about three fourths, to a depth of about eight inches, with a double plow, and then followed with a subsoil plow, loosening but not turning up the soil, to a depth of eight inches more. This tract, as well as the neighboring one, had never been plowed to a depth exceeding six or seven inches. In 1851, the neighboring pieces were plowed the usual depth, and the planting completed on the 7th of May; Mr. Gill completed the planting on the 10th.

In the course of three weeks the corn in the neighboring tracts appeared as forward and thrifty as usual, while that of Mr. Gill appeared pale and rather dwarfed; this, to say the least, was rather discouraging. But in the month of July, that in the neighboring fields appeared to have come to a 'stand still;' the leaves curled and drooped, and gave unmistakable manifestations of sufferings from drought, while Mr. Gill's was growing vigorously, and indicated no lack of moisture. The result was that Mr. Gill obtained 120 bushels per acre, while the adjoining fields yielded less than forty bushels per acre. This fact is well authenticated, and the field was witnessed in July and August by thousands of persons.

While the stalks in Mr. Gill's tract presented a pale and sickly appearance, the roots were pushing downward in search of moisture and nourishment; finding abundance of this, a sufficient supply was stored for the growth of the plant to resist all effects of drought. That in the neighboring fields exhausted the supply at first, and when the drought set in it had no store of supply to fall back upon.

If then deep plowing will secure a good crop in a period of drought, how much more may not be secured by underdraining?

The reason why drainage prevents injury from drought is to be found in the fact that draining "deepens the soil," and "lengthens the season." It is a well-known fact that a deep and mellow soil retains moisture much better than a shallow and compact one.

"'Water is held in the soil between the minute particles of earth. If these particles be pressed together compactly, there is no space left between them for water.' Compact subsoils are but little permeable to water, compared with the same when broken up, pulverized and mellowed. The one is porous and drinks in moisture like a sponge; the other absorbs it but in small quantities, and readily parts with the same on the application of heat. The one takes it from the air, which passes freely through it; the other, impervious to the air, or any slightly powerful influences, remains unchanged. Undrained soils, as we have shown, become compact after heavy rains, by the evaporation of the water with which they are saturated; drained soils, on the contrary, become more porous from the filtration of the same amount of moisture into the drains below.

"Draining prevents injury from drought, by giving a better growth to plants in the early summer. Seed sown on any soil containing stagnant water, sends no roots below that water line, but may for a while grow well from roots near the surface. But let drought come, the water line sinks rapidly, the roots having no depth to seek moisture below, are parched and burned, and without rain, the crop is irreparably injured. On a drained and deepened soil the roots go down without obstruction, and are thus prepared to withstand the effects of the long-continued dry weather so often experienced. That they will do so, a thousand facts in the experience of the farmer will prove to him that observes them."

A correspondent from Pittsburg, Pennsylvania, writing over the signature of B. B., to the *Country Gentleman*, in July, 1854, says:

"There are many portions of high ground in the neighborhood of Pittsburg, Pennsylvania, and along the Monongahela river, remarkable for its productive qualities. For many years past, in has been observed that those high hills, with ordinary cultivation, produce better crops of every kind, and grow superior fruit, to the bottom land in the same region. Many of the farmers would smile if told that the rich qualities of their land might be attributed to underdraining. The idea of draining hills from one hundred to three hundred feet elevation, they would consider ridiculous, from the fact that no swampy or moist land can there exist, and instead of attempting to drain it, some invention should be had to retain the moisture. This very invention they have in the most superior kind of underdraining."

"These hills comprise a portion of the coal region of Pennsylvania, and cover most generally two strata of bituminous coal. The first, about from thirty to sixty to sixty feet from the upper surface, from four to five feet in thickness; and the second, at the distance of about sixty feet beneath the first, of from five to seven feet in thickness. The first strata, upon account of its depth, as well as its quality, is but little worked at the present time, where the second is accessible; and in the immediate neighborhood of Pittsburg, where the first 'crops out,' the second alone is worked. From the quality of this coal, and the great demand for it in all parts of the country, an immense number of tuns are annually extracted—completely undermining many acres of surface, forming mammoth underdrains; and, as a number of acres are taken out, the whole hill is let down—not together in one mass, but broken and mangled by the pillars and supports left by the miners. So that, when the coal from any one hill is extracted and the pits abandoned, the soil upon its surface will have all the advantages of the best underdraining; and not draining of two or three feet in depth, but of from ten to an hundred feet; and, the ground being loosened to such a depth, it is almost impossible that it should suffer from drought. I have no doubt but this is one of the causes of the great crops on some of our hills.

"The drought at this time (July 17) is truly excessive, not a particle of moisture apparent in the ground to the depth of eighteen inches; and the summer thus far has been so dry as to almost check the entire growth of all kinds of spring crops. The farm I cultivate consists of about forty acres, all of which, excepting about ten acres, is undermined and underdrained by the taking out of the coal to the depth of from ten to one hundred feet. My crop of hay above the undermining has averaged over two tuns to the acre, while a 'rich bottom' of one of my neighbors did not produce one half the quantity.

"Again, I have planted upon the underdrained portion about three acres of corn, and on the same place, below the draining, in a rich garden, deeply spaded, there is planted a bed of the same kind of corn. The latter has received careful garden culture, and the former, planted on clover sod, the common field culture. The first looks as if it wanted rain badly, but still has a good color and healthy appearance; but the leaves of the latter look more like torches or fancy cigars, so closely have they wrapped themselves up, than any growing vegetable. The product of hay, as well as the present appearance of the corn, can be, partly at least, attributed to the underdraining.

"These advantages are still more apparent upon the growing of fruit. Formerly the bottom land was always sought after for gardens and orchards. A few years since an enterprising man fixed upon the top of one of our highest hills (Mount Washington.) He now brings the first, largest and best fruits to market, and gets the highest price. *His land is undermined*, and I understand he attributes his success greatly to this fact."

Another correspondent says:

* * "Last spring I was induced to undertake a trial of underdraining in my garden. The soil was originally pretty hard clay, though it has been made lighter and more friable of late years by additions of muck, chip-dirt, manure, etc. The time for making garden being close at hand, and the materials not being conveniently to be had, I was able to drain only about half of my garden, and I was thus, though unintentionally, provided with the means of comparing contiguous portions of similar land, one portion being drained and the other not. The portion that was drained was obviously superior to the other in several respects. 1. It was sooner dry or in a condition to be worked than the undrained. In this respect the draining has, both last spring and this one, made my clay-soil garden almost as early as those on sandy soils. This I consider a great advantage, as it enables me to get in seeds a week or two earlier. 2. During the long term of dry weather last summer, the

things growing upon the drained portion did not suffer so much, in the way of being wilted, pale and stunted in growth, as the plants did on the undrained section. I had thus occular demonstration that drained land will suffer less from drought than undrained. 3. In the case of a few crops which I raised on both portions, for the sake of comparing them, I made myself quite sure that on the drained portion the peas, etc., ripened a few days earlier, and were a little plumper or better than the same crops from the same kind of seed, and with the same kind of treatment, on the undrained."

"In the long drought of 1854, in New England, a pertinent case is mentioned, where two neighbors farmed adjoining fields precisely alike, with the exception of depth of plowing. One plowed four inches deep, and grew oats weighing but seventeen pounds per bushel; while the other, plowing nine inches deep, raised oats weighing thirty pounds per bushel. The proper depth of plowing depends, we think, considerably upon the character of the subsoil, and the condition of the land as to drainage. A porous subsoil would admit of the rising of moisture from below, while a hard pan or clayey soil would need to be plowed to a greater depth, so as to prepare it for taking all possible aid from slight rains, the dews, and the moisture of the air. A well-drained soil would present the same general characteristics of one with a porous subsoil."

"At a Legislative Agricultural Meeting, held in Albany, New York, January 25, 1855, 'the great drought of 1854' being the subject, the secretary stated that 'the experience of the past season has abundantly proved that thorough drainage upon soils requiring it, has proved a very great relief to the farmer;' that 'the crops upon such lands have been far been better, generally, than those upon undrained lands in the same locality;' and that, 'in many instances, the increased crop has been sufficient to defray the expenses of the improvement in a single year.'"

"A committee of the New York Farmers' Club, which visited the farm of Prof. Mapes, in New Jersey, in the time of the severe drought, in 1855, reported that the professor's fences were the boundaries of the drought, all the lands outside being affected by it, while his remained free from injury. This was attributed, both by the committee and by Prof. Mapes himself, to thorough drainage and deep tillage with the subsoil plow.

CHAPTER XII.

DRAINAGE IMPROVES THE QUANTITY AND QUALITY OF THE CROPS.

THAT drainage increases the quantity of the crops, is confirmed by every one who has practiced it to any extent. The writer of "*Talpa, or the Chronicles of a Clay Farm*," states that underdrainage increased the products of a heavy clay farm 27 per cent. Almost all English writers on drainage agree that thorough drainage increases the products to such an extent that the net increase alone, in two or three years, is sufficient to pay all the expense incurred in underdraining. The instances on record demonstrating the increase in crops by underdraining, are sufficiently numerous to make an ordinary-sized volume. From the mass of them we select the following, from England, Germany, and several states in the union, in order to show that climate and culture have not played so important a part in connection with drainage as some have intimated. Much, very much, is due to culture, but more, in many instances, is due to underdrainage; because the best of culture on a stiff clay soil will not produce as great a result as will underdraining.

In the first volume of the *Royal Agricultural Journal*, p. 31, Sir James Graham, bart., states that "a field which he took into his own management was let at 4s. 6d. per acre; it was pasture of the coarsest description, overrun with rushes and other aquatic plants. After draining and subsoil plowing, at an outlay of £6 18s. 4d. per acre, it was let to the incoming tenant, on a fourteen-years lease, at 20s. per acre—yielding an annual interest of rather more than 11 per cent. on the outlay." In vol. 2, p. 276, Mr. J. Burke states that "Mr. Denison, of Kilnwick Percy, purchased about 400 acres of rabbit warren, of an apparently sterile sand, with a heavy ferruginous subsoil; the hills covered with heather, and the hollows with a bed of marshy aquatic plants; and of which the cultivation had been abandoned, as it was found, although pared and burnt, not to produce more than three quarters per acre of oats, and was let at 2s. 6d. per acre. After having been subsoiled, plowed, and drained with tiles and soles, at a cost of £5 4s. 8d. per acre, exclusive of the carriage, and manured in the common way, it produced ten and a half quarters of Tartarian oats per acre, and now bears wheat and oats, on a property which was formerly considered useless."

It is also stated in the same page, "that some land belonging to Rev. Mr. Croft, of Hatton Bushell, which was not worth 5s. per acre, is now let at 21s., evidently from the effect of drainage, and by the breaking up of the moor pan."

In the succeeding pages he continues: "I have, moreover, the authority of the Marquis of Tweeddale for stating that the increased product of his home farm at Yester, in Scotland, has been nearly two thirds on most of the crops, and in some cases much more, upon all the land which has been subsoiled and drained. One field, indeed, which his lordship declares to have formerly carried only 17 bushels of oats per acre, has given 67 bushels of barley, after having been trench-plowed and drained." He goes on to state: "These improvements, by means of drainage, although clearly evincing its importance, both to the landlord in the increased value of his property, and to the farmer in the production of his crops, are yet less decisive than what I shall here briefly attempt to describe. The extra-parochial place of Teddesley Hay, in Staffordshire, is the residence of Lord Hatherton, and contains 2,586 acres. It was originally part of the forest of Cannock, and, with the exception of two anciently inclosed parks, it continued uninclosed until 1820, when the whole became, either by allotment or purchase, the property of his lordship. The extent of the farm lands is 1,832 acres, comprising a range of high and dry hills to the east, adjoining Cauk Chase, which hills were formerly a rabbit warren, covered with heath or fern. Having heard this tract of land below the hills mentioned as exhibiting in a striking manner the results of judicious draining, and employment of the water so obtained, I visited the place, in the latter end of May, 1841. I was conducted over it by Mr. Bright, the respected land steward, who gave me the following statement; and in riding through the farm, which then presented the appearance of the most luxuriant vegetation, described to me the condition of the land in 1820. The larger park, which had been long divided into fields, was ill cultivated, and the lesser park might be fairly viewed as one bed of rushes, and in the lower parts alder; the whole consisted generally of a light soil, rather inclined to peat, the subsoil being chiefly a stiff clay.

Some very deep drains were made in the larger park, which was effectually drained, and from which large volumes of water now issue; as soon as the inclosure was completed, other deep drains were made, and for the most part with excellent effect; things were in this state when Mr. Bright became agent to Lord Hatherton; he immediately conceived the notion of putting a portion of the waste allotments, and the whole of the lesser park, containing a surface of nearly 600 acres, through a regular course of thorough drainage, and afterward collecting the whole of the drain water into two main channels, with the double intention of conducting one of them through the farm-yard, for the purpose of obtaining by it a water power for various objects connected with the estate, and then employing it, in conjunction with the other stream, in making an extensive tract of upland water meadow. It must, however, be acknowledged to have been a bold attempt, which could only have been conceived by a comprehensive mind, and a man of great practical knowledge; but it was liberally seconded by his noble employer, and has been accomplished with admirable success, as the following statement of the improvement by drainage, and the expenditure during ten years preceding 1841, upon such parts of the estate as have been drained, will sufficiently explain. The original value of 467A. 0R. 9P. was £254 10s. 9d.; expenditure, £1,508 17s. 4d.; improved value, £689 13s. 1d.; showing an improved annual value of £435 2s. 4d. These lands having been effectually drained, Mr. Bright's next object was to collect so much of the drain water as the levels permitted into two main carriers, for the purpose of employing them as a power to turn a mill-wheel, and afterward to be employed in irrigation. For the former object, a small reservoir has been constructed, at a favorable level, about half a mile distant from the farm; here, at the farm-yard, a mill has been built, which does infinite credit to Mr. Bright; the stream of water was, of course, not sufficiently powerful to turn an undershot wheel, and to enable it to act with force, it was necessary to bring it out to the upper part of a wheel of thirty feet diameter; this wheel has been placed in the rock, thirty-five feet deep, and the headway has been carried from the bottom through the rock, and comes out in a valley below, at the distance of five hundred yards. The mill and this channel for the water, costs very little more than £1,000; it works a threshing machine, cuts hay and straw, and kibbles oats and barley for the stock, consisting of about two hundred and fifty-horses and cattle, grinds malt, and also turns a circular saw, which does a great part of the sawing for a large estate. The annual saving by this machine has been estimated at about £400, and it is still intended to apply the power to other purposes. From this wheel, and from another small carrier, which is made to pass immediately under the farm-yard (where all the urine and moisture that runs from the manure is carefully collected in a reservoir which overflows into the carrier), the water has been conducted over lands, principally uplands, containing altogether eighty-nine acres, at an expenditure of only £224 4s. 10d., by which an improvement of £2 per acre has been effected, or £178 per annum. This is Mr. Bright's calculation, but it is difficult to estimate the importance of such an acquisition as eighty-nine acres of productive water meadow to a large farm like this, on which there is (especially on the upper part of it) a great quantity of very dry and thin soil. I know no other place in which drain water has been turned to such good account. Luckily the water is all soft, and good for irrigation:

TOTAL INCREASE IN VALUE COLLECTED.

	£	s.	d.	£	s.	d.
Lands underdrained, present value, - -	689	13	1			
Original value, - - - - - -	254	10	9			
				435	2	4
Estimated saving by mill, - - -				400	0	0
Increase in value of water meadows, -				178	0	0
Being an increased value of - -				£1,013	2	4

Resulting only from draining 467 acres, and employment of the drain water over 89 acres of land; affording a clear annual interest on the outlay of full thirty-seven per cent., as will be seen the following

SUMMARY OF TOTAL EXPENDITURE.

	£	s.	d.
Underdraining, as per statement, - - - - -	1,508	17	4
Erecting wheel and machinery, - - - - -	1,000	0	0
Irrigation, - - - - - - - - - - -	224	4	10
	£2,733	2	2

Mr. Herman Wauer, a draining engineer in Prussia, says, in his work on drainage: "Two years ago I underdrained a plat of 37 acres of sandy clay, at an expense of 324 thalers ($216 U. S. currency). The two years preceding, the potato crop was so badly rotted that it did not pay the expense of planting and harvesting. The year preceding the draining, it was put in rye and produced the miserable amount of 5 bushels per acre, and half of that was chess and cockle. After it was drained it was sowed in oats, and produced 900 bushels of oats, which were sold for 500 thalers ($333 33). The next year it produced 5,400 bushels of potatoes, which were sold for 1,500 thalers ($1,000). The present year (1859) the crop of barley which it produced was excellent, but as it is not yet threshed I can not give the figures. The clover which is now appearing on it gives promise of a very heavy crop."

But the most remarkable example of the increase of crops by drainage is that of a domain in Hanover. A tenant leased it in 1844. The tract contained an area of 3,000 acres of heavy wheat land—all of it in an arable condition—and, notwithstanding the rent appeared to be very low, yet several successive tenants became bankrupt on it. But the last, or new tenant, was an intelligent agriculturist, who had thoroughly studied and learned how to drain in England, and he saw at a glance what was necessary to produce good crops. He employed a drainer, and in the course of several years underdrained the entire tract, and, as he held the lease for some years, accumulated an ample fortune on it. There was one tract on this domain of 82½ acres (110 morgen) which gave the following results :

In 1842 it lay fallow, because it was too wet to work in seeding time.

In 1843 it was sowed in vetches, but, as the excessive moisture destroyed most of these, they were plowed down, the land manured, and rape was sowed. This crop, in 1844, scarcely paid for seeding and harvesting, having been badly "winter-killed" and soured. It was then drained, and produced the following increased crops, as the direct result of drainage:

In 1845	it produced	1,944	bush. wheat,	worth	4,860	thalers	($3,240)
1846	"	1,008	" peas,	"	1,400	"	(933 33)
1847	"	1,872	" rye,	"	4,992	"	(3,328 00)
1848	"	2,304	" oats,	"	896	"	(597 00)
1849	"	8,568	" potatoes,	"	3,570	"	(2,380 00)

I have been unable to procure returns of the subsequent crops. The tile were brought from England, and this, of course, enhanced their cost. They cost, delivered on the domain, 25 thalers for morgen ($22 21 per acre), in the aggregate $1,833. It will be seen that the increased amount of the first crop was almost double the entire expense incurred in underdraining.

This is, perhaps, the most remarkable case on record of increased productiveness in consequence of underdraining.

The following was communicated by Mr. James M. Trimble, of Highland county, Ohio, to the *Ohio Farmer*. The farm on which the mole plow, or ditcher, was used is situated in Fayette county, Ohio :

"Mr. Johnston's answer to my letter of inquiry, published in the *Farmer*, did not reach me until I had purchased the implement, with the right to use it; else I should have hesitated, and, perhaps, not bought it. Having witnessed the operation of the ditcher, drawn by a pair of cattle, cutting at the rate of 125 rods of ditch, 3 feet 4 inches deep, in a single day; comparing this work with friend Johnston's statement of a 20 horse power engine being required to operate it, I came to the conclusion that my friend, in his great zeal, as the advocate of tile drainage, could not appreciate the mole plow as a substitute. This, coupled with the cost of tile drains, on a farm of over 1,700 acres, four fifths of which requires underdraining, deterred me from the use of tile, and induced me to give the mole plow, at least, a fair trial, before throwing it aside. To accomplish this, I purchased an accurate instrument to begin with—an engineer's level—and with it ascertained the level of the land to be drained.

"The farm lies on Rattlesnake, the creek running through it from north to south, parallel with and at 75 rods from the east, and 350 to 400 rods from the west line of the survey. There being but little fall to the creek, and the banks low, I had some difficulty in procuring the necessary fall to my open drains, to give a free outlet to underdrains, confining my operations to some 230 acres of prairie land on the west bank of the creek. I laid out my open ditches from the creek west, staking them off at every six rods, and marking the depth of cut and width of ditch on each stake. In this way I laid off 685 rods of open ditch, at 80 rods apart, varying in depth from four to six feet, and in width from six to eight feet, allowing for slope of banks one and three fourths feet to one foot in hight, which was let by contract at 65 cents per rod, and finished in October, 1858. The underdrains were cut in March, April and May. My son superintending the work, he laid off his drains with the level, staking them off more with the view of tapping the wettest portions of land between the open ditches, than a regard to straight lines or thorough underdraining. In this way, with the ditcher, two yoke of cattle and two men, in 16 days, he put in 1,500 rods of underdrain, at a depth of three feet four inches, and a cost of $65.

"At the time of running the mole plow, the surface of the ground was covered with water, from one to six inches deep. The surface soil, to the depth of from one to two and a half feet, is a black clay, or loam, rather a compact, tenacious soil; the subsoil is a close, compact, yellow clay, to the depth of from three to five feet. We followed the ditcher, with a large Illinois sod plow, a steel plow on wheels, drawn by three yoke of cattle, one of Garrett and Cotman's steel sod plows, a No. 8, drawn by three horses, and four steel sod plows, same make as No. 4, with a pair of horses each, and broke the sod up from six to eight inches deep, turning the furrows flat, which was first harrowed—200 acres of it—with the furrows, and then crosswise. It was then marked out, four feet apart, and with Brown's Illinois corn-planter planted in corn, checkered so as to be cultivated both ways. During the time of planting we broke some 30 acres which had been partially underdrained; the sod being tough and the ground very dry, it broke up rough and uneven—so much so that it was planted (without harrowing or marking out) about the 2d to the 4th of June.

"Our first planting was finished the 23d of May. On the 4th of June it was up (with the exception of what the cutworms destroyed), and from six to sixteen inches high. The frost on the morning of the 5th laid it all level with the ground. The largest corn seemed to be most injured; and on the 6th the work of plowing up and replanting was commenced and continued, until the 200 acres were all replanted. The crop was worked three times over with double-shovel plows; the fourth and last time with single shovels. The 30 acres last planted were not cultivated in any way. The weather, from May 23 to September 10, was warm and dry—not to exceed half an inch of rain fell during that time. The corn was all cut up and put in shocks twelve hills square, making about 23 shocks to the acre. We have husked over 100 shocks, and feel confident that the entire crop on the 200 acres will average 60 bushels per acre; and the 30 acres not cultivated will yield 40 bushels per acre. The underdrains all performed their work well up to the middle of July, when they began to fail, and by the 1st of August were perfectly dry. I have been on the farm from the 3d up to the 25th of November, during which time we have had several hard rains; and I have examined the outlets to all of the underdrains, which, without a single exception, are passing off large quantities of water. From a close observation during the summer, I am satisfied that the underdrains were quite as important to the growing crop during the drought, from May to September, as they were in carrying off the surplus water in the spring; and I am equally certain that the increase of crop, resulting from draining, is all of 20 bushels per acre, which would leave the account stand thus: 685 rods open ditch, at 65 cents per rod, $445 25; 1,500 rods of underdrain cost $65; use of ditcher, wear and tear, $25 75; entire cost, $536. Cr., by 20 bushels of corn, on 230 acres, give 4,600 bushels, at 25 cents, $1,150; showing a profit of $614 in favor of the mole plow, in a single year. It would seem superfluous to give the details of so plain an operation, as I have done; yet I am aware of the fact that, in many instances, in the immediate neighborhood of my farm, the use of the mole plow has been condemned, from the fact of improper use, not procuring sufficient outlet, running the ditches too shallow, and failing to reach the clay subsoil with the mole. I have no faith in the use of the implement without a clay subsoil for the mole to operate in; otherwise, the aperture made by the mole will cave and fill up. I have purchased an additional ditcher, and intend to carry on my operations until I have underdrained my farm—at least, all that portion requiring drainage."

Mr. Nathaniel Spalding, of Vermont, purchased a small farm, consisting of twenty-five acres, brook meadow, of clayey soil; some part of it approaching to swamp muck; and 17 acres upland, of cobble stone surface, in wood and pasture—42 acres in all. Mr. Spalding said that he bought it at auction, and moved on to it in 1853—price $460. "An old shell of a house and barn" (to use his own expression) was then upon it, "and some parts of the meadow so wet that a team could not be driven over it to get what little poor hay grew upon it." There was but little of it that could be plowed to advantage. From eight to ten tuns water grass of poor quality was the produce of the first year's hay crop. Mr. Spalding says he has made over 600 rods of drain. Main ditches, three feet wide, and from three to six feet deep; the bottom of the drains are boards; space 12 inches square, covered with flat stones, with shavings from the lumber with which he was erecting new buildings, and hemlock brush, thrown into the drain upon the covering stones, and then filled with earth. The cost of these main ditches averages 62½ cents per rod. His cross drains, leading into the main ones, are four rods apart, 15 inches wide, stones (cobbles) thrown in loose, covered with brush, and filled with earth. The cost of these cross drains, 30 cents per rod.

Mr. Spalding thinks the increase of production for the two years following the draining, paid the whole expense of making these drains. He is undoubtedly correct in his estimates, for this work was performed by himself and boys. Had he employed other labor, or contracted it out, at the high prices farm labor has commanded of late, it would hardly have done it; but he is a man that never puts his hand to the plow and looks back. He is emphatically a practical man, carrying out whatever he undertakes with an energy and skill known to those only of like determination. Above these drains where clover and timothy now grow so heavy as to lodge, a poor miserable water grass grew, scarcely worth the cutting and housing.

Mr. Spalding says the production of these 25 acres in 1857, only four years from the time he commenced on the farm, was 30 tuns English hay, 350 bushels of corn, and 250 bushels oats. And this from a soil, though not exhausted, but so located as to be kept saturated and filled with cold spring water, to such a degree as to discourage and forbid cultivation only on the driest parts and in the driest season.

We found the following in "a paper," without credit, but presume it was written by Luther Tucker, of the *Country Gentleman:*

"We wish to give additional evidence to the value of underdraining, by reporting all accurately stated experiments. Having recently made some on a small scale, we add them to the list. The land is a strong loam in Cayuga county, a medium between a heavy clay and a light loam. The drains were cut two feet nine inches to three feet deep, two rods apart, and completed with tubular tile two inches in diameter. The work being done where the proprietor could not oversee it, cost 40 cents a rod, or $32 per acre.

"The crops on this drained land, the present season, were corn and spring wheat—and being cultivated by a tenant, did not, of course, receive the best treatment. A portion of the cornfield was on a strip of undrained land. The season proving unusually favorable for the latter, but little difference could be perceived till the ears had set. It is now found, however, that while the corn on the drained land is a least forty bushels of sound shelled corn per acre, the undrained portion yields scarcely thirty bushels, and of poorer quality.

"With the spring wheat (China Tea), however, the disparity is greater. Before draining, fifteen bushels per acre was regarded a good crop, and uncertain at that. Three scant acres were sown last spring on the tile-drained land, and yielded *eighty-one* bushels—equal to twenty-seven bushels per acre. The wheat sold promptly for a dollar per bushel—and would probably have brought more as seed, as it was unusually fine, weighing 62 lbs. to the measured bushel.

"The time required to repay the cost of draining would, therefore, be as follows: For the corn, the increase being ten bushels per acre, at seventy-five cents per bushel, four years would be required, if all the seasons were like this. But they are commonly more unfavorable—making a greater difference in favor of the drains; the best cultivation would doubtless place the time for full repayment within three years. The increase of spring wheat being twelve bushels per acre, at a dollar per bushel, repays the cost in less than three years."

It is the unanimous opinion of all who have observed closely, that the plants and fruits grown upon underdrained soil are more fully developed, and of much better quality than those grown on undrained soil.

CHAPTER XIII.

DRAINAGE INCREASES THE EFFECTS OF MANURES.

It has been demonstrated that dew, rain and snow carry with them certain fertilizing agents of great importance to vegetation, such as carbonic acid, nitric acid, and ammonia, or these combined, as carbonate or nitrate of ammonia. When the soil is in a condition to receive all the water that falls upon it in the form of dew, rain or melting snow, these fertilizing agents are carried into the soil and immediately absorbed by it, or at once appropriated by the growing crop. When the soil is already saturated by water, or of a close, impervious character, or when the surface is sufficiently inclined, the water is compelled to run off, and carry with it, whatever elements of fertility it contains. Sandy soils readily receive water, but do not as readily absorb gases as soils containing clay or peat. Clay lands thoroughly drained and deeply tilled, will receive almost any amount of water, and absorb and hold for the future use of plants, all the gaseous fertilizers the water contains. The amount of these fertilizers brought down by the rain, differs greatly under different circumstances. The quantity of ammonia is found to be much greater near cities than in the open country. The amount of nitric acid is greater after thunder storms, and in seasons when thunder storms are frequent, than at other times.

It has been asserted (but at present appears to be a controverted point) that the elements of manure act upon plants only in a state of solution; hence it is of the greatest importance that they be so applied, and that the soil be so prepared that they may not only be readily dissolved by the rain, but that the rain may freely pass through the soil, which, acting as a filter, arrests and holds these elements where they best serve as food for vegetation. On undrained lands the rains dissolve the essential portions of the manure and carry them off, or if lands are more than ordinarily wet, it prevents the rotting of the manure. Herman Wauer mentions an instance where sheep droppings were kept from rotting by moisture for the space of five years. This is one great reason why manures produce such trifling results on heavy lands, especially in seasons of abundant moisture. In very dry weather but little more effect follows their application, from the want of a solvent, such as is ever supplied by the water retained in mellow, porous earth.

"'Draining renders the land penetrable to water,' says a writer on the subject, 'enabling the rain to descend freely through it, carrying to the roots the fertilizing elements with which rain water is always charged,' as well as those it takes in solution from manures. The effect of manures is also much increased by an intimate mixture with the soil. Such mixture can be but imperfectly obtained in the case of hard and shallow land, either in a wet or dry state. It will always be found that mellow and friable soils receive most benefit from manures, and that clayey soils, if made mellow by draining, possess the greatest absorbent powers, and are of the most productive character, compared with sandy and light or mucky loams.

"The true policy of the farmer is to use every means in his power for rendering his labor more effectual, and his farm more fertile, and in no way can this be better accomplished in the case of wet and retentive lands, than by *draining*, and thus deepening and increasing the productive powers of the soil."

Water from drains has repeatedly been collected and analyzed, and that under all imaginable differences of condition. These examinations have been made for the purpose of determining to what extent the water of drainage does bear away with it the fertility of the soil. It is found that drainage water does carry off, in solution, in appreciable quantities, the mineral constituents of soils, that it would be desirable to retain. As might be expected, the amount of loss varies greatly in different circumstances; from sterile lands, the amount of nitric acid, or ammonia, is less than what is furnished in the rain and snow; while on highly manured lands the amount of loss will exceed what is obtained from the atmosphere. From lands well tilled, and in a perfectly friable condition, the loss is greater than from lands imperfectly tilled. Where a crop is growing upon the soil, ready to appropriate whatever is presented in the water passing through the soil, less of these gases escapes than where the ground is fallow. The amount of loss is also found to be much less where the drains are deep, than where they are shallow. Some of the conclusions arrived at, on this subject, are: That there need be no fear that underdraining will rob the soil of its fertility, because the rain, which would run from the underdrained lands and be lost, will either wholly or partially compensate for any loss that occurs through the drains—that there is no method, except by drainage and deep culture, by which stiff, clayey lands can be made to appropriate all the elements of fertility furnished by the atmosphere—that it is better not to manure excessively, at long intervals, because a part of the unappropriated manure will probably be washed through the soil and lost, and, therefore, it is better to apply manure as it is required to meet the present demand. Manure is better applied in a liquid state, for if the soil be dry and deep, and therefore in a good condition for absorbing manure, it will combine with the elements of the soil at once, and the surplus of water will run from the deep drains perfectly clear and inodorous. It is better never to permit naked fallows on such lands, because the soil will be losing more through the drain than it gains from the atmosphere; and much more than it will lose, if covered by a growing crop; but on poor, clayey soils, the case is the reverse; and it is possible for such soils, while undergoing the comminution and exposure of fallowing, to gain more from the atmosphere than they would probably lose from the drains. The loss of any fertilizing agent by drainage is wholly avoided, however, in countries where drainage and irrigation are properly and systematically combined. The waters from manured and tilled lands, being conducted over the meadows below, yield up whatever of fertility they have brought with them, and thus nothing is lost.

CHAPTER XIV.

DRAINAGE PREVENTS RUST IN WHEAT AND ROT IN POTATOES.

The wheat growers of Ohio have often had the misfortune to see that which promised a bountiful harvest suddenly blighted by mildew or rust. In regions where a gravelly subsoil is found, the wheat crop seldom suffers from rust; but the wheat is frequently "rusted" on gravelly soils which rest upon *hard pan*, or impervious clays. Rust or mildew also most frequently attacks wheat on bottom lands, where considerable moisture prevails.

In all our reading and observation we have never heard nor seen a well-underdrained field of wheat attacked by rust, and therefore infer that drainage acts as a preventive of this very undesirable phenomenon.

A series of experiments made in 1857, by H. B. Spencer, of Rockport, Cuyahoga county, proves almost conclusively that the rot in potatoes is due to excessive moisture. We know numerous instances where potatoes, grown on ground having a northern exposure, were sound on the most elevated portions of the field, but badly rotted on the lower or most moist. One instance we remember more particularly, where the potatoes on the hillside were all sound, and on the bottom or swale they were not worth digging. No case of potato rot on well-underdrained ground has come to our knowledge. From this fact, and from Mr. Spencer's experiments, we are inclined to believe that underdraining will prevent rot in potatoes.

The fact that drainage lengthens the seasons, will permit wheat to be sown later in the fall, and thus avoid the ravages of the Hessian fly (*cecidomyia destructor*), and as the wheat will vegetate more rapidly and ripen earlier in spring or summer on underdrained ground, therefore the ravages of the "midge," "fly," or "weevil" (*cecidomyia tritici*), will be greatly lessened, if, indeed, not entirely prevented.

CHAPTER XV.

OTHER ADVANTAGES OF DRAINAGE.

DRAINAGE is of great advantage in many other respects; among these it may be stated that

Drainage facilitates Pulverization.—One object of plowing land is to pulverize it, and render it workable. Every one knows that a wet soil can never be pulverized, and plowing a clayey or loamy soil, when wet, does, perhaps, more injury than if it were not plowed at all, because, if plowed when wet, the soil is pressed together, and is turned over by the plow in almost unbroken slices, which become very hard clods when dry, and render it difficult of culture. Pulverization of the soil is so essential that, more than a hundred years ago, Jethro Tull advocated the idea that complete comminution or pulverization of the soil was a complete substitute for manure. In fact, the little book recently published by a London house, entitled "*Tillage a Substitute for Manure*," is made up mainly from the writings of Jethro Tull. The *Lois Weedon* system of culture, by which more than a dozen successive good crops of wheat were harvested from the same piece of ground, is simply another application of the principle advocated by Tull; and, while this system is not drainage in a direct sense, it undoubtedly partially answers the purpose of drainage. Cultivating to the depth of three feet, as the Lois Weedon system requires, must certainly lower the water line, and thus consummate *one* of the objects of drainage. The deeper any soil is cultivated, the better will it produce.

If the water is withdrawn from the soil, teams can pass

over it with less injury to the soil than on that which is not underdrained. The undrained clay, when tramped by cattle pasturing, or by being frequently hauled over, acquires a consistency to hold water, from which underdrained land is exempt. It is a common practice to haul manure either in the winter or early in the spring, and, in many instances, as much injury is done to the land in hauling as the manure benefits it.

Drainage prevents Surface Washing.—Many plowed fields, especially where the land is rolling, suffer greatly in spring and fall time, from "*washing*" by heavy rains. On drained lands, the rain is at once absorbed, and washing is thus prevented.

CHAPTER XVI.

WILL DRAINAGE PAY.

1ST. *For the Garden.*—With regard to lands designed for garden uses, that have a compact subsoil, there can not be a doubt of the economy of underdraining. Earliness and depth of soil are essential to a good garden; and in many localities these conditions can not otherwise be secured. Drained lands freeze to a greater depth than the undrained, but they are much sooner dry and fit for working, or for seed, in the spring. And during the summer, however wet the season, or recent the rain, the underdrained land may be worked so soon that the weeds do not necessarily get a start.

Ground that is made dry underneath may be cultivated to any desired depth, and may then be brought to any degree of richness, without the bad effects that sometimes follow excessive manuring on shallow soils; and the deeper the soil is stirred, the less injury is sustained from drought. The expense of draining a garden thoroughly is, therefore, a mere trifle, compared with the benefits that may be obtained from the outlay.

2d. *For Nursery Uses*, the soil must be susceptible of deep, early and frequent tillage. These conditions can only be secured on lands having a loose subsoil, or such as have been well drained. When drainage is necessary, the outlay of $20 or $25 an acre will be more than returned in a single season.

3d. *The Orchard* will pay as well for drainage as the garden. The necessity of dry land for the orchard is so generally admitted that the highest and driest parts of the farm are almost everywhere selected for this purpose. Orchards planted on river bottoms, in preference to clayey uplands, are no exception to this; for the bottoms, beside having the deepest soil have the loosest subsoil, and are consequently driest underneath. Apple trees, planted over a subsoil that is for a large portion of the year saturated with moisture, are never thrifty, productive or long-lived. Of the various expedients that may be employed to secure the growth of an orchard on wet land, the cheapest and most reliable is underdrainage. The drains should be about three feet below the surface, and midway between the rows of trees; if they are more directly under the trees, the roots find their way into the drains and ultimately close them. In preparing for an orchard, it is desirable to subsoil the ground as deeply as possible across the drains before planting. The whole expense of such preparation is so inconsiderable, compared with the value of one year's produce of a good orchard, that, even without taking into account the increased longevity of the trees, there is no question about the profitableness of underdraining.

4th. *Tilled Lands.*—There are two principal advantages derived from the thorough drainage of tilled lands. The first is, the lengthening out of the time in which work may be done, because the drained lands may be plowed so much sooner than the undrained. But the chief benefit is the increase of the crop. In Old England the average wheat crop has more than doubled since draining was undertaken in earnest. Results equally favorable, though on a smaller scale, have been obtained in the state of New York, and also in Ohio. This increase depends not so much on larger crops than were ever grown without drainage, but in lessening greatly the causes of failure, so that a fair crop is much more certain. Where the expense of drainage is $20 or even $25 an acre, an increase of four or five bushels of wheat to the acre on every crop, or of ten or fifteen bushels of corn, would make the drainage an excellent investment—far better, indeed, than money loaned at ten per cent. per annum. But this is not the principal advantage; for on drained lands a good crop of grain is often grown, while on adjoining lands, precisely similar and with the same tillage, the crop is a failure, so that the difference in one year has exceeded the whole expense of the drainage. There is another fact, also, worthy of mention: the quality of wheat and other grains is greatly improved by the steady growth which good drainage secures, the grain being uniformly plumper, thinner skinned, and therefore heavier.

5th. *Grass Lands.*—It is desirable to have pasture lands sound and dry, and fit for the tread of animals as soon as the feed starts in the spring. It is equally desirable to have grasses root deeply, so as to escape the influence of summer droughts. It is also advantageous to have lands in such a condition that they will produce a variety of grasses, which, by their different periods of ripening, will keep the pastures fresh through the entire season. Orchard grass and red clover will not prosper unless the soil be dry and loose. In meadows that are too wet, the red-top will gradually take the place of timothy, and what is still worse, wild and innutritious grasses will take the place of all the cultivated kinds.

It is doubtless true that grass will grow upon land too wet for any other purpose, but it is a great mistake to suppose that land can not be too wet for grass. The best varieties of grass, the heaviest crops of hay, and the most uniformly fresh pastures, are to be found on soil properly drained. But will it pay to incur an expense of $20 an acre for these advantages? The dairy farmer can readily see that it will pay, if, by draining a piece of wet clayey land, and afterward subsoiling and seeding with orchard grass and clover, he can thereby secure a month's pasturage in the spring, before the grass has started elsewhere, and fresh green feed through the months of July and August, while other pastures are all dried up. Good pasturage at such times is certainly worth more to the dairyman or any stock farmer than the annual interest on the money expended for the improvement.

It is cheaper to increase crops by drainage than by the purchase and cultivation of additional acres. Drained lands pay no more tax, cost no more fencing, and require no more labor than the undrained. When the cost of drainage has once been paid, the increase of crops involves no new expense, as would necessarily be the case if the same increase were obtained from the cultivation of more land.

Some may be inclined to defer this work of drainage until tiles can be obtained at lower rates. It is probable that tiles will be cheaper and more readily obtained in a few years, but this will only happen if a good demand for them is established. The true way to have tiles cheaper is to begin to use them wherever it will pay at present prices. An increased demand will probably secure a better supply and at lower rates.

CHAPTER XVII.

WHAT LANDS NEED DRAINING.

1. *Low Places, Swamps, etc.*—Where the surface is depressed, and water is received from the surrounding lands, the necessity for drainage of some kind is obvious enough. This may be effected by open ditches; and these, perhaps, are the most economical, where the quantity of water to be disposed of is very great. But where ditches would be inconvenient, or gradually fill in by frost or the treading of cattle, or prove an eye-sore, underdraining may be substituted, and, if properly done, with the effect of converting a low place or swamp into a garden, while a single open ditch up the center, which is the usual course, would have left the ground wet and cold; for if the water, in its descent from the higher ground, be but arrested at the edge of the low lands, by ditches or drains, it is compelled to traverse the low land to the center ditch, and does its mischief before it can escape. Wet and swampy lands, when thoroughly drained, are found to be among the most productive, and hence their improvement by drainage is most marked and satisfactory.

2. *Springy Places.*—At the foot of hills, ridges and highlands, the water if often found even in a dry time, oozing out, not, perhaps, at a single point, or in sufficient quantity to make a useful spring, but enough to make the land for rods or acres around, wet and cold, and worthless. In such situations a single drain, in the right place, is often sufficient to put an end to the mischief, and change worthless into fertile land. But what is oftentimes, and in many places, still a greater benefit,

the water which before evaporated injuriously on the surface is collected by the drain, and made available at a convenient point for stock purposes, forming an artificial spring as durable and often more useful than those formed naturally. On farms as poorly supplied with stock water as many in Ohio, the drainage and improvement of all springy places should be effected without delay.

3. *Sandy or Porous Soils with Clayey Sabsoils.*—Sandy soils, as every one must have observed, are not always warm and dry. There is sometimes found at the depth of a foot, or it may be of two or three feet below the surface, a layer of impervious clay, through which no water can pass, but on the top of which it must flow, if there be an inclination in any direction, with the effect of keeping the surface constantly damp and cold. In all parts of the state such lands may be found; they appear mellow and rich, but are always cold and weedy, and produce no valuable crop. They are much more easily and cheaply underdrained than clayey lands, because a different system may be pursued; and when drained, they soon lose all their disagreeable and unproductive qualities.

5. *Clayey and Impervious Soils.*—Clayey soils transmit water downward, but slowly; and consequently, in a wet time, the surface soon becomes perfectly saturated with moisture. It is too wet for crops, too wet to till, too wet to bear the tread of animals; in short, it is too wet for anything. In drying, it sets hard, and becomes more unmanageable than ever; the roots of grasses or other plants can not penetrate to any considerable depth, and clayey lands are therefore the first to suffer from excessive drought, as well as from excessive moisture; there is scarcely a season that exactly suits them, and only a limited portion of the best of seasons that they can be comfortably worked. When such lands are thoroughly

drained, and at sufficient depth, the surface never becomes saturated with moisture; and in drying it never sets hard. A deeper tillage becomes possible, and is indeed required to secure the full benefits of the drainage. With the deeper tillage, the roots of plants enter the earth to a greater depth, and suffer less from drought. Clayey lands, when drained, can be worked at almost any time; they become more friable in texture, cost less in their cultivation, are suited to almost any crop, and retain their fertility longer than lands of almost any other description.

The writers on the continent of Europe on drainage attach great importance to plants, as being the best exponents of the quality of the soil, and of its condition. Herman Wauer, in his work on Drainage, has compiled a list of plants, occupying nearly ten pages of his book. He states very positively that whenever any of the plants named in the catalogue occur in a field, in observable quantities, that that field requires drainage. Almost all the German works on drainage contain similar catalogues; these lists of plants have very little or no practical value in this country, from the fact that either we have many plants which do not appear in Germany, and which are equally as good indices as are those named by them, which do here; but upon the whole, not one fourth of the plants named by these writers occur here, or else the nomenclature is so different that we have failed to recognize our plants in the lists.

We translate the following from Barrall's (France) great work (3 vols.) on Drainage:

" *External Signs of the want of Drainage.*—The aspect of the soil after heavy rains, or great protracted heat, the mode of culture and the nature of the vegetation are very conspicuous characteristic signs, by the help of which we can easily tell that a ground needs to be drained.

"Whenever after a rain, water stays in the furrows; wherever stiff and plastic earth adheres to the shoes; wherever the foot of either man or horse makes cavities that retain water, like so many little cisterns; wherever cattle are unable to penetrate, without sinking into a kind of mud; wherever the sun forms on the earth a hard crust, slightly cracked, and compressing the roots of the plants as into a vice; wherever three or four days after rain, slight depressions in the ground show more moisture than other parts; wherever a stick, forced into the ground one foot and a half deep, forms a hole like a little well, having water standing at its bottom; wherever tradition consecrated, as advantageous, the cultivation of lands by means of convex, high, large ridges; one may affirm that drainage will produce good effects.

"When water stands on the surface, after rain, or when it oozes from the inside, from below as farmers say, there is no doubt that drainage will be the best improvement that can be made.

"In all the above cases, vegetation can not easily take place; crops are scanty and often amount to nothing; the species of plants which find that kind of lands hospitable, signalize them spontaneously to the exercised eyes of an observing visitor; those parasitical plants are in possession of wet lands, and often expel therefrom productive vegetation; weeding is of no avail, drainage only can effect the cure and restore wholesomeness to the ground, and life to the crops.

"Upon the ground which was drained at the Agricultural Institute at Versailles (farm of the menagerie), and which is composed of green clay, of plastic and somewhat calcareous nature, our great botanist, M. Boitel, determined the nature of the indigenous growth which covered it; this nomenclature may be used as a pattern, and therefore we reproduce it. The number 100, in the following table, shows the most common kind; the other species have figures lower and lower, in proportion as they become more scarce:

Proportional fig.	Latin name of the species.	Vulgar name.
100	Juncus communis,	Common rush.
83	Plantago lanceolata,	Plantain.
67	Colchicum autumnale,	
50	Equisetum arvense,	
50	Ranunculus acris; R. bulbosus,	
50	Carex riparia,	
50	Hypericum tetrapterum,	
33	Ajuga genevensis,	
33	Cirsium palustre,	

Proportional fig.	Latin name of the species.	Vulgar name.
33	Cardamine pratensis,	
33	Agrimonia eupatoria,	
17	Valeriana dioica,	
17	Caltha palustris,	
17	Rumex acetosa; R. crispus,	
1.2	Trifolium pratense; T. repens,	Clover.
0.8	Orchis latifolia,	
0.4	Anthoxanthum odoratum.	

Mr. Boitel adds: "The animals will readily eat the clover and the Anthoxanthum only. One sees in what proportions they are found in wet meadows. The other species are mostly of a nature not suited for forage and characterize wet lands. The colchicum autumnale is known to everybody; from afar its leaves look like those of the large leek; its blossoms, of soft, lilac hue, are about three and a half inches long, and make their appearance during autumn, after the fall of the leaves; its fruit winters in the ground; in the springtime the fruit stretches out and sprouts, surrounded with large, compressed leaves; it is a plant very poisonous, which cattle are careful not to touch; they eat it, nevertheless, at the stable, when mingled with hay; a very small portion will then poison and kill them. This noxious plant is very common in wet meadows; it is dangerous, and occupies the place of a great many others that would be profitable. In order to destroy it, dig out the bulbs or onions, and thereby prevent the seeds from disseminating themselves over the whole meadow. The bulbs, being sunk about eight inches into the ground, would cause their extraction to be difficult, but the produce of an abundant and better vegetation will shortly compensate trouble and expenses.

"Together with this plant, rushes, ranunculacea, sorrel, etc., are certain indications of the utility of drainage; they are fond of moisture; it is, therefore, obvious that drainage will cause them to languish and to die, to be soon replaced by species of better quality. It is only by thorough ditching, or rather by underdrainage—the effects of which are still more efficient—that one will succeed in obtaining so fortunate a transformation."

We neglected to state in the proper place that all lands whose indigenous growth of timber was beech, maple, ash, elm, or any other kinds of timber or shrubs requiring wet soil, is seldom tillable, and never profitably so, until it is underdrained.

Some years since, Congress proposed to donate to each state the amount of swamp lands which they respectively contained. In Ohio every county was authorized to commission a surveyor, or other competent person, to ascertain the number of acres of swamp land in each county, and report to the auditor of state. Not more than five or six counties availed themselves of this opportunity to bring the remnants of public lands into market, and the total number of acres reported amounted to 28,000 only. Governor S. P. Chase expressed the opinion that Ohio was entitled to the entire amount of public lands within her territory as swamp lands—at that time about 250,000 acres. But even this amount is less than the actual amount of swamp lands in the state.

It is almost unnecessary to state that open ditches are not required on underdrained grounds, except as main drains, leading to a creek or river; neither is the furrow necessary between lands for the purpose of surface drainage. By discarding the furrows, considerable area for the growth of plants is added to the field; and in this particular, by increasing the superficial area, drainage is a twofold benefit.

We observed, while traveling on the cars over the Dayton and Michigan railroad, throughout the "Black Swamp" region, through which this road passes, that ditches were made in removing the earth in constructing the road, which answered an admirable purpose for draining. In consequence of these ditches, the timber, being of that class which flourishes best in a moist or wet soil, for several rods on each side of the road, was either dead or dying—the ditches evidently drained to the extent of several rods in every direction, and the trees, finding themselves deprived of their accustomed supply of moisture, could no longer vegetate or exist. Not only were the trees dying, but the succulent plants, which require a wet soil, had released their claim, in consequence of man's improvement, and yielded their place to plants requiring less moisture. Hence, the cheapest and most effectual method of ridding meadows of "sour grasses" (*carices*) is to underdrain.

Annexed is a list of plants whose presence is always an unmistakable evidence of the necessity of drainage—because they flourish only in very moist or wet soil. At the same time we are well aware that every practical farmer understands the condition of his soil better perhaps than he does botany, but there are others who may wish to engage in agricultural operations who understand botany better than they do the character or condition of soils. In the "*Wheat Plant*" we devoted a chapter to an examination of the characteristics and qualities of soil, as indicated by the indigenous forest trees which it produced, and the following list is a further demonstration of the same idea. As soon as the soil is properly underdrained, all the plants named in the list will disappear, because their accustomed supply of moisture will then be withdrawn, and they, of course, will perish.

Botanical Name.	Common Name.
Ranunculus alismaefolius,	Water plantain, Spearwort.
" sceleratus,	Cursed crowfoot.
" Pennsylvanicus,	Bristly crowfoot.
Caltha palustris,	Marsh marigold.
Nasturtium officinale,	Water cress.
" palustre,	Marsh cress.
Cardamine pratensis,	Cuckoo flower.
Impatiens pallida,	Pale touch-me-not.
" fulva,	Spotted touch-me-not.
Flœrkea proserpinacoides,	False mermaid.
Rhus venenata,	Poison sumach, Dogwood.
Sanguisorba Canadensis,	Canadian burnet.
Geum strictum,	Avens.
" rivale,	Water, or purple avens.

Botanical Name.	Common Name.
Rosa Carolina,	Swamp rose.
Rhexia Virginica,	Deer grass.
Lythrum alatum,	Loosestrife.
Nesaea verticillata,	
Epilobium coloratum,	Willow herb.
Ludwigia palustris,	Water purslane.
Penthorum sedoides,	Ditch stone crop.
Saxifraga Pennsylvanica,	Swamp saxifrage.
Heracleum lanatum,	Cow parsnip.
Archemora rigida,	Cow bane.
Cicuta maculata,	Water hemlock.
" bulbifera,	Hemlock.
Conium maculatum,	Poison hemlock.
Cornus sericea,	Silky cornel.
" stolonifera,	Red osier dogwood.
" stricta,	Stiff cornel.
Cephalanthus occidentalis,	Button bush.
Solidago Ohioensis,	Golden rod.
" Riddellii,	
" patula,	
" lanceolata,	
Helianthus giganteus,	Sunflower.
Coreopsis trichosperma,	Tick seed sunflower.
Bidens cernua,	Burr marigold.
" chrysanthemoides,	"
Helenium autumnale,	Sneezeweed.
Cacalia tuberosa,	Tuberous Indian plantain.
Cirsium muticum,	Swamp thistle.
Lobelia cardinalis,	Cardinal flower.
" syphilitica,	Great lobelia.
" Kalmii,	
Plantago major,	Rib grass.
Lysimachia ciliata,	Loosestrife.
" radicans,	"
" lanceolata,	"
Chelone glabra,	Snakehead.
Mimulus ringens,	Monkey flower.
" alatus,	
Veronica anagallis,	Water speedwell.
" Americana,	Brooklime.
" scutellata,	Marsh speedwell.
Gerardia purpurea,	
Pedicularis Canadensis,	Lousewort.
" lanceolata,	
Dianthera Americana,	Water willow.

Botanical Name.	Common Name.
Lippia lanceolata,	Fog fruit.
Physostegia Virginiana,	False dragon head.
Scutellaria lateriflora,	Skullcap.
Myosotis palustris,	Forget-me-not.
Asclepias incarnata,	
Polygonum amphibium,	Knotweed.
" Pennsylvanicum,	
" hydropiper,	
" acre,	
" hydropiperoides,	
Rumex verticillatus,	Swamp dock.
" conglomeratus,	Green dock.
Quercus aquatica,	Swamp oak.
" palustris,	Water oak.
Symplocarpus fœtidus,	Skunk cabbage.
Acorus calamus,	Sweet flag, Calamus.
Typha latifolia,	Cat-tail flag.
Triglochin palustre,	Arrow grass.
Alisma plantago,	Water plantain.
Sagittaria variabilis,	Arrow-head.
Platanthera peramœna,	Great purple orchis.
Spiranthes latifolia,	Ladies' tresses.
Cypripedium spectabile,	Ladies' slipper.
Iris Virginica,	Blue flag.
Sisyrinchium Bermudiana,	Blue-eyed grass.
Scilla Fraserii,	Squill, White hyacinth.
Lilium Canadense,	Wild yellow lily.
Melanthium Virginicum.	
Veratrum viride,	False hellebore.
Juncus effusus,	Bog rush.
" scirpoides.	
" militaris.	
" tenuis.	
Cyperus diandrus,	Galingale.
" strigosus.	
Eleocharis obtusa,	Spike rush.
" palustris.	
" tenuis.	
" compressa.	
Scirpus sylvaticus,	Club rush.
" lineatus.	
" eriophorum,	Wool grass.
Eriophorum polystachyon,	Cotton grass.
Almost all Sedges.	
Leersia oryzoides,	White grass.

Botanical Name.	Common Name.
Leersia Virginica.	
Alopecurus aristulatus,	Wild water-foxtail.
Cinna arundinacea,	Wood-reed grass.
Calamagrostis Canadensis,	Blue-joint grass.
Spartina cynosuroides,	Freshwater cord grass.
Glyceria elongata,	Manna grass.
" nervata.	
" fluitans.	
Phragmites communis,	Reed.
Holcus lanatus,	Meadow soft grass.
Hierochloa borealis,	Vanilla.
Phalaris arundinacea,	Reed canary grass.
Milium effusum,	Millet grass.
Sorghum nutans,	Indian grass.

CHAPTER XVIII.

ON THE ABSORBING QUALITIES OF SOIL AND ANALYSIS OF DRAIN WATER.

It has been urged in some very intelligent circles that drainage would, in course of time, impoverish the soil drained, by the drain water carrying off nutritive substances held in solution. At first view this hypothesis, startling as it was, appeared *rational*, to say the least. Way, Liebig, and other eminent and celebrated chemists, determined to ascertain what proportion, as well as what kinds of nutritive elements were borne away by drainage water, when, to their astonishment, they found that the soils at once fixed, and held all the elements necessary for the growth and maturity of the plant, and that the amount escaping by the drains was in an infinitesimal degree only.

Believing that the views and experiments of these chemists on this subject are eminently proper in this place, we here give them in detail:

In 1850, J. Thomas Way published in the *Journal of the Royal Agricultural Society of England*,[1] an essay "On the Power of Soils to Absorb Manure," detailing a series of most remarkable experiments, which will prove of great importance in modifying the theory, and in confirming or disproving the practice of many agricultural operations. These experiments prove that certain manuring ingredients, when brought (in soluble condition) in contact with soil, lose their soluble form, and combine in a peculiar manner with the soil.

[1] Vol XI, page 313.

Way's experiments were induced by observations made by H. S. Thompson and Huxtable, who had found that liquid manure, when brought in contact with loamy soil, loses its color and odor; and, according to the statement of H. S. Thompson, soils have the faculty of separating ammonia from its combinations by withdrawing it from water.

Way proved that the soil affects caustic, carbonate, sulphate, nitrate, and chlorate of ammonia in this manner; the ammonia is arrested, while the acids remain in the solution. He extended his experiments to the salts of potash, natron, lime and magnesia. He found, moreover, that if a solution of phosphate of natron or of guano, in diluted sulphuric acid (containing phosphoric acid and phosphate of lime), is filtered through soil, the phosphoric acid likewise disappears from the solution, and is arrested by the soil. He finally determined the quantities of ammonia and potash, absorbed and retained in this manner, by given weights of various soils.

He likewise showed that when putrid urine, water from the London sewers, and flax water, are filtered through white clay, and a soil rich in clay (on Pusey's estate), the putrid urine loses its odor and all ammonia; and that the rest lose all their potash and phosphoric acid.

One of the important conclusions for practical agriculture deduced by Way from these experiments, was that the soluble ingredients of manure—in whatever form and dilution they are conveyed to the soil—are retained by the soil for the use of the plants. An English acre of soil (of the quality he used for his experiments) ten inches in depth, weight about 1,000 tuns, would absorb three tuns of ammonia. From this he infers: When the combination has once taken place, there appears to be no power in water to distribute this manure in the soil. It follows that if in the application of manure we are not careful to make an equal distribution, we compel the roots of the plants to seek their food at a distance.

The experiments of Way were mostly made in clay soil—white clay and pipe clay; and the comparison of their absorbing qualities with those of sand, induced him to ascribe the absorbing power of soil to the clay (silicate of alumina). He afterward endeavored to confirm this view, by the discovery of the effects of silicate of clay and lime, artificially produced. This latter view, according to which the absorbing qualities of soil were to be ascribed to a cause purely chemical, can not claim general assent. The pure hydrate of clay soil possesses the power of absorbing potash and ammonia in a higher degree than the soils. But the facts discovered by him are entirely independent of his explanation of them. And if it can be proved, that the absorbing power of the soil belongs to the ground, or arable soil in general—whatever be its composition—these facts will establish a new view on the nutrition of plants, and on the manner in which they receive their non-gaseous substances from the soil.

Mr. Way's observations and conclusions refer to certain soluble salts and ingredients of manure only. But as the manure applied to the fields in practical agriculture, does not cause the fertility of soil, but merely contributes to its preservation, it is obvious that the nutrition of plants, existing in the soil and identical with the ingredients of manure, must operate in a manner similar to the latter. And if the soluble manurial elements applied to the field are separated from the solution, as soon as they come in contact with the soil, and form an insoluble combination with the soil, we must infer that the nutritious substances identical with those elements and existing in the soil, can likewise not be conveyed to the plants in a solution, but that the roots of plants appropriate these sub stances in a manner as yet not ascertained.

The roots of plants receive—according to the views of vegetable physiologists and chemists—the elements for their nutrition from a solution. Rain water, of itself, or aided by carbonic acid, dissolves silicic acid, potash, lime, magnesia, phosphate of lime, phosphate of magnesia, and oxyd of iron. This solution spreads in the ground, and is absorbed by the roots of the plants. The plant acts like a sponge, one half of which is in the air and the other in the soil. The water contained in it evaporates through the agency of leaves, while the roots re-absorb the water thus expelled. The quantity of mineral elements conveyed to the roots depends upon the quantity of fluid absorbed and evaporated, and the substances contained in solution in it.

This view evidently must be abandoned, if it can be proved that rain water of itself, or combined with carbonic acid, does not dissolve the mineral elements serving for the nutrition of plants in so perceptible a quantity, that a certain proportion of vegetation can be ascribed to the quantity conveyed in such a solution. Their absorption must, in this case, be ascribed to an active cause co-operating in the roots of the plants, imparting to the water surrounding the root the power of dissolving certain mineral elements, which by itself it does not dissolve. We must furthermore infer, that the quantity of absorbed mineral elements must be in proportion to the root-surface of the plants, and the aggregate of efficient mineral elements contained in those parts of the earth which are in contact with the root-surface.

In order to obtain more definite results regarding these questions, experiments have been made[1] to determine the relations of salts of potash, silicate of potash, and solu-

[1] By Prof. Leibig.

tions of earthy phosphates, to a large number of earths of various regions and different composition, among which there were clayey soils from Hungary, six limy soils from Havana, limy loam soil from Weihenstephan and Bogenhausen, near Munich, three varieties of lime soil from the neighberhood of Munich of Schleissheim. Particular care had been taken to select such soils as were influenced by salts of ammonia, in exactly the same manner as those which Way used in his experiments.

A syphon having the capacity of 300 cubic centimeters of water, was filled with this earth, and a double volume of the solution of salts of potash was filtered through. The contents of the solution in the salts of potash were known, those of the filtrate were quantitatively determined.

Experiments with Sulphate of Potash.—The solution contained in each cubic centimetre: 1 mgrm. of salt. 260 centimetres of the liquid were filtered through loam soil from Bogenhausen, evaporated to dryness, and treated with choride of platinum, they yielded: 0·0325 of chloride of platinum and potassium, which corresponds to 6·2 mgrms. of potash; 518 mgrms. of potash had consequently been absorbed from 1000 CC. [cubic centimeters] of the solution, or 541 mgrms. of potash.

Soil from Hungary (clay soil) treated with the same solution, yielded a filtrate, 420 of which contained 4·6 mgrms. of potash; 535·4 mgrms. of potash had consequently been absorbed from 1000 CC. of the solution.

Garden mold (rich in lime) yielded a filtrate, 1000 CC. of which retained but 16 mgrms. of solved potash.

It scarcely needs to be mentioned that when filtrates still containing perceptible quantities of potash were again brought in contact with earth, they lost it all.

The same quantity of earth absorbed potash from a diluted solution of nitrate and choride of potash to such an extent that the quantity remaining in the liquid after filtration could not be quantitatively determined.

The experiments with chloride of potash proved also that the soil's power of absorbing was limited to potash, excluding the chloride.

Soils are not indifferent to salts of natron; but their power of absorbing natron from its combinations in solution is much less when compared with the power with which they retain potash.

300 CC. of Bogenhausen lime soil were treated in the manner described, with a solution of nitrate of natron (2000 mgrms. in one litre of water), and the 220 CC. of the filtrate, yielded 204 mgrms. of nitrate of natron; the earth had consequently retained but 54 per cent. of the natron in solution. The same quantity of the same earth was treated with an equally strong solution of nitrate of potash (2 grms. per litre) and left no definable quantity of potash in 220 CC. of the filtrate.

A solution of sulphate of natron (2 grms. per litre) filtered through the same soil retained, in 250 CC. of filtrate, 237 mgrms. of sulphate of natron.

The effect of common salt upon soil is like that of chloride of potash; the entire amount of chloride in the liquid is found again in the filtrate, a certain quantity of the base of natron is retained, and we find in its stead in the filtrate a corresponding quantity of lime and magnesia.

Loamy soil, yielding only a trace of lime to pure water, was brought in contact with a solution of common salt, containing three grms. of salt in one litre; its filtrate showed a very considerable amount of lime and a total absence of sulphuric acid.

Specimens of soil were finally treated with a mixture of liquid manure and water, containing, beside, carbonate of ammonia, salts of potash and natron. The amount of the last two had previously been fixed by the analysis of liquid manure: it contained in 125 CC. 86·7 milligrams of potash and 16·8 mgrms. of natron. The liquid manure was filtered through 300 CC. of earth, and 125 CC. were employed for a new analysis. The potash was, in this filtrate, diminished to 5·6 mgrms; of the 16·8 mgrms. of natron, 5 only had been absorbed. The carbonate of ammonia of the liquid manure had been completely arrested by the earth, so that it could not be traced in the filtrate.

These, and a long series of similar experiments with most various soils, prove that the relation of soil to salts of potash (discovered by Way) is altogether a general quality of arable soil. The inferences with regard to the soluble ingredients of *manure* are thus completely confirmed. The facts ascertained by Way establish, therefore, the law, that the plants do not absorb the manurial substances applied to fields in a soluble state directly, and in the form in which they are contained in manure, but that they previously combine with certain ingredients of the soil, whereby they lose their solubility in the water.

Meadow and wild plants receive manure. Although it seems probable that they, too, do not receive their incombustible substances from a solution of the same, but that their roots must, like those of the cultivated plants, absorb their nutritious elements directly from the soil. The experiments of Way, with respect to the manner of nutrition of plants, do not warrant a general application of his inferences. As to water plants floating on the water's surface, the roots of which do not reach the ground, their mineral ingredients must necessarily have been conveyed in a solution.

J. v. Liebig has made some experiments respecting these questions, and from them he is led to believe that the manner in which the land and water plants receive their nutritive elements may be demonstrated.

The uncultivated plants receive the alkalies of their ashes from the silicates, and the phosphoric acid from phosphate of lime, or phosphate of magnesia.

The relation of silicates of alkalies and of a solution of the above-named phosphates of alkalies in carburetted water to the different soils was examined, and it was found that silicate of potash operates precisely like all salts of potash. The determination of the quantity of potash absorbed by the soils is, by the use of this salt, far easier and less laborious than with the other salts of potash, since it has a strong alkaline reaction, and the decrease of potash in its solution can safely be observed with a good reagent (paper).

If a diluted solution of silicate of potash be brought in contact with soil, it instantly loses its alkaline reaction. The quantity of alkali absorbed by a given weight or volume of earth may thus be readily ascertained. The soils for these experiments were measured in uniform powder by means of a vessel divided into cubic centrimetres and brought into a glass bottle; portions of the solution of silicate of alkali were then added, and they were shaken till the fluid manifested a feeble alkaline reaction.

The solution of the silicate contained, according to a previous analysis, in 1000 CC. 1·166 grains of potash free from water, and 2·78 of silicic acid.

SOILS FROM THE NEIGHBORHOOD OF MUNICH.

I. 400 CC. of garden mold (containing 31·8 per cent. of arbonate of lime) neutralized the alkaline reaction of 810 CC. of the above-named solution of silicate of potash; 1000 CC. of earth absorbed consequently 2·344 grms. of potash.

II. 1000 CC. of the same earth mixed with another solution of silicate of potash, containing 1·183 of potash in 1000 CC., absorbed the potash of 1940 CC. of this solution=2·294 grms. of potash.

III. 1000 CC. of soil (loam) absorbed the potash from 2200 CC. of the same solution=2·601 grms. of potash.

IV. 1000 CC. of soil (loam) absorbed the potash from 2·000 CC. of the same solution=2·366 grms. of potash.

V. 1000 CC. of loam (3·77 per c. of lime) absorbod the potash from 1906 CC. of solution=2·206 grms. of potash.

CLAY SOIL FROM HUNGARY.

This soil is of a brownish gray color, and possesses a quality rarely noticed in other soils in Germany. This earth, with water, forms a plastic mass; when rubbed between the fingers it is imperceptibly fine; when decanted off, no sand remains, at least only a few grains, which are partly dissolved, effervescing with acids. The kneaded mass does not, when dried, fall to pieces, and yields, when burnt, a pale ochry-yellow, inwardly black, porous mass, melting in stronger fire. There were three specimens of earth:

I. Cucuritza Batrin. II. *Alba dolina*; and III. Funtmular. 1000 CC. of these earths weighed, on an average, 1232 grammes. They stood in the following relations to a solution of silicate of potash, 1·183 mgrms. of potash in 1 litre:

1000 CC. of Hungarian soil, I., absorbed the potash of 2855 CC. of solution=3·377 grms. of potash.

1000 CC. of Hungarian soil, II., absorbed the potash of 2785 CC. of solution=3·294 grms. of potash.

1000 CC. of Hungarian soil, III., absorbed the potash of 2685 CC. of solution=3·177 grms. of potash.

HAVANA SOILS.

No. I, of gray color; 1000 CC. of earth absorbed the potash from 1526 CC. of solution=1·805 grms. of potash.

No. II, of yellow color; 1000 CC. of earth absorbed the potash from 1058 CC. of solution=1·251 grms. of potash.

No. III, of red color; 1000 CC. of earth absorbed the potash from 1916 CC. of solution=2·266 grms. of potash.

No. IV, of red color; 1000 CC. of earth absorbed the potash from 1769 CC. of solution=2·092 grms. of potash.

No. V, of gray color; 1000 CC. of earth absorbed the potash from 1210 CC. of solution=1·431 grms. of potash.

No. VI, of gray color; 1000 CC. of earth absorbed the potash from 1150 CC. of solution=1·360 grms. of potash.

The nature and quality of these earths prove that their power of absorbing potash does not belong to a certain composition, and that this quality is chemical, and depends upon a certain mechanical quality or porosity.

The chemical relations are obvious in the relation of the salts of potash to soils, and of their conversion into combinations of lime and magnesia. The soils do not absorb the salts, but potash or the base; and the absorption of alkali would not be likely to occur if the acid did not come in contact with a body representing potash and neutralizing the acid.

If the affinity of soil for potash were chemical, the former would depend upon a chemical combination existing in the soil, and the quantity of alkali absorbed would be in proportion to the quantity of this combination.

All the earths examined were mixtures of clay and

lime, and contained a certain amount of sand in mechanical admixture.

If the absorbing quality depended upon the silicate of clay, it would increase with the quantity of lime, or decrease with that of clay. But there is, in this respect, hardly any difference in the earths examined, with exception of the Hungarian, as will be seen in the following synopsis:

1000 cubic centrimetres of soil from

	Bogenhausen,	Garden mold,	Havana, No. III.
Contain - -	6·6 per cent.	32·2 per cent.	57 per cent. of carbon. of lime.
Absorbed -	2366 mgrms.	2344 mgrms.	2266 mgrms. of potash.

These experiments exhibit no special relation of the absorbing power to the clayey contents of these earths. The Bogenhausen loam is so rich in clay that it is used for manufacturing tiles. The Havana earth, No. III, is a dry, poor lime soil, of a red color, due to its oxyd of iron. Both differ exceedingly in their composition, and have, notwithstanding, the same power of absorbing potash.

As to carbonate of lime, we know that a piece of chalk or a porous limestone, placed in a diluted solution of potash "*water-glass*," becomes a stony mass, almost bearing a polish, and but slightly porous; that carbonate of lime—as Fuchs discovered—is not decomposed, but combines with a certain quantity of the silicate of potash contained in the fluid. If we pulverize chalk very finely, wash it, and bring it in contact with silicate of potash, this latter substance is very sparingly absorbed by the fluid. 10 CC. of the solution of silicate of potash, containing 11·8 mgrms. of alkali, were mixed with chalk, and its alkaline reaction was not perceptible until 115 CC. of powdered chalk had been added; the reaction was caused by the addition of water and the subsequent dilution of the alkaline solution, rather than by the absorption of alkali.

This filtered liquid was concentrated by evaporation, and reassumed the alkaline reaction that had become imperceptible, in consequence of dilution.

Pure hydrate of clay soil was found to separate the greatest amount of silicate of potash from its solutions, so that the latter lose their alkaline reaction.

In one experiment, a quantity of hydrate of clay soil, corresponding=2·696 grms. of burnt clay soil, absorbed silicate of potash from 150 CC. of a solution, containing in 1000 CC. 1·185 grms. of potash and 3·000 grms. of silicic acid. If we suppose that one kilogramme of clay soil occupies, as a dry hydrate, the space of one cubic decimetre, so that it is as heavy as one litre of soil, one litre of this hydrate of clay soil would have absorbed the potash and silicic acid from 4600 CC. of this solution, *i. e.*, 26 grms. of potash. This is about seven times as much as the Hungarian earth No. I may absorb from the same solution. We must, therefore, presume that hydrate of clay, in admixture with silicates of clay, partakes also of the soil's power of absorbing silicate of alkali. We can readily perceive that this quality is very complicated.

The soil is not influenced by silicic acid combined in solution with alkali in the same manner as by an alkali alone.

A solution of the silicate of potash, filtered through forest soil, yielded a brown-colored filtrate of a feeble acid reaction, in which the potash and silicic acid were fixed.

Garden mold, loam soil and Hungarian earth were treated in the same manner, and 250 to 500 CC. of the filtrate (which did not react) were employed to determine the silicic acid and potash contained in it. The following are the results obtained:

1000 CC. of a solution of potash, "*water-glass*," filtered through forest soil, retained in solution 215 mgrms. of potash and 2765 mgrms. of silicic acid.

1000 CC. of the same solution, filtered through garden mold, retained in solution 111 mgrms. of potash and 1699 mgrms. of silicic acid.

1000 CC. of the same solution, filtered through loam soil (Bogenhausen), retained in solution 18 mgrms. of potash and 773 mgrms. of silicic acid.

1000 CC. of the same solution, filtered through garden mold, retained in solution 18 mgrms. of potash and 353 mgrms. of silicic acid.

1000 CC. of the same solution, filtered through Hungarian soil II, retained in solution 14 mgrms. of potash and 136 mgrms. of silicic acid.

The applied ("water-glass") solution contained per litre, 1166 mgrms. of potash and 2780 mgrms. of silicic acid; there remained, after filtering through forest earth, 215 mgrms. of potash and 2965 mgrms. of silicic acid; the forest earth had consequently absorbed 957 mgrms. of potash and 15 mgrms. of silicic acid.

The garden mold I absorbed in a similar manner from the fluid, 1055 mgrms. of potash and 1081 mgrms. of silicic acid.

Bogenhausen loam soil absorbed in a similar manner from the fluid, 1148 mgrms. of potash and 2007 mgrms. of silicic acid.

Garden soil II absorbed in a similar manner from the fluid (amount potash not defined), 2425 mgrms. of silicic acid.

Hungarian soil II absorbed in a similar manner from the fluid, 1142 mgrms. of potash and 2611 mgrms. of silicic acid.

The forest earth and Hungarian soil represent, in these experiments, the utmost limits of affinity for absorbing silicic acid. The former had absorbed, from the solution of the silicate of potash, three fourths of the potash and almost no silicic acid; while the latter had pretty much absorbed all the potash and silicic acid in solution.

The liquids filtered through forest, garden and loam soil were so rich in silicic acid that they gelatinized, when evaporated, like the solution of a silicate in an acid. The forest soil, having absorbed the smallest quantity of silicic acid, yielded, in the beginning, a light brown-colored filtrate of feeble acid reaction. The filtrates of garden soil I and of the loam soil were likewise dark colored, and the slight power of absorbing silicic acid may be explained as having its cause in the organic matter or the decaying vegetable substances, as they contained these earths in larger quantities than the others, which had absorbed more silicic acid from the same solution.

The following experiments will, perhaps, confirm this conclusion. The earths examined were dried at a high temperature, and exposed to a red heat in the air. The forest soil lost in combustible substances 30·9 per cent.; the garden soil I, 18 per cent.; the loam soil, 8·7; the Hungarian soil, 9·84; the Havana soil III contained 5·5 per cent. only, the least quantity of organic substances.

Equally large volumes of the last two earths were mixed with the same solution ("water-glass") until the fluids showed a very feeble but perfectly uniform alkaline reaction; they were then filtered, and the silicic acid of the fluid was determined. In these experiments, the earths came in contact with a very insignificant excess of the solution of silicic potash.

1000 CC. of the filtrate of the Hungarian soil retained in solution 1010 mgrms. of silicic acid; the filtrate of the Havana soil contained in the same volume 580 mgrms. of silicic acid. The organic substances existing in the soil—humus—possess the character of an acid, or the quality of combining with alkaline bases, in a higher degree than the silicic acid, and seem to a certain extent to neutralize this power of entering into insoluble combinations with the silicates of lime and clay soil. The chemical nature of the soil has, however, a great influence in this particular.

Thus, the two garden soils stood in varying relations to the silicic acid of the silicate of potash, although they contained very near the same amount of combustible substances. A certain quantity of earth absorbs 1081 mgrms. of a solution of silicate of potash, while another equally large quantity of soil had absorbed 2425 mgrms. of silicic acid; the latter soil contained a considerable amount of carbonate of lime, while the former contained a large portion of silicious sand. The filtrates of both were perfectly neutral, but differed widely in their coloring. The percolated liquid (of the solution of silicate of potash) of the garden soil, rich in lime, was very slightly brown; that of the soil rich in sand and destitute of lime was of a deep brown color.

The forest soil, which absorbed scarcely any silicic acid, yielded, when calcined, a residuum which did not effervesce with acids, and consisted for the greater part of silicious sand. This soil was mixed with about 10 per cent. of washed chalk, dried, and afterward a solution of silicate of potash was filtered through it. The filtrate was neutral and much less colored than it was before (without the chalk). 95 CC. of filtrate yielded 199 mgrms. of silicic acid; 100 CC. 21 mgrms. of potash; 1000 CC. gave, therefore, 2090 mgrms. of silicic acid and 210 mgrms. of potash. The solution (of "water-glass") contained, before its contact with the soil, per litre: 1277 mgrms. of potash and 3230 mgrms. of silicic acid. The same earth which previously absorbed 15 mgrms. only of silicic acid, and 951 mgrms.

of potash—with a large amount of organic substances and a lack of alkaline bases—from one litre of solution (of "water-glass"), containing 1167 mgrms. of potash and 2765 mgrms. of silicic acid—the same earth now absorbed, from the same volume of solution, 1140 mgrms. of silicic acid and 1060 mgrms. of potash. Washed chalk absorbs by itself, under these circumstances, no definable quantity of alkali and silicic acid.

The same forest soil was finally incorporated with lime water into a paste, to which a sufficient quantity of lime water was added from time to time, until it exhibited a feeble alkaline reaction; the latter was neutralized by a supply of earth. This earth had thus lost its acid reaction without the presence of an excess of lime; it was subsequently dried, and combined with a solution of silicate of potash. The filtrate exhibited a feeble alkaline reaction of lime; but the silicic acid had decreased from 3230 mgrms. per litre to 61 mgrms., and the potash from 1277 to 290 mgrms., which had remained in the solution.

Soils, by burning, undergo a remarkable modification in their power of absorbing silicic acid. Loam soil (Bogenhausen) was burnt in the air (in order to destroy the organic substance), and combined with the solution of potash (water-glass).

No diminution of alkaline reaction had taken place in the filtrate; but the silicic acid had been completely separated from the solution. 20 CC. of the filtrate required 24 CC. of a solution of oxalic acid to neutralize it. There existed more alkali in the filtrate than in the liquid employed, and a close investigation proved that a certain quantity of caustic lime had been dissolved.

It results from these experiments that vegetable remains in the soil exert an influence on the distribution of the hydrate of silica to the roots. This may, perhaps, explain the influence of a certain amount of humus in the soil—or of the organic remains of plants with widely-spread roots, as clover—upon the growth of the subsequent plants; as well as the occurrence of plants abounding in silicic acid in stagnant waters and swamps, upon the soil of which great quantities of decaying vegetable matter are accumulating.

These facts warrant the conclusion that potash is afforded to the plants in one relation only, or that they separate it from one combination only.

Chloride of potassium and sulphate or nitrate of potash do not operate in the soil in the form in which they are applied; but the base separated from the acid, the latter forming, with lime and magnesia, salts of another chemical nature.

The plant does not absorb those substances in consequence of a decomposing process in its organism after their reception. The soil accomplishes this decomposition previous to absorption, inasmuch as it separates the potash from the acid with which it was combined, and renders it insoluble in water.

Any soil possesses a certain power of absorbing potash; this power can be represented by a figure; and it is not unlikely that the quality of a soil may be determined by this illustration.

Ammonia, either pure or in the form of salts, acts precisely like potash.

A manufacturer on the Rhine, being desirous of extracting oxide of copper from bituminous marl-slate, in which it existed in the form of malachite and lapis lazuli, hit upon the idea of using ammonia for this purpose, as it had, in experiments on a smaller scale, furnished satisfactory results. He constructed, at considerable expense, an extracting apparatus on a large scale, consisting of two basins, connected with each other by a very wide pipe. One of them was used for the ammoniacal liquid; the pipe was filled with bituminous marl-slate; the second basin served as condenser. Ammonia and water vapor were, according to this arrangement, to be driven with the copper ore through the pipe, to condense there and to dissolve the oxide of copper; the solution was to flow into the second basin. The pipe was afterward to be filled again with copper ore, and the ammonia of the satiated solution was to be driven out by boiling, and to serve again to extract another portion of the copper ore. As the apparatus was hermetically closed, it was hoped that the same ammonia could be employed without loss to extract large quantities of copper ore. One of the two basins served alternately as condenser. The first experiment was satisfactory, inasmuch as a solution of oxide of copper was collected in one of the basins; but when the ammonia was driven through another portion of bituminous marl-slate, it disappeared in an unaccountable manner to the manufacturer, so that the proceeding was ultimately abandoned. The disappearance of ammonia in these operations had been undoubtedly caused by being absorbed by the bituminous marl-slate. This fact may be regarded as a proof of the powerful affinity between them, which does not appear to be neutralized even by the influence of a high temperature.

The power of certain soils to absorb ammonia may be sufficient, perhaps, to separate—in manufacturing artificial manure—ammonia from very diluted ammoniacal liquids, putrid urine and other liquids, and to combine them instead of an acid.

Urine substances, which in putrid state are converted into carbonate ammonia, are not separated from their solutions by the soil. A solution of urine (2000 mgrms. per litre), before or after filtration through soil, required an equally large portion of nitrous oxide of mercury to precipitate it, so that not the smallest portion of it seemed to have been absorbed by the soil.

The relation of a solution of phosphate of lime, phosphate of magnesia, or phosphate of ammoniated magnesia to soil, is similar to that of a solution of salts of potash or ammonia. This seems to prove that the effect produced by soil upon these solutions, is based partially upon the formation of chemical combinations.

While, in the salts of potash and ammonia only the alkali is extracted and retained by the soil, this affinity extends to the phosphates, and more especially to phosphoric acid.

Liebig mixed lime water with diluted phosphoric acid, so that neither alkaline nor acid reaction could be perceived. The precipitate thus produced was dissolved in water holding carbonic acid in excess. Similar solutions were subsequently made (by him) of phosphate of ammoniated magnesia in carburetted water.

Measured quantities of these solutions were then brought into contact with different soils until specimens of the filtered liquids gave manifest indications of the presence of phosphoric acid by a distinct reaction of molybdate. Thus, it was approximatively ascertained that from the solution of phosphate of lime which contained 610 mgrms. of phosphate of lime per litre:

1000 CC. of loam soil absorbed 1098 mgrms. of phosphate of lime.
" " garden soil " 976 " " "
" " soil of Weihenstephan 976 " " "
" " " Schleissheim 976 " " "

These experiments show that equal volumes of these may differ very little in their affinity for phosphoric acid.

The liquids filtered through these soils were almost as

limy as before, and appeared to have lost the phosphoric acid only; but these soils contained a considerable quantity of carbonate of lime. While, therefore, the phosphate of lime contained in the soil was separated from its solution in carburetted water, the latter retained its power of dissolving the carbonate of lime. An acetate was formed from the carbonate of lime, which filtered through without being decomposed.

An addition of chalk decanted off will not cause phosphate of lime to be separated from a solution of phosphate of lime in carburetted water; the fluid retains its reaction on phosphoric acid.

The relation of soils to phosphate of ammonia and phosphate of magnesia is similar to that of phosphate of lime. They exhibited in this salt, too, very slight difference in their power of absorption.

Equal volumes of Bogenhausen loam soil, garden soil, and soils from Weihenstephan and Schleissheim, absorbed the same quantity of phosphate of magnesia-ammonia—that is, they separated this salt from an equal volume of its solution in carburetted water. 1000 CC. of these soils required, in order to determine the presence of phosphoric acid in the filtrate, 1800 CC. of a solution containing 1425 mgrms. of salt of magnesia per litre. The phosphoric acid, magnesia and ammonia disappeared simultaneously from the solution, and the filtrate received an abundant quantity of lime in their place. The filtrate of the Schleissheim soil contained 884 mgrms. of lime in 1800 CC.; the filtrate of the garden soil, 524 mgrms.; that of the soil from Weihenstephan, 402 mgrms.; that of the loam soil from Bogenhausen, 456 mgrms. of lime. These quantities of lime are evidently in no relation to each other or to the salt previously dissolved.

The salt of magnesia is not precipitated from a solution

of phosphate of magnesia and ammonia in carburetted water, when brought in contact with decanted chalk; lime does not supplant magnesia. From the relation of soil to lime, ammonia and phosphoric acid, we may infer that the majority of our cultivated plants do not receive their most important mineral substances from a solution. For if the potash and ammonia are so completely separated from the acids with which they are combined, and from water, that, after the percolation of their solutions through strata that are not deeper than tillable soil, chemical analysis can hardly discover any traces of these substances; it can not be supposed that rain water in itself, or with the aid of a small per cent. of carbonic acid, should be able to separate these substances from the soil, and to form a solution capable of spreading in the soil without losing the substances held in solution. The same remark will apply to phosphoric acid and the phosphates generally. Water, holding carbonic acid in excess, will dissolve this salt wherever it meets in connection with phosphate of lime; but this means of solution can only cause the distribution of phosphates in the soil; the solution can not leave the place where it was formed without its soluble salt being separated again from soil not saturated with it.

These substances exist in the soil in a condition ready to be absorbed by means of the roots; but they are not soluble in themselves by rain water, and can not be separated by means of this solution till the soil holds it in excess.

The composition of our common river water, of springs and of drain water upon fields, serves to support these inferences.

A number of excellent analyses of river and spring water have been made by Graham, Miller and Hofmann, from which it appears that 10,000 gallons, or 500 tuns, of Thames water, taken from five different places of the Thames, contained:

Pounds.	Thames, Dillen.	Kew.	Barnes.	Redhouse, Battersea.	Lambeth.
Potash,	7.3	4.71	3.55	10.	7.3

The following spring waters contained in 100,000 gallons = 1,000,000 pounds:

Pounds.	Whitley.	Cutshmere.	Vellwood.	Hindhead.	Barford.	Cosfordhouse.
Potash,	2.71	2.5	3.	0.7	1.8	6.

Thomas Way found in drain water, *i. e.* in rain water filtrated through soil in a natural manner, the following ingredients in specimens of seven different fields:

	Grains in 1 Gallon=70.000 Grains of Water.						
	1.	2.	3.	4.	5.	6.	7.
Potash,	trace.	trace.	0.02	0.05	trace.	0.22	trace.
Natron,	1.00	2.17	2.26	0.87	1.42	1.40	3.20
Lime,	4.85	7.19	6.05	2.26	2.52	5.82	13.00
Magnesia,	0.68	2.32	2.48	0.41	0.21	0.93	2.50
Oxyd of iron and clay soil,	0.40	0.05	0.10	—	1.30	0.35	0.50
Silicic acid,	0.95	0.45	0.55	1.20	1.80	0.65	0.85
Chlorine,	6.70	1.10	1.27	0.81	1.26	1.21	2.62
Sulphuric acid,	1.65	5.15	4.40	1.71	1.29	3.12	9.51
Phosphoric acid,	trace.	0.12	trace.	trace.	0.08	0.06	0.12
Ammonia,	0.018	0.018	0.018	0.012	0.018	0.018	0.006

Dr. Krocker obtained quite similar results in his analysis of drain water at Proskau (Liebig and Kopp's Jahresb. f. 1853, 742). See table on following page.

These drain waters contain all the substances which rain water can dissolve in the soil, and their composition gives an idea of the quantity which a plant can possibly obtain from this solution during its period of vegetation.

We will suppose that twelve millions of pounds of rain water fall upon an acre of ground during a year, and that the third part of this water has dissolved or holds in

	Drain Water (in 10,000 Parts.)					
	a.	b.	c.	d.	e.	f.
Organic substance,	0.25	0.24	0.16	0.06	0.63	0.56
Carbonate of lime,	0.84	0.84	1.27	0.79	0.71	0.84
Sulphate of lime,	2.08	2.10	1.14	0.17	0.77	0.72
Nitrate of lime,	0.02	0.02	0.01	0.02	0.02	0.02
Carbonate of magnesia,	0.70	0.69	0.47	0.27	0.27	0.16
Carbonate of iron,	0.04	0.04	0.04	0.02	0.02	0.01
Potash,	0.02	0.02	0.02	0.02	0.04	0.06
Natron,	0.11	0.15	0.13	0.10	0.05	0.04
Chloride of the base of natron (natrium),	0.08	0.08	0.07	0.03	0.01	0.01
Siliceous earth,	0.07	0.07	0.06	0.05	0.06	0.05
Total of solid substances,	4.21	4.25	3.37	1.53	2.58	2.47

excess all the substances like the above-mentioned drain waters; that these four millions of pounds are completely absorbed in the months of June, July, August and September, by the roots of the potato plants cultivated in this soil, and are evaporated through their leaves. All the potato plants together would, in this case, not receive a single pound of this solution from the first four fields of an acre each; they would receive somewhat more than one pound from the two other fields (Nos. 5 and 6), and two pounds of potash from the seventh (No. 7).

Now an acre of ground produces an average crop of potatoes containing 408 pounds of ashes, in which there are 200 pounds of potash.

Supposing the fields—whose drain water was analyzed by Dr. Krocker—to be planted with beets, and that four millions of pounds of rain water, holding the mineral substances from the soil in excess, had been conveyed to the plant during the period of its vegetation; this rain water could have conveyed to the beet plants of the four fields of an acre each, only 8 pounds, of another sixteen pounds, and of a third twenty-four pounds of potash.

The average crop of beets on an acre amounts, together

with the leaves, to 100,000 pounds, containing 1,144 pounds of ashes, in which there are 495 pounds of potash!

The amount of ammonia in the drain water analyzed by Way is extraordinarily small. It can scarcely be imagined that one pound of ammonia dissolved in three and a half millions of pounds of water should exert any perceptible influence upon vegetation.

Its quantity could not be determined in a gallon (70,000 grains), of Thames water taken from four places to that amount, and there are in the water taken from the Thames near Redhouse Battersea, 3 parts in 7 million parts of water. The Thames would, when used for irrigation, undoubtedly produce a considerable increase of the hay crop on many meadows; but, assuredly, not by the supply of ammonia, of which this water, as well as river and brook water in general, is so destitute.

The amount of phosphoric acid in the drain, river and common spring water is = nought. Krocker did not find any in drain water; Way found only traces of it in the water from three drains.

It appears from the relations of soil that the plant itself must be active in absorbing its nourishment; its existence as an organic being does not entirely depend upon exterior causes.

If the land plants received their nourishment from a solution, they could (according to time and proportion) absorb only as much as evaporates through their leaves; they could only absorb what the solution contains and conveys. It is certain that the water of the soil, and the evaporation by means of the leaves co-operate in the process of assimilation as necessary agents of conveyance; but there exists in the soil a police protecting the plant from conveying injurious materials, and selecting only what the latter needs. Whatever the soil affords can be conveyed into its organism only by the co-operation of an active cause in the root.

The greatest number of cultivated plants are compelled to receive their mineral nutrition directly from the soil, so that their subsistence is endangered, and they are stunted and die away, if these substances are conveyed to them in a solution.

It is very difficult to imagine in what manner the plants bring about the solution of mineral ingredients. As a matter of course, water is indispensable for its transition.

There are frequently found in meadows smooth lime stones, the surface of which is covered with fine net-like furrows; if the stone is taken fresh from the ground, each deepened line or furrow is seen to correspond to some fiber, as if it had eaten its way into the stone.

The difficulty of explanation should not prevent us from investigating the facts in all directions, and to ascertain the full extent of their influence. There are always exceptions enough.

There must, of course, be other laws for the absorption of mineral nutrition by those water plants whose roots do not reach the ground. They must, like the sea plants, receive it from the surrounding medium; for wherever a plant is growing, it must find the conditions of its existence.

The examinations of waterworts (Lemna trisulea), gave rise to some interesting observations in this respect. This plant grows in stagnant waters, ponds and bogs, and floats on the surface of the water, so that its roots are in no connection whatever with the ground.

A number of these plants were collected (by Liebig), dried, burnt, and the ashes analyzed. Ten to fifteen litres of the swamp water from which they had been taken, and which was slightly green, were filtered and evaporated to dryness. The ashes and salt residuum of the water were subsequently subjected to an analysis.

The large quantity of mineral ingredients contained in this plant was really surprising; still more so was the quantity and quality of the elements of the swamp water, which indicated, by analysis, a very unexpected composition. We will put their analysis in juxtaposition, in order to facilitate comparison.

	Ashes of Waterworts. 100 parts of dried worts yielded 16.6 parts of ashes. There are in 100 parts of the burnt ashes:	Salt residuum. 1 litre contains 0.415 grms of residuum (slightly burnt). There are in 100 parts of salt:
Lime, - - - -	16.82	35.00
Magnesia, - - -	5.08	12.264
Common salt, - -	5.897	10.10
Chloride of potassium, - -	1.45	—
Potash, - - -	13.16	3.97
Natron, - - - - -	—	0.471
Oxyd of iron, with traces of clay soil, - - - -	7.36	0.721
Phosphoric acid, - -	8.730	2.619
Sulphuric acid, - - -	6.09	8.271
Silicic acid, - - -	12.35	3.24

The comparison of the composition of water with the ingredients of ashes, shows that all the mineral substances of the former are to be found in the plant, with the exception of natron, but in a relation very much changed. The water contains 45 per cent. of lime and magnesia, the plant only 21 per cent.; the water contains 0·72 per cent. of oxide of iron, the plant ten times more. The differences between phosphoric acid, potash, etc., are not less remarkable. A selection had obviously taken place: the plant absorbed the mineral substances in the proper proportions for its growth, and by no means in the relations offered by the fluid.

The great amount of mineral elements in the water is very remarkable; for it more than ten times exceeds that in drain water, and from twenty-five to thirty times that in spring water. This water represents, therefore, in its qualitative analysis, a mineral water nowhere else to be found.

The accession of the amount of potash, phosphoric acid, sulphuric acid, silicic acid and iron can be explained without difficulty. There are a vast number of decaying generations of plants gradually gathering in a swamp, the roots of which have taken up from the soil a great quantity of mineral substances. These remains of plants rot upon the bottom of the swamp ground, *i. e.*, they are burnt, and their inorganic elements (or their elements of ashes) are dissolved under the co-operation of carbonic acid, and, perhaps, of organic acids in the water; they remain dissolved when the surrounding mud and earth have been saturated with this solution.

It has been ascertained that this boggy water, when filtered through soil taken up about a foot from the margin of the water basin, does not lose its potash, while the potash of any other soil is rapidly separated from the same water.

Mud of ponds (*muck*), stagnant waters and bogs, is often highly valued as an excellent means of improving the fields and increasing their fertility. This mud operates evidently like soil.

Water percolating through a soil in which remains of plants are accumulating and decaying, dissolves, of course, many mineral substances otherwise not found in those soils.

Verdeil and *Risler's* investigations as to the quantity of potash and phosphoric acid separated from soil by tepid water, are unfortunately not conclusive. They extracted about 40 pounds of soil by means of tepid distilled water, dried the clear yellowish extract, burnt it, and analysed the remains. They found, in the majority of cases, not more than 4, in others 6 to 8 and 9 per cent. of phosphate of lime. Another sediment contained 11, and another 18 per cent. of phosphate of lime. Chloride of potassium and natron amounted together from 3 to 9, in other cases to 14 and 18 per cent. The potash and natron of the silicates together in no case reached 8 per cent.

This investigation did not determine what per cent. of ashes was left by the extract, and how much of these substances had consequently been dissolved by the water. And this was evidently the principal point of this analysis. Had the watery extract of soil been 40 pounds with $\frac{1}{20}$ per cent. of ingredient of ashes, the mineral substances dissolved from the 40 pounds of earth would have amounted to 20 grms., and only 31 mgrms. of potash and 40 mgrms. of phosphate of lime would, according to the analysis, have been separated from 1000 grms. of soil. If the water extracted one fortieth of one per cent. of these substances, the analysis would, of course, indicate only one half of the quantity of potash and phosphate of lime mentioned.

DRAIN WATER CONTAINS—ACCORDING TO KROCKER'S AND WAY'S ANALYSIS—$\frac{2}{10000}$ TO $\frac{4}{10000}$ OF DISSOLVED MINERAL SUBSTANCES.

The relation of arable soil to mineral nutritive substances, which exist in the soil or are conveyed to it in manure, result in some conclusions and applications of great importance to practical agriculture. The first inference is: that the quantity of ingredients absorbed in such cultivated plants as derive their nourishment directly from the soil, in a given time and under otherwise equal conditions, increases in proportion to the extent of the surface of roots, and that the fertility of soil is limited by its contents of nutritive matter in each part of the intersection of the soil downward as far as the roots reach. Liebig observes, in his *Chemical Letters:*

"The principal effect of manure on our fields seems to consist, in many cases, in the circumstance that the upper crust of the fields is more abundantly supplied with nourishment; the plants shoot out, therefore, during the period of their development, a tenfold, perhaps a hundred and a thousand-fold of (root) fibers; their subsequent growth is in proportion to this number of organs, by which they are enabled to find and to appropriate the less copious nutritious matter of deeper strata. This may be the reason why a comparatively small quantity of ammonia, alkalies and phosphates increases the fertility to such a high degree."

PART II.

CHAPTER I.

PRACTICAL DRAINAGE.

Before commencing drainage operations, many things are to be taken into account, the most important of which, in all probability, to the farmer, is, what kind of drains shall be made. Where lands can be purchased from $5 to $15 per acre, it would, perhaps, not be advisable to underdrain with tile, at a cost from $15 to $25 or $30 per acre.

Drainage is designed to be a permanent improvement; as much so as building a house or barn. In all farm improvements, the farmer in the West is proverbial for "cutting the coat according to the cloth." The western farmer is emphatically a practicable man, makes use of such means and materials as he can command, whether it be in accordance with any system "found in books," or not; and to this fact, perhaps, as much as to anything else, do we owe the amount of progress made in agriculture in the state of Ohio, and in the West generally. If the farmer had withheld all improvements, until they could have been made in the most approved manner, we possibly might yet be in the full enjoyment of log cabins and "gar skin" plows throughout the state. Instead of pursuing that policy, however, they have more generally adapted themselves to the circumstances by which they were surrounded, and made such improvements as their means warranted, and it is, perhaps, best for the permanent progress and improvement in agriculture, that the same policy should be pursued with regard to underdraining.

As there are several ways of underdraining, and different materials, or no materials at all, employed in keeping open the water-courses, we will in a synoptical manner enumerate the various kinds of drains, and then devote more space to giving the details of each kind of drain.

MATERIALS FOR KEEPING OPEN THE WATER-COURSES.

1. *Wood.*—This material has long been used, in various forms, for making drains. In swamps, where a general outlet is secured by an open ditch, the side drains leading into it, as well as drains made for the cure of springy places, are often kept open by the brush that is usually found growing in such places. It is doubtless bad economy to use brush, when a better material is at hand; but as this is not always the case, it will be found that a brush drain is much better than none. Saplings, or round sticks, or split wood, are frequently used, cut into equal lengths of three or four feet, and put in the drains at an angle, in the same manner as brush. Or a different plan may be adopted. Straight sticks, of any length, may be used, by laying one on each side of the drain, leaving the necessary space between; then a third pole, or piece, is laid upon them, so as to cover the space, and prevent the side pieces from crowding together. Timber is sometimes split into thin, wide pieces, resembling staves, or the shakes formerly used for the covering log houses, and a water-course is obtained by laying one edge of each piece on the bottom, on one side of the drain, and letting the other edge lean against the other side, some inches from the bottom. In this case, the drains must be dug narrow, or

the stuff split sufficiently wide, so that it can not be forced down flat into the bottom of the drain. For short distances, lumber is occasionally used. A narrow board forms the bottom of the drain; a piece of scantling forms each side, and another board makes the top. This is an expensive method; and although the drain is good while it lasts, it is not especially durable. The choice of these various forms of wood drains must depend on the kind most readily obtained.

Turf Drains.—Almost everybody, perhaps, has heard of turf drains, and, therefore, the question may naturally arise, if turf drains will answer, why incur the expense of tiles? Before tiles were as cheap in the British Isles as they are at present, millions of acres were drained with turf. One method was to dig the drain wedge-shaped, or much the narrowest at the bottom; then the turf, which had been taken from the top, was cut of such a width and shape with the spade, that when inverted and laid in carefully, it would rest eight or ten inches from the bottom, and support all the earth thrown upon it, in filling in the drain, leaving a small wedge-shaped channel at the bottom, which lasted many years. Another plan consisted in cutting down the sides perpendicularly, to within eight or ten inches of the intended bottom; then, with a narrow spade, or one made with one edge turned up about two inches, a narrow channel was dug, and good shoulders left, on which the turf could be firmly laid. These turf drains, in clayey land, and where the work was well done, often lasted a lifetime. Of late years, however, tile have superseded turf in all kinds of soil. Turf drains are, perhaps, more familiarly known as wedge and shoulder drains.

Mole Plows.—These were extensively used in various parts of Europe, some fifty years ago. They seem to be attracting some attention in this country, at the present time. On lands where the subsoil is a tolerably soft and plastic clay, without stones, and where the surface has a regular inclination, they do good service; and the water-courses opened in this way, often continue for many years.

Stone.—Drains of stone are formed either by placing them so as to secure a clear water-course, or the drain is partially filled up with small stones thrown in, and the water is left to find its way between them. A good water-way may be secured either by placing a stone on each side of the bottom, and laying a flat stone upon them, or by setting a flat stone upon the edge, on one side of the drain, and leaning another flat stone against it from the other side. In either way, care must be taken to cover well, with more stones, all the spaces through which sand or earth might pass to obstruct the drain. When small stones are thrown in to form a drain, a great many troublesome little stones may be disposed of, and a tolerable drain made, but it is very liable to become choked with sand or earth, and so made useless. In short, almost all drains, made with timber or stones, are liable to be injured by lobsters or moles, or be otherwise destroyed and rendered unsatisfactory.

Tiles.—Where good tiles can be obtained, at a reasonable rate, no other material should be used, under any circumstances, because no other material makes so perfect a drain, is so durable, or so cheap. The value of tiles depend upon their form, the quality of the clay of which they are made, and the perfection of the burning. Horse-shoe tiles—that is, those of which the end represents a semicircle, with the sides compressed a little, were, many years since, extensively used in England; but this form has, by everybody in that country, been abandoned for better. The water running through them, softens the floor on which they stand, and consequently one, or both sides sink down, and the drain is obstructed. This difficulty is obviated by the use of soles, of the same or some other suitable material; but this adds to the expense and trouble of laying. Narrow boards are used in this country instead of soles. This increases the cost, and the drains, when finished, have only the durability of the wood that is used. Sole tiles are, in general, nearly of the same form, but with the sole added in the manufacture. These are far better than horse-shoe tiles, but are still liable to some objections. Being widest at the bottom, the stream of water, when but little is flowing, is spread out over a wide surface, or it makes for itself a narrow channel, which turns from side to side of the tiles, and deposits in its course the sand which always finds its way into the drains, sometimes stopping them altogether; while, if the tiles were contracted at the bottom, the water would flow along the center in a straight line, and carry the sand out of the drain. Another objection grows out of the necessity of laying them all the same side up; for, if warped in drying or burning, as tiles are liable to be, it is impossible, at all times, to make perfect joints. A form of sole tile became very popular for a time in England, having the sole very narrow, and wholly on the outside, while the inside was contracted at the bottom, so that the opening was egg-shaped, and stood the small end down. This form perfectly obviated the first objection, but was open to the second, and this in practice was found so serious that this form of tile has been abandoned. *Pipe tile*, as perfectly round as they can be made, are now the most approved by experienced drainers. The principal advantages of this form are, the ease with which they are laid, for as all sides are alike, and the tiles are warped in drying or burning, there is no difficulty in so turning them as to secure a perfectly level water-course, and at the same time make perfect joints; they also confine the stream to the center of the channel, and, therefore, leave in the tile no deposit of sand; it is also easier to prepare the bottom of the drain for their reception. The advantages are so decisive, that every one about to purchase a tile machine, of whatever pattern, should order pipe dies; and any one who has the choice of different forms, should select pipes, in preference to any other, for laying. But more on this subject in the proper place.

Therefore, in regions where stone or tile are scarce, or would prove more expensive than in districts where they are more abundant, and especially in sections where land is cheap, it would be as well to commence with

BRUSH DRAINS.

On lands where stones are scarce and tile dear, but where a good descent for the drains may be had, brush drains will answer a good purpose. Being nearly excluded from air, the brush will not decay so rapidly as where more exposed. We have read accounts of some brush drains doing good service for fifteen years; other accounts are to the effect that they cave in and become useless in the course of three or four years. Much, of course, depends upon the character of the soil in which they are made. In situations where the drains will be required to carry off a large amount of water, it must be expected that the sides and bottom will *wash* more or less—the more rapidly it washes, as a matter of course, the sooner will the drains become useless. At best they act upon the same principle as the filter in the ley leach or vat.

"The drain for brush is dug like any other drain, but is best if a foot or more wide. The brush may be cut a few feet in length, and should not be more than an inch or two in diameter. If the branches are straight and nearly parallel, they may be larger and longer than if crooked and spreading. In the latter instance they must be cut quite short, or they will not lie well. Commence always at the *upper* end, and let the butts rest on the bottom of the drain, with the tops pointing upward, or *from* the descent. This position tends constantly to throw the descending water to the bottom or lowest part of the drain. If a sufficient quantity of brush be laid in to fill the ditch, it will occupy, after being trodden down and the earth filled in, only about one third of the ditch. Inverted turf forms a good cover for the brush, before throwing the earth in. The sides should be nearly perpendicular, or the brush will not settle well.

"Being nearly excluded from air, the brush will last many years. Some kinds, as for example, cedar, will last much longer than others. But even when quite decayed, there will still be a good channel for the escape of the water, in the many veins left among the decayed branches, the earth having become compact and well settled above, especially in soils of some tenacity."[1]

Mr. French says:

"Open the trench to the depth required, and about twelve inches wide at the bottom. Lay into this poles of four or five inches diameter at the butt, leaving an open passage between. Then lay in brush of any size, the coarsest at the bottom, filling the drain to within a foot of the surface, and covering with pine, or hemlock, or spruce boughs. Upon these lay turf, carefully cut, as close as possible. The brush should be laid butt-end up stream, as it obstructs the water less in this way. Fill up with soil a foot above the surface, and tread it in as hard as possible. The weight of earth will compress the brush, and the surface will settle very much. We have tried placing boards at the sides, and upon the top of the brush, to prevent the caving in, but with no great success. Although our drains thus laid, have generally continued to discharge some water, yet they have, upon upland, been dangerous traps and pitfalls for our horses and cattle, and have cost much labor to fill up the holes, where they have fallen through by washing away below."

The brush should be laid, as Mr. Thomas says, so as to "let the butts rest on the bottom of the drain," and at

[1] J. J. Thomas, in *Rural Register.*

the same time have the butts laid "down stream," and not up stream, as directed by Mr. French. Where the butts are laid up stream, from some cause or other, the drains choke much sooner than when laid the contrary way. If the water were to pass *over* the brush instead of under or through it, Mr. French's directions would be correct.

The French system (or system practiced in France) of brush drains is perhaps the best. At the bottom of the drain short pieces of wood are driven in to the depth of

FIG. 13.

several inches, as at *a a*, Fig. 13, on which the brush is laid as indicated in the cut. But unless the drains are in a stiff clay, the *wash* will be so great that in a few years the whole drain will be useless.

A very practical gentleman, who is an occasional correspondent of the *Country Gentleman*, says:

"We have at different times within the past fifteen years, made use of small poles for filters in our drains—have used various kinds, such as hemlock, spruce, birch, maple, and recently, black alder poles. These last were from one to three inches in diameter at the stump, and from ten to twenty feet long. The drains were two and a half feet deep, and about ten inches wide at the bottom. Commence at the upper end of the ditch; lay in from four to six poles, according to size, and so on to the end of the ditch, lapping the poles, as directed in filling in brush. Have ready a supply of hemlock, cedar, or spruce boughs, and immediately cover the poles, to prevent the soil from the sides of the ditch falling in and clogging. After the boughs are nicely *shingled* over the poles, step into the ditch, drawing in with a hoe a few inches of soil, treading it solid;

working backward, so as to press the covering firm upon the poles. The ditch can then be finally filled with the shovel or plow.

"Drains thus made fifteen years ago, and at many times since, are this day running as freely as any tile or stone drain would discharge the water. A few years since, I drained a wet, flat, frost-heaving piece of land; before it was drained it was nearly worthless, now it will annually pay the net interest of more than one hundred dollars per acre. It was sowed with winter wheat the first of last September; early in February, the snow disappeared, since which time the surface of the soil has been frozen and thawed more than twenty times, yet none of the wheat plants are thrown out or winter-killed, and the field is as green as when the snow came last November. Without drainage, we think wheat on this land could not have lived at all through such a severe trial. In thorough underdraining there is much hard work and expense, but as far as our experience goes, it is a thing *that will pay.*"

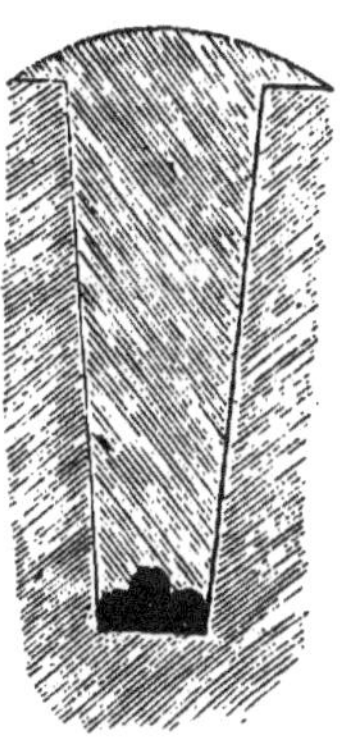

Fig. 14.

In many portions of the country, drains are made as follows: Two poles or saplings are laid on opposite sides at the bottom of the drain; then a third pole or sapling, somewhat larger in diameter, is laid over the two, as represented in Fig. 14; when the poles are laid down, the ditch is then filled with the material which was dug out of it. Drains of this kind, particularly in wet, swampy or mucky land, answer a good purpose for ten or fifteen years. Many such drains are to be found in Northwestern Ohio, where they have given general satisfaction.

In constructing drains of this kind, the poles should be covered with turf, or some other material, to prevent the earth from being admitted between or under the poles. In some instances straw, small stones, and even brush have been placed on the poles in order to make the drains "*draw*," as it is called; but this is simply material and labor lost, because the water will very readily find its way into the drains, and wash out the bottom and destroy the whole drain, if great care is not exercised in constructing them. In some parts of the county, fence rails are used instead of poles. But neither brush, stone, poles, nor rails should be used, if tile can be obtained at reasonable prices. The digging and filling up of the drain cost about the same, whether brush, poles or tiles are used, and since tile will last so much longer, we have cited an instance where tile were laid in 1620, and has made the ground more fertile for all subsequent time, until their removal; it is but reasonable to conclude that tile are in the end much the cheapest. Underdraining at once produces a marked effect upon the crops, whether the conduits are made of brush, poles or tile; the owner of the land is not obliged to wait for years for a remunerative result, as in the case of planting an orchard; therefore, where the farmer can command the means, it is by all means advisable to make the *best* kinds of drains.

PLUG DRAINING, OR SUBSOIL DRAINING.

This system of underdraining does not require the use of any foreign materials, the channel for the water being wholly formed of clay, to which this kind of drain is alone suited. It was the invention of Mr. Lumbert, a highly talented agriculturist, at that time living at Wick Rissington, Gloucestershire (England), where he made the first experiment about the year 1803;[1] in 1845, the tenant (Wm. Bliss), wrote to Mr. Newman, as follows:

"In answer to your letter I have the pleasure of stating that the drains in the field you named are as perfect as when you last wrote me,

[1] Charles Newman. *Hints on Practical Land Drainage*. London, 1845.

and as likely to last as when first made; and my opinion is that if drains are well rammed, and not made when the weather is frosty, the clay draining will last as long as any other drain that can be made. What I have ever seen fail in this neighborhood, has been in a year or two after being made, and in my opinion resulted from not being properly rammed down, or allowing the work to be done in frost, which has the effect of causing the clay to crumble into the drain."

This method of draining requires a particular set of tools for its execution; consisting of, first, a common spade, by means of which, the first spit is removed, and laid on one side; second, a smaller sized spade, by means of which the second spit is taken out, and laid on the opposite side of the trench thus formed; third, a peculiar instrument called a bitting iron, consisting of a narrow spade three and a half feet in length, and one and a half inches wide at the mouth, and sharpened like a chisel—the mouth, or blade, being half an inch in thickness, in order to give the necessary strength to so slender an implement. From the mouth, on the right hand side, a wing of steel, six inches long, and two and a half broad, projects at right angles; and on the left, at fourteen inches from the mouth, a tread, three inches long, is fitted.

The method of using this tool is as follows: When the first and second spits have been removed, the bitting iron is pushed down into the soft clay to the required depth of the drain; it is then withdrawn, and, after being turned round, is again pushed down to the same depth as before, but six inches further back in the trench. By these two cuts, a piece of clay, six inches in length and of the depth to which the tool had been pushed, is separated on all sides, withdrawn by the tool, and deposited beside the second spit. These operations are repeated until a neatly formed trench is completed, from which any crumbs are removed by a narrow scoop.

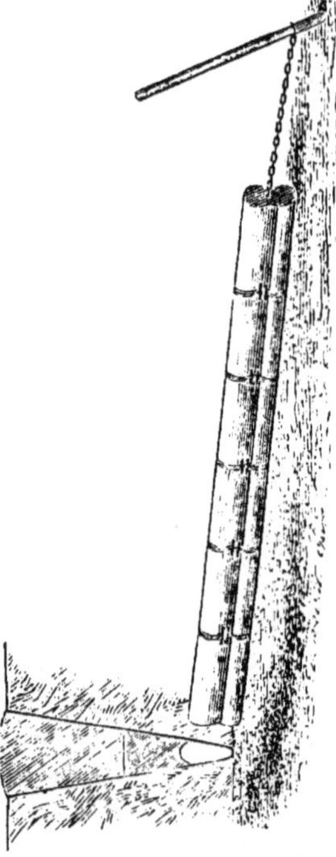

FIG. 15.

A number of blocks of wood (see Fig. 15), each one foot long, six inches high, and two inches thick at the bottom, and two and a half at the top, are next required. From four to six of these are joined together by pieces of hoop-iron let into their sides by a saw draught; a small space being left between their ends, so that when completed, the whole forms a somewhat flexible bar, as shown in the cut; to one end of which a stout chain is attached. These blocks are welted, and placed with the narrow end undermost in the bottom of the ditch, which should be cut so as to fit them closely; the clay which has been dug out is then to be returned by degrees upon the blocks and rammed down with wooden rammers three inches wide. As soon as the portion of the trench above the blocks, or plugs, has been filled, they are drawn forward, by means of a lever thrust through a link of the chain, and into the bottom of the drain for a fulcrum, until they are all again exposed, except the last one. The further portion of the trench, above the blocks, is now filled in and rammed, and so on, the operations proceed until the whole is finished.

Plug draining should never be used when there is a want of fall in the drains, or when there is any risk of flooding, for the tubes formed in the clay are rapidly destroyed when any water remains standing in the drain.

Plug draining, as may readily be supposed, can not be executed very cheaply. The nicety required in all the operations connected with it demands the services of skillful workmen, so that it sometimes exceeds the cost of tile draining. It can only be carried on on lands which yields the material for making pipes; and now that (thanks to railways) coals are so much at the command of most districts, it can not be recommended; and is mentioned here rather as a method which has been used than with any view to encourage its adoption.

WEDGE AND SHOULDER DRAINS.

These were made to a considerable extent, in former times in England, even after the mole plow was laid aside, although they are of the same general character of the plug and mole plow drains, that is, no foreign material is required to form a water channel. Figs. 16 and 17 present a sectional view of the wedge and shoulder drains respectively. The description of them we copy from *Morton's Cyclopædia of Agriculture.*

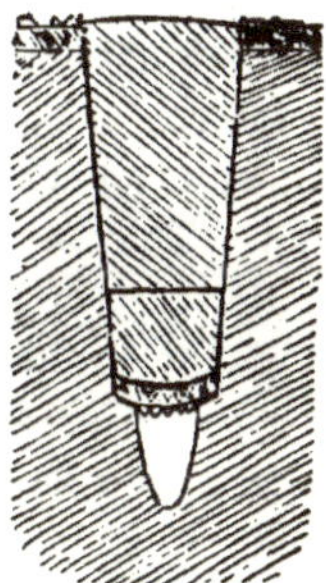

Fig. 16.—Wedge Drain.

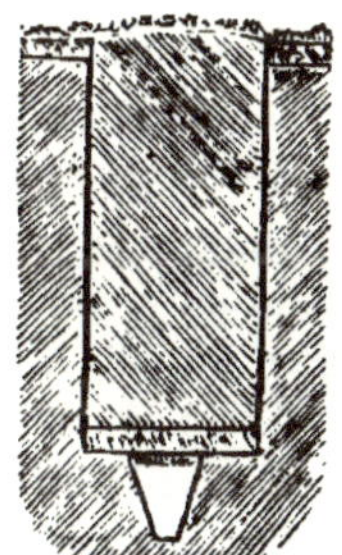

Fig. 17.—Shoulder Drain.

Wedge and Shoulder Drains.—These, like the last-mentioned drains, are mere channels formed in the subsoil. They have, therefore, the same fault of want of durability, and are totally unfit for land under the plow.

In forming wedge drains, the first spit, with the turf attached, is laid on one side, and the earth, removed from the remainder of the trench, is laid on the other. The last spade used is very narrow, and tapers rapidly, so as to form a narrow wedge-shaped cavity for the bottom of the trench. The turf first removed, is then cut into a wedge so much larger than the size of the lower part of the drain, that when rammed into it with the grassy side undermost, it leaves a vacant space in the bottom, of six or eight inches in depth.

The Shoulder Drain does not differ materially from the wedge drain. Instead of the whole trench, forming a gradually tapering wedge, the upper portion of the shoulder drain has the sides of the trench nearly perpendicular, and of considerable width, the last spit only being taken out with a narrow tapering spade, by which means a shoulder is left on either side, from which it takes its name. After the trench has been finished, the first spit, having the grassy side downward, as in the former case, is placed in the trench and pushed down till it rests upon the shoulders already mentioned, so that a narrow wedge-shaped channel is again left for the water.

These drains may be formed in almost any kind of land which is not a loose gravel or sand. They are a very cheap kind of a drain; for neither the cost of cutting, nor filling in, much exceeds that of the ordinary tile drain; while the expense of tiles, or other materials, is altogether saved; still such drains can not be recommended, for they are very liable to injury, and even can only last a very limited time.

MOLE PLOW.

After the advantages consequent upon underdraining became apparent to English farmers, they conceived the idea of underdraining by machinery. Several plows were invented and patented in England, the object of which was to make surface drains of a few inches depth only. The first account of a mole plow which we have succeeded in finding is in the "*Repertory of Arts and Sciences*," vol. 8, a serial London publication, commencing about the year 1796. This is the first record we could find of an implement or machine with which covered or underground drains were successfully made. It was pretty generally used throughout England during a few years, but was soon laid aside—at least, we find no reference made to it as being in general use after about 1805. The following, from Mr. Newman's work on drainage, indicates that greater confidence was reposed in plug drains than in the drains made by the mole plow:

"I should state that the mole plow, worked by a windlass, was a favorite machine of Mr. Lumbert [the inventor of plug draining], for which he had a patent. After his invention of the subsoil system, the mole plow was laid aside—a great proof of the superiority of the former. Although it must be admitted that the windlass mole plow, on soils suited for its purpose, is a very useful machine, it is only calculated for strong clay land; and even on such land it has been frequently found that there is a degree of uncertainty arising from some sorts of clay being too soft, and consequently filling up the orifice and spoiling the drain. It may, however, be considered useful as a temporary and cheap method for the tenant, but it can not be called an effectual measure."

We intended at first merely to mention the mole plow as one of the means devised years ago, and then abandoned, for making drains. But the many recent successes with it in the state of Ohio, in Fayette, Clinton, Madison and Union counties, make it worthy of more than a mere passing notice. We, therefore, copy the account of the *first* mole plow (Fig. 18) from the *Repertory of Arts and Sciences:*

THE PIONEER MOLE PLOW.

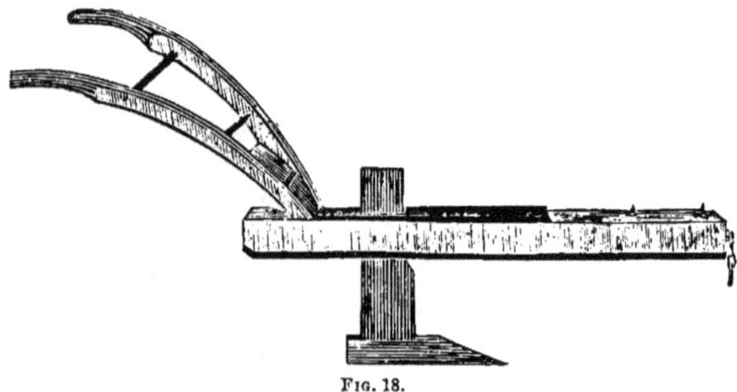

FIG. 18.

DESCRIPTION OF A MACHINE FOR DRAINING LAND, CALLED A MOLE PLOW.[1]

"For this invention a bounty of thirty guineas was voted to Mr. Scott[2], who has described the manner of using the machine in the following letter:

"The bounty above mentioned was given to Mr. Scott in the spring of the year 1797, and, in the month of October following, a patent was taken out by Mr. Henry Watts, for an implement for draining land, the similarity of which to Mr. Scott's mole plow it is unnecessary for us to point out; but what we think highly important to inform the public is, that Mr. Scott, who sold his mole plow for two guineas and a half (indeed, Mr. Welton's letter says, 'the price of the plow complete is about two guineas'), is now the agent

1 By Mr. Adam Scott, of Guildford Survey.—[From the Transactions of the Society for the Encouragement of Arts, Manufactures and Commerce.]

2 When bounties for machines, etc., are given by the Society, it is always upon the condition that the machine, or a model thereof, shall be deposited in the Society's collection, for the use of the public; it is also expressly stated, that "no person shall receive any premium, bounty or encouragement from the Society, for any matter for which he has obtained, or proposes to obtain, a patent."

for the sale of Mr. Watts' patent improvement, at the enormous price of ten guineas. Such of our readers as desire a further account of this matter, will find a long letter concerning it in the *Gentleman's Magazine* for February:

"The mole plow has been used in Sutton Park, for John Webbe Weston, Esq., these three years past, and is found to answer every purpose of underground draining, without breaking the surface any more than by a thin coulter being drawn along, the mark of which disappears in a few days. A man and boy, with four horses, may drain thirty acres in a day, provided there is an open gripe or ditch cut at the lower side of the ground to be thus drained, in order to receive the water from those small cavities which the plow forms in the ground, at the depth of twelve inches or more. The method of using it is, to go down and up, at the distance of fifteen, twenty or thirty feet, as the land may require. This alludes to grass lands; but it is equally good for turnips, when it is too wet for sheep to feed them off, or on any land that is too wet to sow; either of which evils it will remedy in a very short time, provided there is some declivity in the ground. The best time for this operation, in grass lands, is in October or November, when the land has received moisture enough for the plow to work, and not so much as to injure the land or render it soft.

"A further account of this plow is given in the following letter from Mr. Weston, dated Sutton Place, December 9, 1795:

"'With respect to the mole plow, I really think too much can not be said in its commendation; for the purpose of temporary draining, where such is useful, as is the case with great part of my land laid down to grass, it being on a declivity, and is too wet (in the autumn and winter only), after great falls of rain or snow. It being free from land springs, I conceived it improper to be underdrained in the usual way, as thereby the moisture necessary for its producing a crop of grass would be carried off equally at all seasons.

"'The soil is very light, but not sandy, to the depth of from nine to eighteen inches, or more; and underneath is a strong clay, which renders the surface absolutely pouchy in winter; but, from the use of this instrument, the ground on which a man could not walk will, in the course of forty-eight hours, be enabled to carry any cattle. From ten to twenty acres may be easily drained in one day, by a single team, which makes the expense trifling, though it should be necessary to be done every year.

"'The drains made by the plow should be in direct lines, at from ten to twenty feet apart, and all vent themselves into an open furrow or gripe at the bottom. I have used this machine for four seasons past, and with great success. The price of the plow complete is about two guineas. The plow, to the best of my knowledge, is the sole invention of my steward, Adam Scott, whose ingenuity on this and many other occasions deserves every encouragement.'

"There are also two letters from Edmund Bochen, Esq.; the first from Burwood Park, dated March 20, 1796, is as follows:

"'Mr. Scott's mole plow is so contrived that it makes the drains from one foot to eighteen inches deep; the bore two inches and a half in diameter; the soil rather a stiff clay. I made use of six horses, but am inclined to think, from the ease with which they worked it, that four would be fully sufficient. I shall have, next season, a better opportunity of coming to a certainty on the subject. Should you wish for further information, I shall then be happy to communicate what may have occurred.

"'I apprehend this plow can only answer in soils where it is not likely to meet with any material obstruction; in mine, I flatter myself, I shall find much benefit result from its use.'

"In his second letter, from Ottershaw Park, dated February 12 1797, Mr. Bochen says:

"'On the first of this month, in light land, my drain being fourteen inches deep, I worked the plow, without difficulty, with two oxen and three horses; but, in the strong clays, found it work enough for four horses and two oxen, although I reduced my depth two inches. The drains I have drawn on low wet lands and clay run instantly after the plow. On these lands I have generally drawn the drains about twenty feet asunder, and find them much firmer and drier. I conceive that, except in very heavy land, four oxen would be sufficient, and fully equal to two oxen and three horses, as the former step and consequently draw much better.

"'The mole plow, in my opinion, fully answers the intents in such lands as it can properly work in; my only objection being to the strength required to work it, which makes it impracticable when a large team is not kept.

"'It may be worthy remark, that the last year's drains, which were in clay, are as entire, and run as freely, as the first moment they were made.'"

Major Dickinson, of Steuben county, New York, appears to have been the first one to introduce the mole plow in this country. Major Dickinson himself, in a recent address, thus speaks of what he calls his

SHANGHAE PLOW.

"I will take tho poorest acre of stubble ground, and, if too wet for corn in the first place, I will thoroughly drain it with a Shanghae plow and four yoke of oxen in three hours.

"I will suppose the acre to be twenty rods long and eight rods wide. To thoroughly drain the worst of your clay subsoil, it may require a drain once in eight feet, and they can be made so cheaply that I can afford to make them at that distance. To do so, will require the team to travel sixteen times over the twenty rods lengthwise, or one mile in three hours; two men to drive, one to hold the plow, one to ride the beam, and one to carry the crowbar, pick up any large stones thrown out by going to the right or left, and to help to carry around the plow, which is too heavy for the other two to do quickly.

"The plow is quite simple in its construction, consisting of a round piece of iron, three and a half or four inches in diameter, drawn down to a point, with a furrow cut in the top one and a half inches deep; a plate, eighteen inches wide and three feet long, with one end welded into the furrow of the round bar, while the other is fastened to the beam. The colter is six inches in width, and is fastened to the beam at one end, and at the other to the point of the round bar. The colter and plate are each three fourths of an inch thick, which is the entire width of the plow above the round iron at the bottom.

"It would require much more power to draw this plow on some soils than on yours. The strength of team depends entirely on the character of the subsoil. Cast iron, with the exception of the colter, for an easy soil would be equally good; and from eighteen to twenty-four inches is sufficiently deep to run the plow. I can as thoroughly drain an acre of ground in this way as any that can be found in Seneca county."

Within the past three or four years, some five or six patents have been granted to persons in Madison county, Ohio, for improvements on the mole plow. These Ohio mole plows, as well as the Marquis and Emerson or Gopher plow of Illinois, are operated by a capstan, as shown in Fig. 19.

Fig. 19 is a view of the sweep power, capstan, cable and plow, in operation. The team is driven around the capstan attached to the lever, by which the cable is wound upon the capstan, and the plow thus drawn forward. F F are anchor stakes, to secure the power in place while being operated. The cable is 100 feet long, and, when the plow is drawn up to the first anchor stakes, the team is hitched to the body of the power, and it is dragged forward 100 feet, and set again for another turn. In the plow, the shank, B, is set to go the desired depth. This machine is the one made by Lane & Loomer, of Lockport, Ill. Fig 20 is a view of the mole and foot of the shank; this mole is in flexible sections.

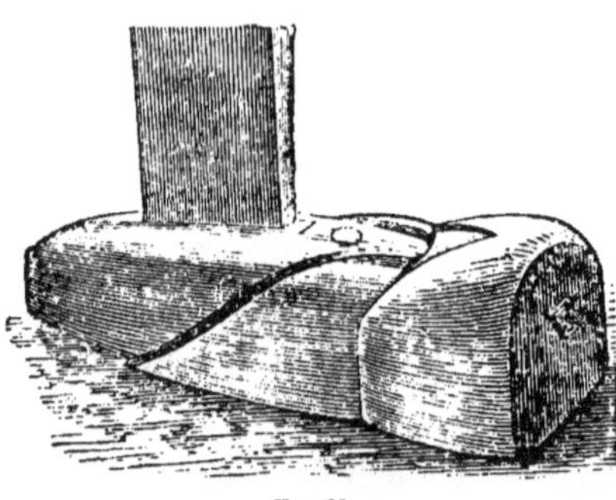

Fig. 20.

Figs. 21 and 22 represent Rowland & Forbis' mole plow, of London, Ohio, patented in 1859; it is also known as the Witherow plow.

The improvements here represented are said to be well adapted to the purposes intended; and the *simplicity* of the adjusting apparatus, in combination with the *strength* of the supports, is certainly theoretically much in its favor. Fig. 21 represents an elevation of the machine; and Fig. 22 is a plan or a view taken when looking down upon it. Similar letters refer to similar parts in both figures.

At A, are the runners for carrying and guiding the

FIG. 19.

front part of the machine. Upon the cross bar of "this sled" rests the end of the beam, B, to which the draft is attached by the rope, and through the sled by the bolt, *l*.

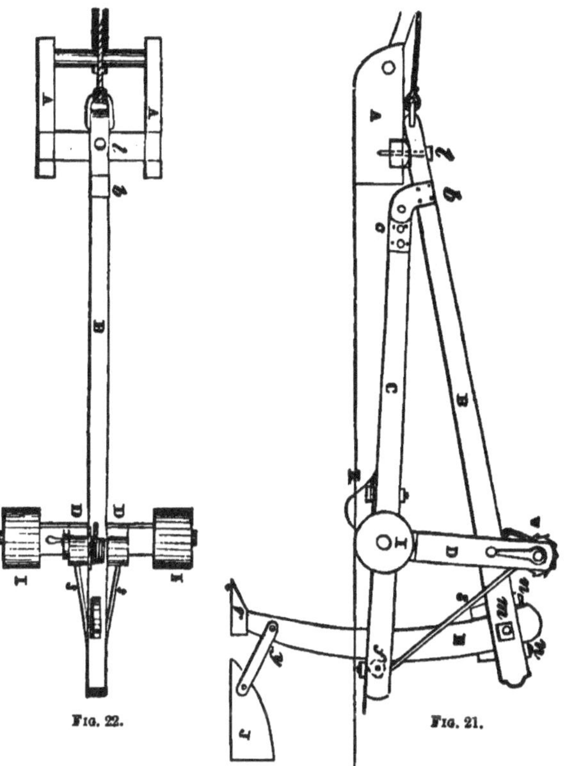

Fig. 22.

Fig. 21.

The rear end of this beam, B, is supported by the cord or rope, and the windlass, by means of which the depth of the mole is regulated. C, is another beam, fastened by iron couplings to the beam, B, as shown at *c* and *b*, and having its rear end resting upon the axis of the trucks, I I.

Upon this axis are placed the supports, D D, which sustain the windlass. These supports are stayed by the rods, *s s*, and hence can be inclined backward, as shown in Fig. 21.

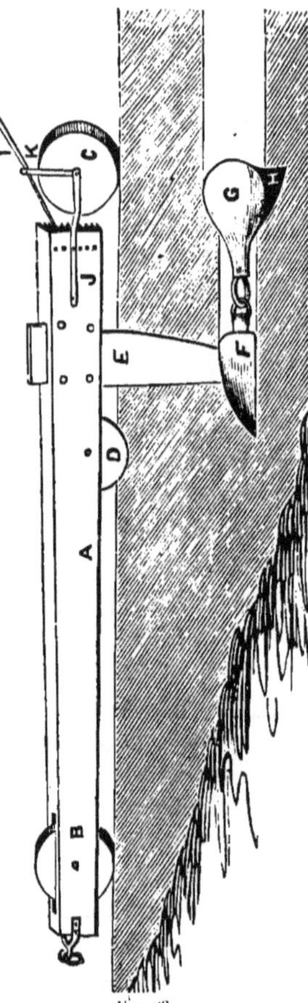

Fig. 23.

E, is the arm to which the moles, *r* and J, are attached. It is fastened securely in the rear end of the beam, B, by the iron bolt, *m*, and keys, *n n*, and passing through a mortise in the beam, C, works up and down against the friction roller, at *f*, and, as is evident, receives great resistance from it. The advantage of having the mole, J, attached as here shown at *y*, is, that it may yield to stones and other objects which would be likely to break it.

There is at K, a cutter bolted underneath the beam, C, which opens the sod and separates roots, etc., before the arm, E.

Letters patent were granted for these valuable arrangements April 19, 1859, to H. W. Rowland and E. Forbis, and the right was assigned to themselves and Washington Witherow.

Fig. 23 represents the Cole & Wall mole plow. A, is the beam, in which a wheel at B, moves in a mortise; at D, is a circular cutter, intended to cut the sod, and thus lessen friction. E, is a cutter bar or arm, to which the moles, F and G, are attached; the mole, G, has a fin, H, attached to the under surface; C, is a wheel, to the axle of which two stout iron bars, J and K, are attached. The bar J, serves as a regulator of the cutter, E, elevating or depressing the mole the depth of the beam, A.

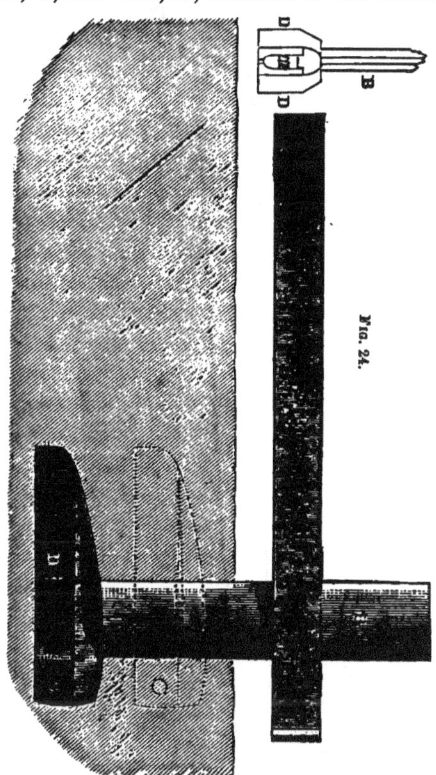

Fig. 24.

Fig. 24 represents the mole and cutter of the Bales' mole plow.

The improvement here represented pertains principally to the mole and cutter—its adjustability to various depths, etc. A, represents the beam; B, the cutter shaft, which is made of cast steel, made light and sharp, and polished, so as to pass through the ground smooth and as easily as possible, that the ground may

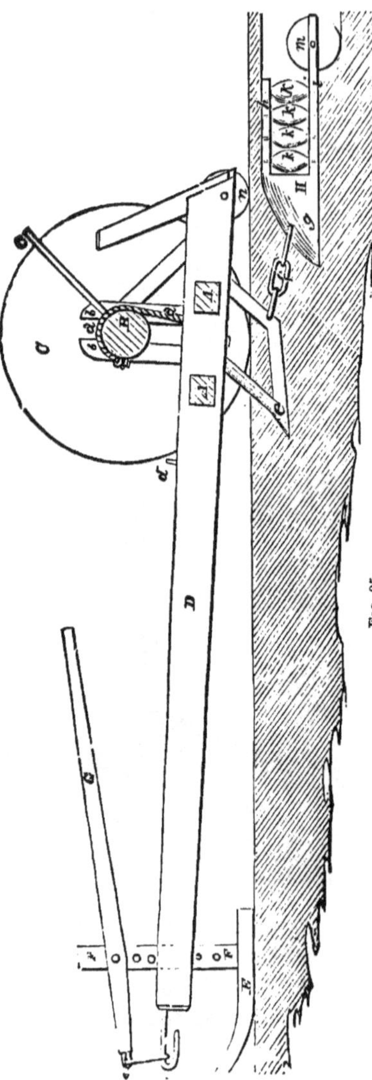

Fig. 25.

readily close behind it. The mole, D, is in the form of a wedge, having a sharp edge in front, and so curved in its upper surface as to form an arch-shaped trench, as shown in section D D, 5 by 7 inches. The mole is hollow in the bottom, so as to prevent its pressing the bottom, permitting the water to rise freely through the bottom, and made of cast steel well polished.

Fig. 25 represents A. Defenbaugh's mole plow. The mole to this plow is attached by a stout link to the lower portion of the cutter bar, or colter, *E*. The mole, H, has a circular fin, *m*, attached to it, and the sides of the mole are furnished with friction blocks or pulleys, *k k k k*, on each side. The colter is fixed and not adjustable, but the beam, *D*, may be elevated or depressed, by means of the windlass, *B*. The forward end of the beam is attached to a *shoe* or *sled*, *E F*.

It would require a very strong team to operate one of these plows if the power were applied directly—that is, if the team were attached to the end of the beam, as in the case of the ordinary plow. But, by employing a capstan, as represented in Fig. 19, two yoke of oxen can operate it with comparative ease. By attaching a dynamometer at the end of the lever or sweep to which the team is hitched, it appears that about 250 pounds is all the draught required to cut a mole 36 to 40 inches in ordinary moist clay; but if the dynamometer is attached where the cable is attached to the beam (D, Fig. 19), the direct draught is indicated, and amounts to about 5,000 pounds. By a simple arithmetical process the direct draught is readily determined, viz: multiply the power applied at the end of the lever or sweep, by the length of the sweep in inches (counting from the center of the capstan or reel), and divide the product by half the diameter of the reel or capstan; the quotient will be the direct amount of power required to operate the implement or machine. For example, the sweep, in Fig. 19, to which the oxen are attached, measures 16½ feet or 198 inches; the capstan or reel measures 16 inches in diameter, and the dynamometer indicates a draught of 250 pounds; what is the actual force or power necessary to operate the plow?

250 pounds×198 inches=49,500÷8 inches=6,187½ pounds.

After the reel or *spool* has been wound full from top to bottom, the doubling of the cable on the reel will cause an increase of power to be applied; the double cord or thread will make the dynamometer indicate an increase of 75 to

22

100 pounds; thus making, in the above-named instance (with a two-inch cable) the actual draught to be 6,435 pounds.

During the first days of July, 1859, a trial of five different patents of the mole plow was had at London, Madison county, at which the writer acted as chairman of the examining and awarding committee. It is not deemed inappropriate to insert the report of that committee in this place—the writer having in the meantime neither seen nor learned anything in relation to these plows to cause him to change a single idea expressed in the report.

"According to previous notice, there were assembled at London, Madison county, O., a large concourse of persons, chiefly farmers, to witness the trial of reapers, mowers, and mole plows or ditching machines. It may not be generally known that within the past twelve months there have been five patents obtained on mole plows, by persons in Madison county. Each one of these plows has special merits, and the agricultural community in that county manifested considerable anxiety to learn, by means of a trial, the comparative merit of each. The entries and description of these plows were as follows:

	Length Beam, ft.	Length Cutter, ft.	Breadth Cutter, in.	Diameter Mole. inches.	Diam. Capstan, in.	Lever Capstan, ft.	Adjustability.	Cost.	Draft.
Witherow & Co.,	16	5	8	5 x6½	16	16½	good.	$100	300 double cord. 225 single "
A. Defenbaugh,	15	5	9	5⅜x7⅜	15	16⅖	good.		250 single " 300 double "
— Bales,	18	4½	7	5 x7	16	16½			250 single " 325 double "
Cole & Wall,	16	5	8x6	4 x5 5 x8	18	18½	not good.	110	250 single " 300 double "
Marquis,	16	4½	8x6	5 x7	18	18½	not good.		225 single " 300 double "

"There being great uniformity in the operation and draft of the plows, the committee found it impossible to take the working qualities as a basis of the award, and therefore took into account cost, adjustability, and the shape of the mole. The adjustability of the Witherow plow, being very convenient in operation, and so graduated that the operator can know at all times the precise depth (by means of a graduated scale) at which the ditch is being made, together with the cost of the plow, determined the committee to award it the first premium. The mole of this plow is an angular ovoid, six and a half inches high, five in horizontal diameter, running down to a flat base of about two inches. The mole might be considerably improved in form.

"The Defenbaugh machine is adjusted with regard to depth by a windlass, attached in the rear of the cutter, or colter, by which a change of eighteen inches may be made in the depth of the ditch, but the operator has no means of knowing precisely at what depth he is cutting. The form of the mole is that of an ellipse, with a flat base, from the center of which proceeds a sharp fin, downward, an inch or more. Upon the whole, the mole is rather better than that of the Witherow plow.

"The Bales plow is not without merit. On the trial he used the capstan of the Witherow machine. The adjustability is more difficult than in either of the preceding ones, while the mole is certainly the most objectionable. The mole is seven inches in perpendicular diameter, and five in horizontal. It is well known that a small quantity of flowing water requires a very limited channel. The mole of this plow presents the same sized channel to a small, that it does to a large quantity of water. When water has a wider channel than absolutely necessary, it forms a zigzag course, and deposits whatever foreign matters, such as sand, roots of vegetables, etc., it may bring with it, at the curves it has made in its course, and in a short time, comparatively, fills up from this cause. But if the channel is so constructed that a small quantity of water has a very narrow channel, and a larger quantity of water a wider channel, the probability is that the channel will be kept clear a much longer period than where a uniformly wide channel is prepared for *all* stages of water.

"Although the Cole & Wall plow is defective in being readily adjusted to different depths, yet, in the opinion of the chairman of the committee, the mole was certainly the best shaped of any presented for competition. Its form is ovoid, and has a fin four inches in depth, extending from the base downward; this fin is about half an inch thick, and makes a deep incision in the earth, in the bottom of the drain, thus making a very narrow channel for the water when at a low stage. When operating, two moles are attached; the first one measures four by five inches, while the second one is five by eight inches. It is claimed that the second mole, being a short distance behind the first one, and being three more inches in perpendicular diameter, completely closes the incision made by the colter, and thus prevents the drain from filling by substances falling in from above, more effectually than the others. On account of the superiority of the mole, the committee awarded to this plow the second premium.

"The Marquis, or Illinois mole plow, is one among the earliest patented in this country. On the trial, it was operated by the Cole & Wall capstan. It is defective in adjustability to different depths, and the shape of the mole was, by the committee, considered to be not superior in form to that of the Bales plow, although evidently more durable in structure, yet objectionable because it makes a drain with a flat bottom of five inches in width.

"Each plow was furnished with one hundred feet of two-inch cable, and each drained or ditched at about the depth of three feet, or forty inches. The length of drain which each is capable of making per day is about the same. The character of the land on which the trial was made, may be said to consist of a stiff clay subsoil, and a rather stiff loamy clay soil. With a good team, any one of these plows can ditch from seventy-five to a hundred rods per day, in the kind of soil in which the trial was made.

"The committee desire it to be distinctly understood that they do not consider these mole plows to be of any considerable utility in any other than level, or very slightly undulating clay lands. For sandy loams, or gravelly soils, or very undulating lands, they can not commend them. In such lands, the only method of securing the advantages of underdraining, is to employ drain pipe tiles.

"The mole plow is useful, inasmuch as it helps to demonstrate the benefits of thorough draining more promptly than it could otherwise be done, although this has never been found the best method of making drains. The fact that work done in this manner is never permanent, and that mole plows are adapted to use on a part only of the lands needing drainage, has always prevented their coming into general use, while tile draining has the advantage of being suited to lands of every character, whatever the nature of the soil or surface, and the further advantage of being permanent, and in most cases actually cheaper than any other method."

We will conclude this notice of the mole plows with the following communication to the *Ohio Farmer*, from the pen of James M. Trimble, of Hillsboro, Highland county, Ohio. Mr. Trimble is engaged in farming on an extensive scale, and is well and favorably known to the Ohio agricultural community:

"Spending some six or eight days on my farm in Fayette, while looking over the farm accounts, I was reminded of my promise made, to give you the result of our ditching and underdraining operations for the year 1860. I have with some care taken from the diary of the farm the ditching and underdraining account. The present account includes the work of 1858 and 1859, which I gave you last year. The creek runs north and southerly across the farm; the work has been confined to the prairie land west of the creek. The open ditches contain in all 2,041 rods, varying from 3½ to 6 feet in depth, and cut at a cost of $910. The land next to, and adjoining the creeks, for some 30 to 50 rods, is from two to five feet higher than it is from 100 to 200 rods west, which required the outlets or open ditches to be some three to five feet deeper across this elevation than those are from 100 to 200 rods west of the creek.

"The drains were put in at from three feet to three feet six inches deep, which received the working of the mole at a sufficient depth in the clay subsoil to make the drains more permanent and lasting. The total amount of underdrains put in is 4,560 rods, at a cost of $190, making the entire cost of open and underdrains $1,100, from which, deducting $536, expended in 1858 and 1859, leaves $564, expended in 1860. Most of the open ditches cut during the last year, were made with the plow and scraper. They are not only a cheaper but a better ditch than those cut with a spade, the dirt being removed so far from the bank as to prevent its washing or falling into the ditch. The fall in the drains and open ditches is barely sufficient to carry off the water. In time of high water, the creek leaches up some of the open ditches a distance of 200 rods. The underdrains were laid so as to receive a regular fall of about one inch to 500 feet, which I think a decided advantage in mole drains, as it secures them permanently from any current or wash, in throwing off the surplus water.

"The land, with the exception of an occasional grove of timber was broken up, planted, and *well* cultivated in corn this year, with the exception of sixty acres of prairie sod, broken late; it was planted in corn with Barnhill's drill, thinned out, but not cultivated.

"Although not accurately measured with compass and chain, yet we have so far measured the ground as to satisfy us, that, excluding groves, we had 400 acres in corn, 200 acres of which have been cut up and put in shock. Of that left standing, we have husked and fed at the rate of 100 bushels per day, for the last sixty days, and have husked out a number of fallen shocks of that shocked up. From the amount husked out, we have no doubt that the entire crop of 400 acres will average over 66 bushels per acre. My son and Mr. Jere Shelton, who have charge of the work and farming the land, concur in the opinion that fifteen bushels of corn per acre is but a fair estimate for the *excess in yield, on account of underdrains.* I would put it at twenty; but taking their estimate, the account will stand thus: Farm Dr. to open and underdrains, $1,100; Cr. by extra yield of crop, 15 bushels per acre on 400 acres, which is 6,000 bushels of corn, at 25 cents per bushel, makes $1,500, from which deduct $564, cost of open and underdrains for 1860, and you have $936 as the profit for 1860.

"Now, this looks like extravagant work on paper, and the question might be asked if 11¾ rods of underdrain, and some two rods of open ditch per acre (about one half of the open ditch, extending over 900 acres of land), at a cost of $2 per acre, or less, well give these results, what will thorough drainage do? To this I would reply, judging from the corn, directly over and within six to eight feet of the underdrains, and comparing it with that beyond the influence of the drains, I would put the crop at 100 bushels in lieu of 66 bushels.

"As to the question of durability of mole drains (a very important item in their economy), I can only say, from present indications, my better impression is, that they may last ten or more years, and that they *will last five years,* I have no doubt.

"The fall rains have started the most of my underdraining; they are throwing off small streams of water, as freely as they did last spring; if there is a single defective drain on the farm, I am not aware of it. Last spring, in constructing new drains, it became necessary to connect with those made a year previous. In digging down to, and boxing up these connections, we found the original drain sound and perfect in every instance. I have no fears of my drains crumbling in, caving in, or filling up, at least for many years to come. Most of the defective mole drains that I have seen or heard of, cave in from the top of the ground to the bottom of the drain, or they fill up. This is owing to two causes: First, too much fall has been given to the drain, and the seam or aperture made by the cutter bar is not permanently closed at the top of the drain, either of which is fatal to a mole drain. Second, the grade of the drain should be regular, and not run so as to make a syphon; lead pipes or tile may answer in such drains; without either the one or the other, the drains will fill up. A mole drain, with a regular, gradual fall of one inch to 1,000 feet, is abundantly better than one with irregular falls and rises, as the inequalities of the ground happen to be, with a fall of three feet in 1,000.

"In the construction of mole drains, my experience has taught me that the great trouble and danger is in the top, the arch and roof of the drain, and not in the bottom, as some suppose. The last ditcher (Emmerson's patent), the mole has been improved—as I think, very much improved in form and shape. The bottom is hollowed out more; the sides are rounded from the bottom to the apex of the cone of the mole, so as to throw the pressure equal from the side and bottom of the drain to the cone or arch, and the *nub* on the end of the mole, back of the cutter, is elevated some three inches above the level of the mole, which effectually closes the aperture made by the cutter, making it as solid and permanent as any other portion of the drain. My examinations in digging down to, and cutting away the arch, or the roof of the drain, has satisfied me that nine tenths of all the water going into the drains, enters at the bottom of the drain, and not through the roof and sides; they are more impervious to water than tile. From the best information I can get, there are at this time not less than 200 miles of mole drain in Fayette and Clinton counties, put in within the past two years, and at an average cost not to exceed five cents per rod (when the owners of the land put it in); in some cases, by contract, ten and fifteen cents per rod was charged. The drains, so far as I can learn, have very generally given satisfaction.

"I may have been tedious in this statement, if so, my apology is in the importance of the subject of underdraining our lands, and the economy in the use of the mole plow in preference to tile or stone."

Mr. J. C. Miller, of Union county, Ohio, is the inventor of a traction engine, propelled or operated by horse power, to which is attached a mole plow, that opens the channel and cements it as it progresses, by letting down a cement ready made of water lime through a flat tube in the rear of the cutter or colter, directly upon the rear of the mole, and which is pressed into shape by a wooden mole trowel following.

For ourself we have no confidence in this matter of cement, for the reason that none of the arches break down from above, but where the arches do fall in, the sides and bottom wash so as to let down the top, and the same causes operating will cause the cemented arch to fall. In making the drain with the mole plow, the top and portions of the sides become very hardly packed; in fact, so much so as almost to exclude water, and the coat of cement will give it no additional support. A sufficient amount of cement we do not think could be introduced to make a substantial arch for a loamy soil.

In addition to the kinds already described, there are yet, aside from pipe tile and stone drains, those described in British works under the head of *Bog drains* and *Sheep drains*. Having never witnessed any of these kinds of drains, and being doubtful whether any exist in the United States, we copy the following description and figures from *Morton's Cyclopædia of Agriculture:*

SHEEP DRAINS.

"In forming sheep drains, the main drains should be first opened in the most suitable places, and the minor drains then led into them. In peaty or boggy places, the workman first proceeds to mark out, on both sides of the drain, with a strong and heavy edging-tool. This tool should be edged with steel, and have a cross handle, which the workman can seize with both hands. This is the tool we have found all workmen to prefer; for by simply raising it a foot or so above the surface, and causing it to descend with sudden force, it cuts through the tough, wiry stems of heath, or other obstacles, and at once makes a cut of considerable depth into the sod also. When the line of drain is thus marked out, the workman proceeds to divide the sod which he has separated, into convenient lengths, by transverse cuts with the same tool; and these he drags out and deposits on the lower side of the line of drains, by means of a light drag, placing the grassy side undermost. The drain is afterward deepened, and finished off with a common spade; the soil or peat dug out is placed upon and behind the sods already removed, after which, a blow or two from the spade gives a finish to the bank and completes the operation, giving the trench and bank the form represented in the following cut. In places where the surface is not of a boggy nature, the common spade must take the place of the edging-tool and drag.

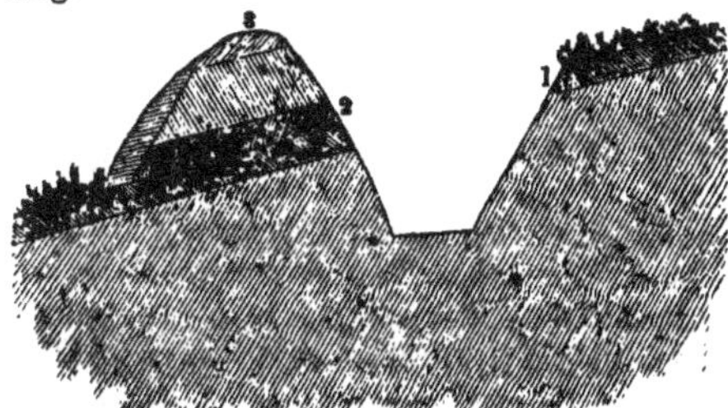

FIG. 26.—SHEEP DRAIN.

"*Bog Draining.*—In draining deep bogs, the removal of water causes a great alteration in the hight of the surface of the bog, which rapidly sinks as the drying process goes on. This constant alteration of the level, and the soft nature of bogs, render the use of heavy materials, for forming drains in them, improper. Where the bog does not exceed six or eight feet in depth, the best plan is to cut quite through it, to the bed of solid material on which it rests, and then to form drains of some of the more permanent materials; but when the bog is so deep as to render this plan impracticable, the proper course to pursue is to divide it into brakes, by means of large, open ditches, into which the subsidiary drains are made to empty. The subsidiary drains may be formed somewhat in the manner of the shoulder drain. They must be made at least eighteen inches wide, and the turf first taken out is to be laid on one side, to be used for forming the roof of the drains. The trench should not be taken out the full depth of the drain at once, but should be left unfinished for a few months, in order that the bog may subside. Before the autumnal rains commence, the drains should be finished, by paring down their sides in a perpendicular direction, to within a foot of the depth they are intended to be made. The bottom spit is then taken out, precisely as in the case of the shoulder drain in grass lands, except that it must be wider, to allow for the sides coming somewhat together, owing to the soft nature of the bog. When the trench is neatly and properly finished, the turf first removed is to be returned into the drain, which it should just fit. The surface portion should be placed undermost, resting upon the shoulders; the rest of the trench should then be filled up with the remainder of the peat which had been removed. If proper care is exercised in cutting the drain, the pieces may be made to fit neatly into the trench again, forming a very complete and efficient drain.

"In bog draining, a conduit has been employed, formed of peat, somewhat in the form of a pipe in two halves, Fig. 27. These pipes are formed at once in the bog, by means of a peculiar kind of tool, invented by Mr. Calderwood. If properly dried, they are very durable, and can be formed at a very low price, as an expert workman can turn out two or three thousand a day, when the peat is of suitable description. From their lightness, they answer well in bog draining, and it has been attempted to introduce them into draining operations in arable land; but the low price at which tiles can now be made, almost at any place, renders such a practice very questionable economy."

Fig. 27.—Peat Tiles.

These, then, are the principal kinds of drains in which the conduit is composed solely of the materials of the ground in which they are formed. The cost of a permanent drain now so little exceeds even the cheapest of those described, that special and weighty reasons alone can justify the employment of any other. We shall now consider the more durable forms of drains.

STONE DRAINS.

Every portion of the country appears to be abundantly supplied with materials of some description for draining. Where timber is scarce, stone is abundant; or, if both are wanting, then there generally is an excellent deposit of clay, from which tile may be manufactured. It is an established fact, that underdraining will pay all reasonable expenses incurred in its construction in the course of three or four years, and not unfrequently the first year alone, by the increased productiveness. It therefore behooves the farmer to consider well what kind of drains his present means will justify him in making. The digging and filling the drains will cost about the same for any kind, except tile—the difference in cost, then, will depend upon the material employed. If stones are to be hauled two or three miles, then, perhaps, wooden drains, as already described, would be cheaper, and will answer the purpose for some five or six years—at the end of which time the farmer will be enabled to redrain, and in a more thorough manner. But, if stones are abundant on the field to be underdrained, or in the adjoining fields, it would, perhaps, be a matter of economy to employ the stone, for two reasons: *first*, stone will make a drain which will secure the object intended; and *second*, the surface of fields will be cleared of a great nuisance and hindrance to a more perfect system of culture.

Stone drains never should be dug *less* than three feet deep, and one foot wide on the bottom. Stone should be filled in to the depth of one foot, at least, and then be covered with brush, straw, leaves, sod with the grassy side down, or some such material, so as to prevent the dirt from falling in and filling up the interstices.

What kind of stone shall be employed, and how should they be placed in the drain?—In many places persons would, perhaps, be obliged to use the rounded little bowlders, found in the beds of streams, or stones of this character which are found on the surface of fields. Flat stones, or fragmentary ones from quarries, are not always accessible or within a proper distance. Where the rounded bowlders are employed, many persons are of opinion that the manner in which they are laid in the drain is a matter of no consequence whatever, and, to use their expression, they "just throw them in 'higglety-pigglety,'" feeling certain that there will at best be ample space for the water to pass through. A drain of this description is represented in Fig. 28. It must be obvious to every one that where the drain is filled with stones, without any regard to forming a continuous channel for the water, fine dirt will be carried down from the sides, or from the stones themselves, and in a very short time the interstices in the bottom layer will be found to be completely filled up, and, as a matter of course, rendered entirely useless. Layer after layer will, year after year, be filled up, until the drain is rendered valueless in the last degree. A better method of using the bowlders is described by C. G. Calkins, of Ashtabula, O.

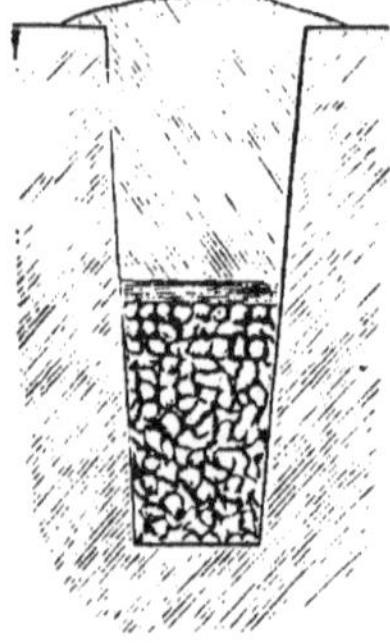

FIG. 28—DRAINS FILLED WITH SMALL STONES.

"Some four years since, an old countryman in my employ informed me that he could lay an effectual 'pipe' of small stones regularly in three courses, one on each side, and one on the 'shoulders' of these, forming the top. The top course must be laid so as to wedge between the others, to keep them apart, and must be covered with turf, straw, or something to keep the earth from filling in, until an enduring crust is formed. We tried 'taking up' a water vein, in a hillside, running along nearly on a level, and forming numerous springs. There is a strata of quicksand, in or at the bottom of which the stones were laid. The trench was dug from two to four feet deep, and no wider at the bottom than was necessary to receive the 'pipe'—say one foot. It was filled rather imperfectly, being on a steep bank, where tilling could not be done. It was fully successful—intercepting all the springs, emptying them in a single and constant stream at the mouth of the drain, and continues as good as at first.

"The amount of stones required in a drain of this kind is not large, and an experienced hand will lay 30 or 40 rods in a day.

"Some days after a light rain, and when all around is dry, this drain is seen discharging water, though dug in a ridge of the hardest clay soil.

"It will avail little to express my faith in the utility or practicability of this or any other mode of underdraining, so long as that faith is not followed by 'works' more extended. It, however, appears providential that, in a region almost destitute of stone, these little bowlders should appear so well dispersed, and at the same time so fitted to this important use—draining the soil they now incumber. However, I am of the opinion that if the manufacture of draining tiles was commenced, there would soon be a good demand for them, as being the most convenient and suitable.

"In building a fence on the side of the garden, we dug a trench some two and a half feet deep, set the posts on the bottom, and filled around them with loose stones to the top of the ground, then filled the spaces between with stones thrown in at random, mainly to the depth of a foot or more, and, after covering with turf, filled up with dirt. This has been a good and useful piece of work, for, beside draining the land, it preserves the fence from the action of frost, and in a measure from decay."

In commenting on this statement of Mr. Calkins, the editor of the *Country Gentleman* says:

"We have practiced this mode for many years, before the introduction of tile. When stones are on the ground and abundant, they may be used to advantage. The objections to their use are two: first, the earth is liable to work down among the stones, or 'cave in,' where streams run across the surface in heavy rains or in thaws, and find their way down through the soil; second, the increased labor of digging a drain wide enough to lay the stones well, will pay for tile, if not very remote from a tile manufactory.

"The mode of laying must vary with the soil. Those soils which approach quicksands in character, render it almost impossible to use stone successfully. When they are saturated with water, they will find their way among the stones through every avenue—at the top, bottom and sides. It is rare that such drains endure many seasons uninjured. The best security for them is to lay, first, flat stones on the bottom (or hard, durable boards or slabs), to prevent the cobble stones from sinking into the earth; to use as small stones as practicable against the sides of the ditch, so that the interstices there may be too small for the soil easily to enter; and to cover the top with very small stone, and then very coarse gravel, or with flat stone, for the same object, before the straw or inverted sods preceding the earth covering are applied. In stiff or clayey soils, the earth rarely falls among the stones, even when little precaution is taken. After practicing underdraining with stone on such lands for many years with entire success, we had occasion to adopt the same mode in another district of country, where the soil was light and much more sandy. The first spring destroyed the value of most of them by the caving in of the soil, and this evil was only prevented effectually, by covering the stone filling either with flat stones, gravel, or hardwood slabs, before applying the earth at the top.

"As a general rule, we would not recommend the use of cobble stone, except in soils of considerable tenacity.

"The importance of a good drain under every post fence, is not generally understood, and we are glad to see the subject alluded to by our correspondent. Wherever post holes retain water, they are sure to be heaved by frost, and the fence thrown out of shape; and the posts can not last long, where they are alternately subjected to water soaking and drying. But if all the water which falls, passes immediately down into the ditch, it can not lie in contact with the posts long enough to soak them, and as a consequence, they must remain perpetually dry, and last for a long period. Robert B. Howland, of Union Springs, New York, who has used Pratt's ditcher with success, found it cheaper to cut ditch with this machine, in which to set the posts for a fence, than simply to dig the post holes by hand, and he thus attained all the advantages of drainage, beside a practice well worth copying.

"A single suggestion on the efficacy of underdraining, on lands that do not at all appear to need it. It is a very good rule for determining its necessity, to observe whether water will *stand* in holes dug two or three feet, for this purpose. If the subsoil is porous, the water will immediately sink away, and ditches would be wholly useless. But if water will stand forty-eight hours in the holes, draining is necessary to relieve the subsoil of this cold and chilling mass which fills it.

"Now, if the surplus water in the soil and subsoil, at the wettest period, is only equal to a depth of two inches, then for a ten acre field it would amount to more then *seven thousand hogsheads.* Suppose, therefore, that this field has a slope, so as to give it what many would suppose a *natural* drainage—'not needing any ditching'—'dry enough already'—then, in getting rid of these seven thousand hogsheads of hurtful water, it must, every gill of it, soak, drop by drop, from one particle of earth to another, until it all passes slowly down, almost imperceptibly, from one side of the field to the other. No wonder that days, and even weeks, are required to complete the process, and to render the land dry enough to become friable and fit to receive seed, and promote the extension of the young roots of crops. No, give this field a smooth, tubular channel of tile, for every two roads of its whole surface, the shortest way down the slope; the water in the soil then has only about *one rod* to soak through the soil before reaching one of these drains, and most of it much less than a rod. When it reaches them, it shoots rapidly down the smooth descending tube, and in a few minutes has passed the boundary of the field, instead of being otherwise compelled to *soak* its weary way the whole forty or fifty rods, or entire breadth of the field. This rapid discharge reduces the soil to dryness in so short a time, as to surprise those who have never before witnessed it, and to lead to the common supposition that the simple statement of the practical advantages of thorough underdraining, by those who have given it a trial, are wild exaggerations."

Where flat stones can readily be procured—in places where bowlder or cobble stone abound—a combination of the two will make the best drain.

The most common way, and usually the best, for filling stone drains, where the stone are nearly *round*, is made by just laying a row of small stones on each side of the bottom, leaving an open channel between them about three inches wide, and then covering this channel with flat stones, and filling the ditch with small ones promiscuously thrown in, to within about 15 or 18 inches of the surface, so as to be below the reach of the plow—and the remainder with earth. It is hardly necessary to remark that the upper surface of the stone must be either covered with coarse gravel or small flat stone, and then with straw or inverted sods, to exclude the earth from the stones; and if the soil is nearly free from clay, more care in this respect will be needful—and perhaps a covering of hardwood slabs will be necessary to keep the earth in its place. If the bottom of the drain inclines to quicksand, a layer of flat stones must be first laid on the bottom. We mention this common mode of constructing stone drains, in order to show the superiority of the flat stones spoken of by our correspondent. The chief objection to the mode just described, is the necessity of cutting a ditch nearly a foot wide at the bottom, to allow laying the channel. The flat stones, when they can be procured in any quantity, on the contrary, obviate the labor of cutting a wide ditch; the channel being constructed by placing three flat stones together, as shown in Fig. 29. The bottom of the ditch is cut with a pointed spade, so as to have an angular trough; flat stones and then selected, all of the same width, and fitted into and meeting each other at the bottom, and then covered by a third flat stone, reaching across them. The ditch above this is partly filled with irregular fragments of stone, and covered as already described.

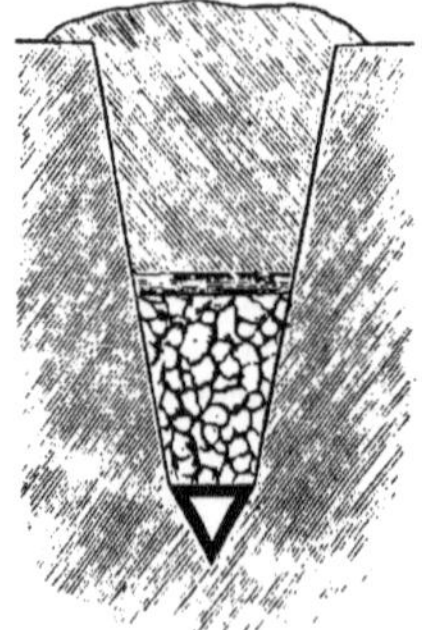

FIG. 29.—THE TRIANGULAR STONE DUCT.

A still better way, where the earth is hard and the quantity of water not large, is as follows: The ditch is cut with the narrowest kind of spade—a mode familiar to English ditchers, and which they execute with great expedition. Flat stone, without regard to their exact width, are placed against the sides, open at the top. Into this opening, one or more thicker flat stones are thrust, as represented in the cut, and the drain then filled as before mentioned. The advantage of this mode is in obviating the necessity of selecting the stones, as almost any width will answer.

The last two modes, if well made, will last as long as

tile drains; as the earth can not fall into them from the sides, nor rise from the bottom, even if of a quicksand nature; and in the last described, the stones being mostly vertical, admit the free descent of the water from above.[1]

A correspondent of the *Country Gentleman*, writing from Monroe county, N. Y., says:

"I have made several hundred rods, and make more or less every year, and have made at all seasons of the year. The best time is in the spring, as soon as the ground is settled, especially where there is hard-pan, as that then works the easiest. My mode is to commence with team and plow—cut two furrows, one from the other—then put the plow in the center, and cut deep as I can—then shovel out and dig from three to five feet deep, and even more where I cross ridges—the bottom ten inches wide. In filling, I take flat stone and set them on the edge on the outer side of the ditch, and let the tops come together, forming Λ—fill in with small stones up to within eighteen inches of the top or surface. Then take litter or straw and cover the stone lightly, and then take the plow and fill up rounding, as it will settle more or less. Some of mine have paid expenses the first crop. I have drains that were made in 1838, and answer their purpose well yet."

But this method of underdraining with stone, will be very much improved by laying a flat stone in the bottom of the drain, as represented in Fig. 30.

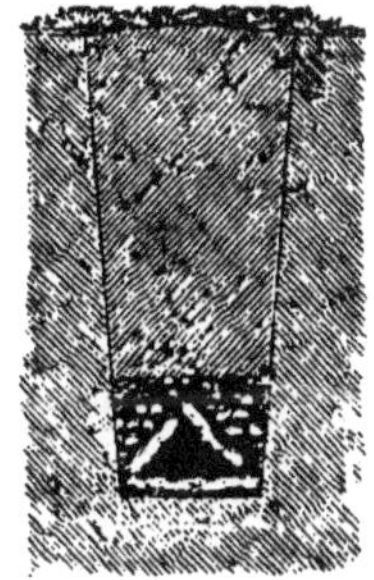

FIG. 30.—THE COUPLED STONE DUCT.

Mr. L. Griswold, of Litchfield, Connecticut, says:

"This last fall I have drained four acres of my eight acre meadow, in this way, viz: We lay out the short drains forty feet apart—though we vary from this rule some; when it comes near a wet hollow we go through that—we cut them three and a half feet deep, two feet wide at the top, and slant down to six inches on the bottom. We scrape

[1] J. J. Thomas, in *Rural Register*.

the mud from the bottom perfectly clean, so that it is hard, like rock. This thoroughly done, we begin to fill with small round stone, taking care that no one stone is large enough to reach across, for the first layer, and so on five or six inches; then the cobble and broken stones may be thrown in with less care, extending up to a hight of eighteen inches; the little slivers from the broken stone, and such like, we scatter alone on the top to fill up the cavities; then place inverted turf on snugly, and press it down with our feet. The dirt dug from the ditch is then filled in, and it is finished.

"The whole cost is about sixty cents per rod, including drawing the stone, which pays by getting them out of the way. I am well convinced that these drains will continue to act well, and I can not see why stone is not quite as good, if not better than tile, and it costs something less here."

Almost anywhere in Ohio, the best of tile drains can be made at a considerably less expense than sixty cents per rod; whatever merit may attach to Mr. Griswold's method, there will be one insuperable objection to it in the West, on account of the cost.

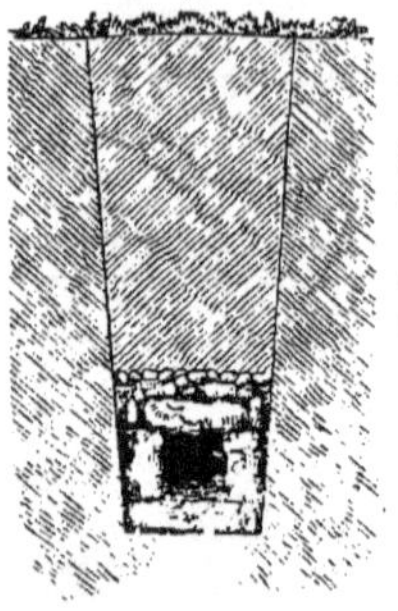

Fig. 31.

In locations where the clay is somewhat destitute of the quality of stiffness, and is inclined to crumble, it may be advantageous to protect the sides, as in the drain represented by Fig. 31. This drain is made by laying a flat stone across the entire bottom; then a flat stone against each side, and another covering the last two—the covering stone should, if practicable, be as wide as the ditch. Rough boards, or "*slabs*" from the sawmill will answer an excellent purpose as covers.

Considerable draining with stone has been done in Ohio, from a belief that it was cheaper and more permanent than almost any other kind of drain. Farmers generally have teams of their own; and there often occur "odd half days," or times when the team and hands, according to their system of farming, could not be profitably employed, and they say that during such times stones may be gathered, drawn to the proper field and distributed along the line of the contemplated drains; and in this way the stones are place in readiness on the ground at no cost, or at most a comparatively small cost. This may be true in certain cases, but surely no man would undertake to drain his neighbor's field, and gather, draw and distribute the stones as mere pastime; and in estimating the cost of drains, no item should be omitted, however trifling. A man purchases a farm in the wilderness, and during the "odd half days," gets out, and draws together, the timber for a new house—in the estimate of the cost of the house would this form *no* item of expense? "*Time is money*," and the time expended in preparatory measures is just so much money expended—that farmer has certainly not adopted the best system of farming, who can command sufficient leisure to gather, draw, and distribute stone for drains, without regarding it as an important item in the cost of drains.

Hon. John Howell, of Clark county, Ohio, has upward of a thousand rods of stone drains on his farm. They are 28 to 30 inches deep, and filled to the depth of 10 to 12 inches with stone. The stone were brought a distance of one and an eighth mile. The cost of the drains was 47 cents per rod, exclusive of the boarding furnished the hands.

Drawing and distributing stone, - - - -	27	cents	per	rod.
Digging and filling drains, - - - - - - -	10	"	"	"
Laying stone in drains, - - - - - - - -	10	"	"	"
Total cost, - - - -	47	"	"	"

Mr. Howell assured us that tile drains—the tile being

furnished at a manufactory within the country—would have cost, completed, 32 cents per rod only, instead of 47, as the stone drains did. In fact, he says the cost should be put down at 30 cents, because he filled the drains mainly by plowing the ground in that was dug out, and he has not taken the cost of plowing into account.

The subjoined experience of the editor of the *N. E. Farmer*, may be of much value to those who are hesitating between two opinions, or which to choose, stone or tile:

"We have plenty of stones for the purpose of drainage, and have constructed many drains of them, in both dry and sandy loams. They operated well for a time, but the first star mole that made his way to one of them left an inviting opening for the next drenching shower to follow. This, of course, was repeated a good many times and in a good many places during the year, and the work of *destruction* was begun. Unless laid very deep, the frost also deranges the upper portion of them, and lets the fine soil down. We have, therefore, great doubt whether it is not best to use tile in the first instance. It costs much less to lay tile than stone, and where the work is once done, and the tile entirely below the frost, a drain is made of great permanence and utility."

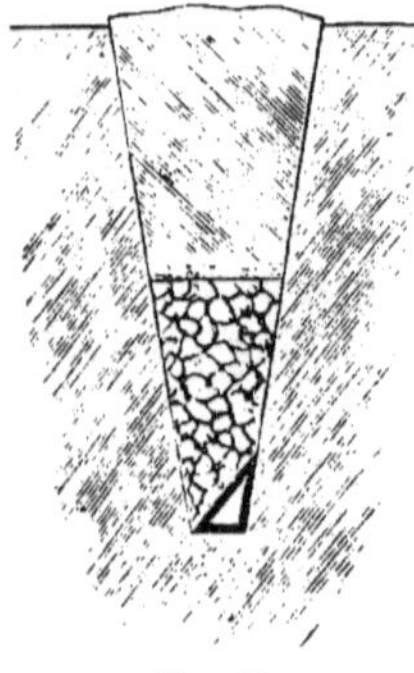

Fig. 32.
Irish Drain.

In some places stone drains are made by placing a flat stone on the bottom, then one on the side, and a third one in a diagonal direction from the bottom, as represented in Fig. 32. This is called an *Irish* drain. They are filled either with cobble or small stones, 10 or 12 inches deep, like the other stone drains, or else may be filled entirely with the material which was thrown out in making the ditch.

Stone drains made according to any of the systems

described are peculiarly liable to be obstructed, because there is no regular water-way, and the flow of the water must, of course, be very slow, impeded as it is by friction at all points with the irregular surfaces.

Sand, and other obstructing substances, which find their way, more or less, into all drains, are deposited among the stones—the water having no force of current sufficient to carry them forward—and the drain is soon filled up at some point, and ruined.

Miles of such drains have been laid on many New England farms, at shoal depths, of two or two and a half feet, and have in a few years failed. For a time, their effect, to those unaccustomed to underdrainage, seems almost miraculous. The wet field becomes dry, the wild grass gives place to clover and herdsgrass, and the experiment is pronounced successful. After a few years, however, the wild grass re-appears, the water again stands on the surface, and it is ascertained, on examination, that the drain is in some place packed solid with earth, and is filled with stagnant water.

The fault is by no means wholly in the material. In clay or hard-pan, such a drain may be made durable, with proper care, but it must be laid deep enough to be beyond the effect of the treading of cattle and of loaded teams, and the common action of frost.[1]

TILE DRAINS.

We have already alluded to the fact[2] that pipe tile was in use, as conduits for underground ducts or causeways, in France, as early as the year 1600. From some cause, which we have failed to discover, in our agricultural literary researches, they were discontinued; and it is exceedingly doubtful if they were used in England previous to

[1] *French's Farm Drainage.* [2] Ante, page 7.

the present century. In Elkington's treatise on land draining, edited by Johnstone, and first published in London, in May, 1797, are described "draining bricks." On page 41 we find: "When flat stones can be got, they are preferable to brick; but there are several kinds of brick, beside the common sort, invented and used solely for the purpose of draining, in several parts of England, where the expense of stone would become greater. Of these, the figures in the annexed plate are some of the best kinds. When small drains are wanted, and when the water is to be conveyed to a house, etc., Fig. 33, is

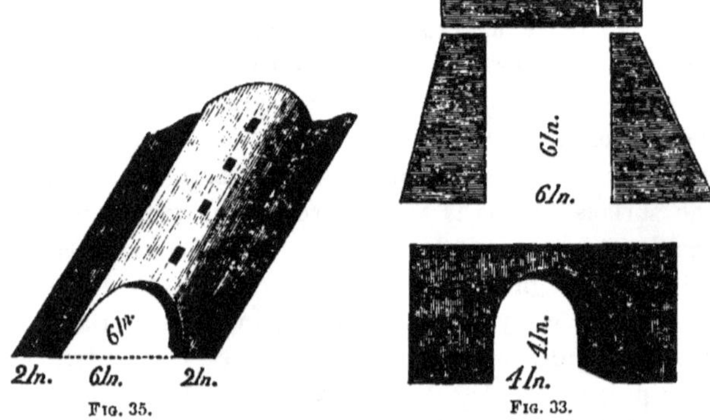

Fig. 35. Fig. 33.

commonly made use of. For larger drains, Figs. 34 and 35 are well adapted, especially, Fig. 35, lately invented by Mr. Couchman, of Bosworth Temple, in Warwickshire, and with which Mr. Elkington has laid several drains. They are laid single, without one reversed under, for when that is done, the water running on the under one, occasions a kind of sludge, which in time becomes so incrusted on it, as totally to obstruct the passage of the water, and render the work useless in a few years. In clay bottoms, they may be laid single, or without anything under; but in soft, sandy bottoms, a common building brick should be laid under each side, to prevent them from sinking down, and should be so laid as to form a regular arch (*i. e.*, the side bricks laid with an equal hight), the better to support the pressure above from breaking them or causing them to slip."

The brick represented by Figs. 33 and 35, were called *soughing* brick. That part of a drain forming the conduit for water, was termed the "*sough*" or "*surf*," and as these brick formed a conduit, they were accordingly termed "soughing brick." The only ones we ever saw were on the farm of Norton S. Townshend (formerly President of the Ohio State Board of Agriculture), in Lorain county, Ohio. That pattern of soughing brick represented by Fig. 35, had a series of "*eyelet*" holes, as they were termed, on the upper portion of the brick, in order to afford a means of ingress for the water to get into the drain.

It is, perhaps, unnecessary to mention, that these brick or tile are made of clay, sun-dried, and burned the same as other brick. The drain brick represented by Figs. 33–4–5, have long since been entirely abandoned by persons draining on a large scale. For a long time, the "*horseshoe*" tile was more in use than any other (See Fig. 36). These tiles were made by hand; the clay was "rolled" out, somewhat after the fashion that housewives roll out dough, and then were pressed by hand over a cylindrical substance, and set away to dry. Of course, they were expensive; at present, they are made by machines. Many very excellent drainers in Ohio and New York, are partial to the horseshoe tile, because less care is required in laying them. We regret to state, that a very large proportion of tile, used at pres-

FIG. 36.—HORSESHOE TILE [resting on a Sole of Tile or Board].

ent, are of this form—a form which we have long since considered almost the worst possible, and have not hesitated to express this opinion on all occasions—in newspaper articles, in lectures, and in ordinary conversation. We had not read Gisborne's *Essays on Agriculture*, until December, 1860, and therefore could not have been influenced in our opinion by his writings, but had based our opinion upon our knowledge of hydraulics and hydrostatics. But as Gisborne presents the views entertained by us, in this respect, based upon experience and observation, we prefer to quote his language:

"We shall shock some and surprise many of our readers, when we state confidently that, in average soils, and, still more, in those which are inclined to be tender, horseshoe tiles form the weakest and most failing conduit which has ever been used for a deep drain. It is so, however; and a little thought, even if we had no experience, will tell us that it must be so. A doggerel song, quite destitute of humor, informs us that tiles of this sort were used in 1760, at Grandesburg Hall, in Suffolk, by Mr. Charles Lawrence, the owner of the estate. The earliest of which we had experience were of large area and of weak form. Constant failures resulted from their use, and the cause was investigated; many of the tiles were found to be choked up with clay, and many to be broken longitudinally through the crown. For the first evil, two remedies were adopted; a sole of slate, of wood, or of its own material, was sometimes placed under the tile, but the more usual practice was to form them with club-feet. To meet the case of longitudinal fracture, the tiles were reduced in size, and very much thickened in proportion to their area. The first of these remedies was founded on an entirely mistaken, and the second on no conception at all of the cause of the evil to which they were respectively applied. The idea was, that this tile, standing on narrow feet, and pressed by the weight of the refilled soil, sank into the floor of the drain; whereas, in fact, the floor of the drain rose into the tile. Any one at all conversant with collieries is aware that when a *strait* work (which is a small subterranean tunnel six feet high and four feet wide, or thereabout) is driven in coal, the rising of the floor is a more usual and far more inconvenient occurrence than the falling of the roof: the weight of the two sides squeezes up the floor. We have seen it formed into a very decided arch without fracture. Exactly a similar operation takes place in the drain. No one had till recently dreamed of forming a tile drain, the bottom of which a man was not to approach personally within twenty inches or two feet. To no one had it then occurred that width at the bottom of a drain was a great evil. For the convenience of the operator the drain was formed with nearly perpendicular sides, of a width in which he could stand and work conveniently, shovel the bottom level with his ordinary spade, and lay the tiles by his hand; the result was a drain with nearly perpendicular sides, and a wide bottom. No sort of clay, particularly when softened by water standing on it or running over it, could fail to rise under such circumstances; and the deeper the drain the greater the pressure and the more certain the rising. A horseshoe tile, which may be a tolerably secure conduit in a drain of two feet, in one of four feet becomes an almost certain failure. As to the longitudinal fracture—not only is the tile subject to be broken by one of those slips which are so troublesome in deep draining, and to which the lightly-filled material, even when the drain is completed, offers an imperfect resistance, but the constant pressure together of the sides, even when it does not produce a fracture of the soil, catches hold of the feet of the tile, and breaks it through the crown. Consider the case of a drain formed in clay when dry, the conduit a horseshoe tile. When the clay expands with moisture, it necessarily presses on the tile, and breaks it through the crown, its weakest part.[1] When the Regent's Park was first drained, large conduits were in fashion, and they were made circular by placing one horseshoe tile upon another. It would be difficult to invent a weaker conduit. On re-drainage, innumerable instances were found in which the upper tile was broken through the crown, and had dropped into the lower. Next came the ∩ form, tile and sole in one, Fig. 37, and much reduced in size—a great advance;

FIG. 37.—HORSESHOE TILE AND SOLE IN ONE.

and when some skillful operator had laid this tile bottom upward, we were evidently on the eve of pipes. For the ∩ tile a round pipe

[1] The tile has been said, by great authorities, to be broken by contraction, under some idea that the clay envelopes the tile and presses it when it contracts. That is nonsense. The contraction would liberate the tile. Drive

molded with a flat bottomed solid sole ⍜ is now generally substituted, and is an improvement; but is not equal to pipes and collars, nor generally cheaper than they are.

"Almost forty years ago, small pipes for land drainage were used concurrently by the following parties, who still had no knowledge of each other's operations:—Sir T. Wichcote, of Asgarby, Lincolnshire (these, we believe, were socket-pipes)—Mr. R. Harvey, at Epping—Mr. Boulton, at Great Tew, in Oxfordshire (these were porcelain 1-inch pipes, made by Wedgwood, at Etruria)—and Mr. John Read, at Horsemonden, in Kent. Most of these pipes were made with eyelet-holes, to admit the water. Pipes for thorough draining were incidentally mentioned in the Journal of the Agricultural Society for May, 1843; but they excited no general attention till they were exhibited by John Read (the inventor of the stomach-pump), at the Agricultural Show at Derby in that year. A medal was awarded to the exhibitor. Mr. Parkes was one of the judges, and brought the pipes to the special notice of the Council, and was instructed by them to investigate their use and merits. From this moment inventions and improvements huddle in upon us faster than we can describe them. Collars to connect the pipes, a new form of drain, tools of new forms—particularly one by which the pipe and collar are laid with wonderful rapidity and precision, by an operator who stands on the top of the drain—and pipe-and-collar-making machines (stimulated by repeated prizes offered by the Royal Agricultural Society), which furnish those articles on a scale of unexampled cheapness. For all these inventions and adaptations we are mainly indebted to Mr. Parkes. The economical result is, a drain 4 feet 6 inches deep, excavated and refilled at from 1¼*d.* to 2*d.* per yard—the workmen earning 12*s.* and upward per week; and 333⅓ yards of collared 1½-inch pipes for 18*s.*—being 12*s.* per thousand for the pipes, and 6*s.* per thousand for the collars; larger sizes at a proportionate advance. We shall best exemplify the improvements to our readers by describing the drain. It is wrought in the shape of a wedge, brought in at the bottom to the narrowest limit which will admit the collar by tools admirably adapted to that purpose. The foot of the operator is never within 20 inches of the floor

a stake into wet clay; and when the clay is dry, observe whether it clips the stake tighter or has released it, and you will no longer have any doubt whether expansion or contraction breaks the tile. Shrink is a better word than contract.

of the drain; his tools are made of iron plated on steel, and never lose their sharpness, even when worn to the stumps; because, as the softer material, the iron, wears away, the sharp steel edge is always prominent. The sloping sides of the drain are self-sustaining, and the pressure on its floor is reduced to a minimum; the circular form of the pipe and collar, see Fig. 38, enables them to sustain any pres-

FIG. 38.—PIPE AND COLLAR.

sure to which they can be subjected; the adaptation of the bed in which they lie, to their size, prevents their wriggling. They form a continuous conduit, and whose continuity can not be broken except by great violence. However steep the drain, the water running in the pipe can never wash up its floor. They offer almost insuperable impediments to the entrance of vermin, roots,[1] or anything except

---

[1] I am afraid that I must materially modify this expression, as far as roots are concerned. The words, "almost insuperable impediments," are not applicable. My own experience, as to roots, in connection with deep pipe draining, is as follows:—I have never known roots to obstruct a pipe through which there was not a perennial stream. The flow of water in summer and early autumn appears to furnish the attraction. I have never discovered that the roots of any esculent vegetable have obstructed a pipe. The trees which, by my own personal observation, I have found to be most dangerous, have been red willow, black Italian poplar, alder, ash, and broad-leaved elm. I have many alders in close contiguity with important drains, and, though I have never convicted one, I can not doubt that they are dangerous. Oak, and black and white thorns, I have not detected, nor do I suspect them. The guilty trees have in every instance been young and free growing; I have never convicted an adult. These remarks apply solely to my own observation, and may of course be much extended by that of other agriculturists. I know an instance in which a perennial spring of very pure and (I believe) soft water is conveyed in socket pipes to a paper mill. Every junction of two pipes is carefully fortified with cement. The only object of cover being protection from superficial injury and from frost, the pipes are laid not far below the sod. Year by year these pipes are stopped by roots. Trees are very capricious in this matter. I was told by the late Sir R. Peel, that he sacrificed two young elm trees in the park at Drayton Manor to a drain which had been repeatedly stopped by roots. The stoppage was nevertheless repeated, and was then traced to an elm tree far more distant than those which had been sacrificed. Early in the autum of 1850 I completed the drainage of the upper part of a boggy valley,

water, and they are more portable both to the field and in the field than any other conduit previously discovered: cheap, light, handy, secure, efficacious."

The ordinary form of *pipe* tile is represented by Fig. 39, but all tile having a tubular form, like those of Fig. 40, 41, are called *pipe* tile.

FIG. 39.—PIPE TILE.

FIG. 40.—PIPE TILE.

FIG. 41.—OCTOGANAL PIPE TILE.

We presume we shall be obliged to rest content with

---

lying, with ramifications, at the foot of marly banks. The main drains converge to a common outlet, to which are brought one 3-inch pipe and three of 4 inches each. They lie side by side, and water flows perennially through each of them. Near to this outlet did grow a red willow. In February, 1852, I found the water breaking out to the surface of the ground about 10 yards above the outlet, and was at no loss for the cause, as the roots of the red willow showed themselves at the orifice of the 3-inch and of two of the 4-inch pipes. On examination I found that a root had entered a joint between two 3-inch pipes, and had traveled 5 yards to the mouth of the drain, and 9 yards up the stream, forming a continuous length of 14 yards. The root which first entered had attained about the size of a lady's little finger; and its ramifications consisted of very fine and almost silky fibers, and would have cut up into half a dozen comfortable boas. The drain was completely stopped. The pipes were not in any degree displaced. Roots from the same willow had passed over the 3-inch pipes, and had entered and entirely stopped the first 4-inch drain, and had partially stopped the second. At the distance of about fifty yards a black Italian poplar, which stood on a bank over a 4-inch drain, had completely stopped it with a bunch of roots. The whole of this had been the work of less than 18 months, including the depth of two winters. A 3-inch branch of the same the pipe tile for some years to come, but we do not consider them the "hight of perfection," although they are a very decided improvement on the horseshoe tile. A better form, in our opinion, is that of which an end view or section is represented by an egg, with the small end down. A conduit of this shape affords a very narrow channel, when a small quantity of water only is to be discharged; but as the quantity of water increases the channel increases also. We are well aware that some may deem this a matter of very small importance, if indeed it be worthy of consideration at all. The importance in the form of the conduit can not be better illustrated than by citing well known facts.

It must be self-evident, and therefore requires no argument to prove, that a body of water confined in a channel one inch high, and one inch wide, will pass off more rapidly than if spread over a surface eight inches wide and one eighth of an inch high. In the former case the water will move rapidly, and carry with it all the particles of sand, clay and other impurities which may find their way into the conduit, while in the latter case the resisting power of the current would not be sufficient to remove them, and they would form a nucleus for a permanent stoppage of the water.

The California gold diggers at first employed the ordinary "box," or "square" form, for a conduit to carry off the water during the process of washing gold; this conduit became clogged daily, and much time was spent in keeping the channel open. One of the miners con-

system runs through a little group of black poplars. This drain conveys a full stream in plashes of wet, and some water generally through the winter months, but has not a perennial flow. I have perceived no indication that roots have interfered with this drain. I draw no general conclusions from these few facts, but they may assist those who have more extensive experience in drawing some, which may be of use to drainers.—T. G.

structed a conduit of two boards, placed together in this shape V, which required no cleansing or further care and never became clogged or choked. Instead then of having the widest of the conduit at the bottom, as in the case of the horseshoe tile, the very narrowest possible should form the base, and for this reason, if none other, pipe tile are to be preferred to the horseshoe.

Pipe tile are more readily laid in the drain than any other kind, for the reason that all tiles are more or less warped in drying and burning, and where it is desired to made perfect work, there is no "wrong-side-up" to them; they can be turned with any side up, so as to make not only better joints, but a straighter run for water—which is very important.

The best authorities on the subject differ widely with respect to the importance of *collars* to connect the pipes in the drains; and as we do not think that they will be used to any extent in the country—at least for some years to come—we will refer our readers to several English authorities for views on this subject. Mr. Gisborne says:

"We were astounded to find, at the conclusion of Mr. Parkes, Newcastle Lecture, this sentence: 'It may be advisable for me to say, that in clays, and other clean-cutting and firm-bottomed soils, I do not find the collars to be indispensably necessary; although I always prefer their use.' This is a barefaced treachery to pipes: an abandonment of the strongest point in their case—the assured continuity of the conduit. Every one may see how very small a disturbance at their point of junction would dissociate two pipes of one inch diameter. One finds a soft place in the bottom of the drain, and dips his nose into it one inch deep, and cocks up his other end. By this simple operation the continuity of the conduit is twice broken. An inch of lateral motion produces the same effect. Pipes of a larger diameter than two inches are generally laid without collars; this is a practice on which we do not look with much complacency; it is the compromise between cost and security, to which the affairs of men are so often compelled. No doubt a conduit from three to six inches in diameter is much less subject to a breach in its continuity than one which is smaller; but when no collars are used, the pipes should be laid with extreme care, and the bed which is prepared for them at the bottom of the drain should be worked to their size and shape with great accuracy.

"To one advantage which is derived from the use of collars we have not yet adverted—the increased facility with which free water existing in the soil can find entrance into the conduit. The collar for a 1½ inch pipe has a circumference of three inches. The whole space between the collar and the pipe on each side of the collar is open, and affords no resistance to the entrance of the water; while at the same time the superincumbent arch of the collar protects the junction of two pipes from the intrusion of particles of soil. We confess to some original misgivings that a pipe resting only on an inch at each end, and lying hollow, might prove weak, and liable to fracture by weight pressing on it from above; but the fear was illusory. Small particles of soil trickle down the sides of every drain, and the first flow of water will deposit them in the vacant space between the two collars. The bottom, if at all soft, will also swell up into any vacancy. Practically, if you re-open a drain well laid with pipes and collars, you will find them reposing in a beautiful nidus, which, when they are carefully removed, looks exactly as if it had been molded for them.

Mr. Denton says:

"The use of collars is by no means general, although those who have used them speak highly of their advantages. Except in sandy soils, and in those that are subject to sudden alteration of character, in some of the deposits of red sandstones, and in the clayey subsoils of the Bagshot sand district, for instance, collars are not found to be essential to good drainage. In the north of England they are used but seldom, and, in my opinion, much less than they ought to be; but this opinion, it is right to state, is opposed, in numerous instances of successful drainage, by men of extensive practice; and as every cause of increased outlay is to be avoided, the value of collars, as general appliances, remains an open question. In all the more porous subsoils, in which collars have been used, the more successful drainers increase the size of the pipes in the minor drains to a minimum size of two inches bore."

CHAPTER II.

SIZE OF TILE, ETC.

The size of tile to be employed in underdraining depends upon, 1, the amount of fall; 2, the length of the drain; 3, the distance between the drains; 4, the depth of the drains.

It is very evident that the greater the amount of fall, the smaller the conduit may be; and the converse of this proposition is equally true, viz.: that the less the amount of fall, the larger the pipe must be. Actual experiment has demonstrated that if a drain of 100 feet in length, having *eight* feet fall, is laid with pipe having a caliber or capacity of 1½ inches, will in 24 hours drain the same quantity of water that 2 inch pipes, having a fall of 2 feet 3 inches, will drain in the same period, or a 3 inch pipe having a fall of only 6 inches, or a 4 inch pipe hav- a fall of less than 3 inches. Hence the size of the tile is determined by the amount of fall.

Again, it may be necessary to discharge 50,000 gallons of water in 24 hours, where only two feet fall in 100 feet in length can be had; a 3 inch tile with one foot fall will effect this, but if 2 inch tile were employed, it would require a fall of 4 feet 6 inches. Hence the length of the drain, or what is the same thing in effect, the amount of water to be discharged, governs the size of the pipe.

If 50,000 gallons are discharged by each of the two drains, A and B (Fig. 42), 100 feet apart, it is evident that if a third drain, C, were placed between them so as to make the distance between the drains 50 feet, then each drain would discharge 33,333 gallons. But if two more

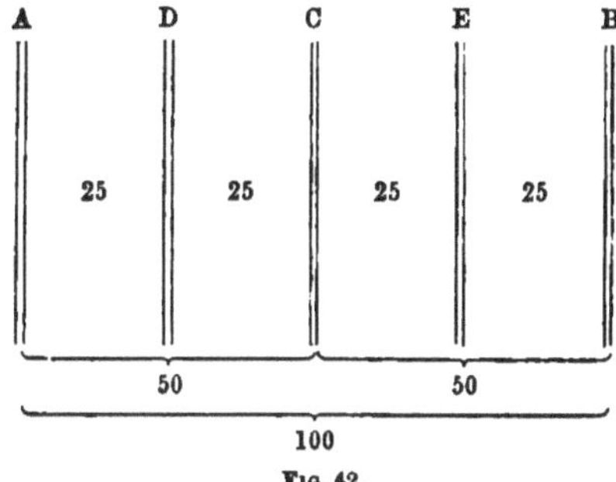

FIG. 42.

drains, D and E, are placed between A and C, and C and B, then the distance between the drains will be 25 feet, and the amount discharged by each drain will be 20,000 gallons only.

If, then, the drains are made 25 feet apart, 2 inch tile, with a fall of 9 inches in 100 feet will drain off as much water as A, B and C would with the same sized tile, having a fall of two feet, or 3 inch tile, having a fall of five inches. Or if the two drains, A and B, only are employed, then if laid with 2 inch tile, they must have a fall of four feet six inches; with 3 inch tile a fall of about one foot; or a fall of five inches if 4 inch tile are employed. Hence the distance between the drains determines the size of the pipes.

From these propositions it is very evident that the depth of the drains exerts a controling influence on the size of the pipes.

Suppose the drain 7, 5, in Fig. 5, page 99, were placed no deeper than the point indicated 8, it is very evident that in such case it would have less than half the amount of water to drain that it has at 7.

As the entire efficiency of drains depends upon a correct understanding and compliance with the principles involved in the four propositions stated at the commencement of this chapter, we shall dwell at some length upon them.

Stone drains require more fall than tile drains, on account of the friction, or the retardation water meets in passing through angular crevices. Friction must not be omitted in our calculations of fall and capacity. Where water can flow in a *straight* direction in a smooth and regular channel, much more water can be discharged in a given time than where the angles and curves occur in the direction; and where the surface is smooth, the flow is more rapid than where it must pass through a channel full of rough points or inequalities.

In some recent English experiments "it was found that with pipes of the same diameter, exactitude of form was of more importance than smoothness of surface; that glass pipes, which had a wavy surface, discharged less water, at the same inclinations, than Staffordshire stoneware clay pipes, which were of perfectly exact construction. By passing pipes of the same clay—the common red clay—under a second pressure, obtained by a machine at an extra expense of about 18 pence per 1000, while the pipe was half dry, very superior exactitude of form was obtained, and by means of this exactitude, and with nearly the same diameters, an increased discharge of water of one fourth was effected within the same time."

"On a large scale, it was found that when equal quantities of water were running direct, at a rate of ninety seconds, with a turn at right angles, the discharge was effected in one hundred and forty seconds; while, with a turn or junction with a gentle curve, the discharge was effected in one hundred seconds."

CALIBER AND MINIMUM FALL OF DRAIN PIPE TILE.

Vincent, an English writer, has adopted Eytelwein's formula,

$$c = 6.42 \sqrt{\frac{50\ d\,h}{1 + 50\ d\,h}}$$

in the computation of the minimum fall and caliber of the pipe tile. In this formula, c = velocity of current per second, d = diameter, or caliber of pipe, and h = the fall in 1 foot, as *data* from which to determine the velocity of the water in a pipe of a given diameter, and certain fall. He determined from this formula, by an inverse process, the requisite caliber of the pipe tile d, for a given distance or length of drain—assuming six inches per second to be the minimum velocity of water discharged from an acre. The calculations in question were, however, based on hypothetic values only; because the co-efficient, 50, occurring in the formula, refers, originally, to metal tubes or pipes; and there was no data at hand for that of clay or earthenware pipes. Owing to the great importance of having the pipe tile of proper caliber, John, Waege, and v. Möllendorf,[1] made a series of experiments.

For this purpose they laid a number of drain tile in a trough making a slight angle with the horizon, and secured the joints by moist clay. Water was then let into the pipe from a reservoir, and the time occupied in flowing through the pipes in seconds and the quantity discharged in cubic feet was very carefully observed and recorded.

The data obtained from repeated observations, with given lengths of pipe and various degrees of inclination

[1] These experiments are quoted in detail in *Zeitschrift fur Deutsche Drainirung*, 1855, pp. 79 and 106; also, in *Dingler's Polytechnische Journal*, 138 257.

or fall, and various capacities or diameters of pipe—and notwithstanding that the extremes of twenty-two experiments varied about 40 per cent. sufficient data was obtained to change the co-efficient 50, and to make the following as nearer the truth, viz.:

$$c = 6.42 \sqrt{\frac{46.5\, d\, h}{1 \div 46.5\, d\, h}}.$$

If we adopt Vincent's data, and assume that the velocity of water, in pipe tile, amounts to 6 inches per second, as a minimum, in order to secure the pipes from being filled by detritus, or particles of earth entering them and being deposited by gravity overcoming velocity, then the following will be the minimum fall (by Vincent's formula) for drains of 120 feet in length.

For drains with tile of	1	inch caliber,	2.33	inches fall,	(4.8)
"	"	1¼	"	1.88	"	
"	"	1½	"	1.58	"	
"	"	2	"	1.20	"	(2.4)
"	"	3	"	0.82	"	(1.8)
"	"	4	"	0.63	"	(1.0)
"	"	5	"	0.52	"	(0.7)
"	"	6	"	0.44		
"	"	7	"	0.39		
"	"	8	"	0.35	"	(0.5)

The figures in parentheses are those adopted by *Vincent.* It must be remembered, however, that these figures are applicable only to such drains as are as good and as carefully laid pipes as those in the experiment.

In calculating the capacity of drain pipe tile, two other contingencies must be taken into account, namely: *the quantity of water to be discharged*, and *the distance between the drains.* Upon the supposition that well-arranged drains must discharge the rainfall of a month in fourteen days (assuming 4 inches as the maximum), Vincent has fixed the amount discharged per second, from an acre, at 0.00625 cubic feet.

In the course of several articles in *Zeitschrift fur Deutsche Drainirung*, John has compared the replies of several writers to the question, "What is the capacity of pipe tile of a given caliber?" Schönermark, the practical draining engineer of the Duchy of Brunswick,[1] has communicated some calculations upon this same subject, and, as the principles assumed by him do not vary materially from those of the other authors, and furthermore, as the measures employed by him are purely Brunswickian, and therefore can not well be compared with those of John, without being reduced to Prussian (almost American) measure, reference must be made by those interested to the "*Zeitung*" itself.

The quantity of water discharged per acre, per second, has been fixed at

0.0095 cubic feet by	Stephens and Leclerc,	
0.0095 " " "	Vincent,	
0.00289 " " "	Stocken,	
0.00276 " " "	Waege and	for heavy soil.
0.00376 " " "	v. Mollendorf	" light soil.
0.00426 " " "	Schonermark.	

Stephens and Leclerc have assumed that a heavy rainfall, in a day, will amount to 0.382 inches, and that this amount is to be removed by the drains in 24 hours; the amount evaporated is not to be taken into account.

Stocken assumes that, after deducting 20 per cent. for evaporation, that the drains will discharge the remaining 80 per cent. of the maximum of eight days of winter rains (which in that latitude would average 1¼ inches for that period) in 9 days.

Waege and v. Möllendorf, as well as Vincent, assume

[1] Agronomische Zeitung, 1855, page 360.

that the drains will discharge the rainfalls of autumn, winter and spring in an aggregate of 14 days, after deducting for evaporation. They, together with Dickinson, have assumed the evaporation for clay to heavy loam soil, to be 45 per cent., and for a lighter loam to be 25 per cent. of the entire rainfall.

Schönermark's calculation is based on the following data: the maximum monthly rain for Brunswick is 4 inches (Brunswick measure); this quantity *may* fall in 8 days; consequently, ½ inch in one day—this latter quantity the drains are to discharge in 48 hours, after deducting 25 per cent. for evaporation.

From the data just given, the comparative capacity of one inch and three inch pipe tile, to carry off a given rainfall, is determined to be as follows, by the various experimenters:

	Fall of 3 6-10 inches in 120 ft.	
	Inch tile.	Three inch tile.
Stephens & Le Clerc, - - - - -	0.33 acre.	——
Vincent, - - - - - -	0.40 "	7 acres.
Stocken, - - - - - -	0.98 "	15 "
V. Mollendorf and Waege } heavy soil, - -	0.86 "	13.4 "
V. Mollendorf and Waege } light " - - -	0.64 "	9.8 "

The experiments demonstrate that, for Ohio or the Middle and Western states generally, one inch tile is entirely too small, while three inch tile is perhaps larger than necessary, especially when it is considered that two inch tile will answer the purpose and cost a great deal less. Alderman Mechi says: "I seldom use any larger than one inch bore, except for large springs. I am practically convinced they are as large as are required. We make some sad mistakes as to water: a rope of water one inch thick, spread eight inches wide, forms a *broad-looking* stream one eighth of an inch thick. It is perfectly ludicrous to see immense six, nine, and twelve inch bore pipes put, in many cases, to carry an insignificant stream that would fold up into a one, two, or three inch coil. We must bear in mind that a two inch pipe will carry as much as four one inch; a three inch is equal to nine one inch."

Presuming that the alderman is correct for England, where there is an average rainfall of twenty-three inches, should we not have at least two inch pipe in Ohio, where the average rainfall is nearly forty-six inches, or double that of England? One and a half inch pipe would be just the proportion, according to the rainfall; but then we must remember that in England it is a perfect *drizzle* from January until December, while here we sometimes have forty days of drought, and then a rainfall of two inches per day!

Judge French obtained from Messrs. Shedd & Edson, of Boston, the following valuable tables, showing the capacity of water pipes, with the accompanying suggestions:

DISCHARGE OF WATER THROUGH DRAINS.

"The following tables of discharge are founded on the experiments made by Mr. Smeaton, and have been compared with those by Henry Law, and with the rules of Weisbach and D'Aubuisson. The conditions under which such experiments are made, may be so essentially different in each case, that few experiments give results coincident with each other, or with the deductions of theory; and in applying these tables to practice, it is quite likely that the discharge of a pipe of a certain area, at a certain inclination, may be quite unlike the discharge found to be due to those conditions by this table, and that difference may be owing partly to greater or less roughness on the inside of the pipe, unequal flow of water through the joints into the pipe, crookedness of the pipes, want of accuracy in their being placed, so that the fall may not be uniform throughout, or the ends of the pipes may be shoved a little to one side, so that the continuity of the channel is partially broken; and, indeed, from various other causes, all of which may occur in any practical case, unless great care is taken to avoid it, and some of which may occur in almost any case.

"We have endeavored to so construct the tables that, in the ordinary practice of draining, the discharge given may approximate to the truth for a well laid drain, subject even to considerable friction. The experiments of Mr. Smeaton, which we have adopted as the basis of these tables, gave a less quantity discharged, under certain conditions, than given under similar conditions by other tables. This result is probably due to a greater amount of friction in the pipes used by Smeaton. The curves of friction resemble, very nearly, parabolic curves, but are not quite so sharp near the origin.

"We propose, during the coming season, to institute some careful experiments, to ascertain the friction due to our own drain pipe. Water can get into the drain pipe very freely at the joints, as may be seen by a simple calculation. It is impossible to place the ends so closely together, in laying, as to make a tight joint, on account of roughness in the clay, twisting in burning, etc.; and the opening thus made will usually average about one tenth of an inch on the whole circumference, which is, on the inside of a two inch pipe, six inches—making six tenths of a square inch opening for the entrance of water at each joint.

"In a lateral drain 200 feet long, the pipes being thirteen inches long, there will be 184 joints, each joint having an opening of six tenth square inch area; in 184 joints there is an aggregate area of 110 square inches; the area of the opening at the end of a two inch pipe is about three inches; 110 square inches inlet to three inches outlet; thirty-seven times as much water can flow in as can flow out. There is, then, no need for the water to go through the pores of the pipe; and the fact is, we think, quite fortunate, for the passage of water through the pores would in no case be sufficient to benefit the land to much extent. We tried an experiment, by stopping one end of an ordinary drain pipe and filling it with water. At the end of sixty-five hours, water still stood in the pipe three fourths of an inch deep. About half the water first put into the pipe had run out at the end of twenty-four hours. If the pipe was stopped at both ends, and plunged four feet deep in water, it would undoubtedly fill in a short time; but such a test is an unfair one, for no drain could be doing service, over which water would collect to the depth of four feet."

1½ INCH DRAIN PIPE.

Area: 1.76709 inches.

Fall in 100 feet. ft. in.	Velocity per second in feet.	Discharge in gallons in 24 hours.	Fall in 100 feet. ft. in.	Velocity per second in feet.	Discharge in gallons in 24 hours.
.3	.71	5630.87	5.3	3.75	29704.51
.6	1.04	8248.03	5.6	3.84	30454.28
.9	1.29	10230.73	5.9	3.93	31168.06
1.	1.52	12054.81	6.	4.	31723.21
1.3	1.74	13799.59	6.3	4.10	32516.36
1.6	1.91	15147.83	6.6	4.18	33150.76
1.9	2.10	16654.68	6.9	4.25	33705.91
2.	2.26	17923.61	7.	4.33	34340.38
2.3	2.41	19113.23	7.3	4.41	34974.85
2.6	2.56	20302.86	7.6	4.49	35609.30
2.9	2.69	21333.86	7.9	4.56	36154.45
3.	2.83	22444.17	8.	4.65	36878.23
3.3	2.94	23150.71	8.3	4.71	37354.08
3.6	3.06	24268.25	8.6	4.79	37988.55
3.9	3.16	25061.34	8.9	4.85	38464.40
4.	3.28	26013.03	9.	4.91	38940.25
4.3	3.38	26806.11	9.3	4.98	39495.39
4.6	3.46	27440.58	9.6	5.04	39971.24
4.9	3.56	28233.66	9.9	5.10	40447.10
5.	3.65	28947.43	10.	5.16	40922.93

2 INCH DRAIN PIPE.

Fall in 100 feet. ft. in.	Velocity per second in feet.	Discharge in gallons in 24 hours.	Fall in 100 feet. ft. in.	Velocity per second in feet.	Discharge in gallons in 24 hours.
.3	.79	10575.4	5.3	4.11	55018.9
.6	1.16	15528.4	5.6	4.22	56491.5
.9	1.50	20079.9	5.9	4.31	57696.3
1.	1.71	22891.1	6.	4.40	58901.1
1.3	1.94	25970.	6.3	4.49	60105.9
1.6	2.16	28915.1	6.6	4.58	61309.7
1.9	2.35	31458.5	6.9	4.66	62381.6
2.	2.53	33868.1	7.	4.74	63452.5
2.3	2.69	36009.9	7.3	4.83	64667.3
2.6	2.83	37884.	7.6	4.91	65728.3
2.9	2.97	39758.2	7.9	4.99	66799.2
3.	3.11	41632.4	8.	5.07	67870.1
3.3	3.24	43372.6	8.3	5.15	68941.
3.6	3.36	44979.	8.6	5.23	70011.9
3.9	3.48	46585.4	8.9	5.31	71082.8
4.	3.59	48057.9	9.	5.38	72019.9
4.3	3.70	49530.5	9.3	5.46	73090.9
4.6	3.80	50869.1	9.6	5.53	74027.9
4.9	3.91	52341.6	9.9	5.60	74965.
5.	4.02	53814.1	10.	5.67	75902.

3 INCH DRAIN PIPE.

Fall in 100 feet. ft. in.	Velocity per second in feet.	Discharge in gallons in 24 hours.	Fall in 100 feet. ft. in.	Velocity per second in feet.	Discharge in gallons in 24 hours.
.3	.90	24687.2	5.3	4.57	125356.2
.6	1.33	36482.2	5.6	4.68	128373.5
.9	1.66	45534.2	5.9	4.78	131116.6
1.	1.94	53214.7	6.	4.89	134133.9
1.3	2.19	60072.2	6.3	4.98	136602.6
1.6	2.43	66655.5	6.6	5.08	139345.6
1.9	2.63	72141.5	6.9	5.18	142088.7
2.	2.83	77627.6	7.	5.27	144557.4
2.3	3.	82290.7	7.3	5.37	147306.4
2.6	3.16	86679.6	7.6	5.46	150069.1
2.9	3.31	90794.1	7.9	5.55	152237.8
3.	3.47	95182.9	8.	5.64	154706.6
3.3	3.60	98748.9	8.3	5.73	157175.3
3.6	3.74	102589.1	8.6	5.82	159644.0
3.9	3.87	106155.	8.9	5.91	162112.7
4.	3.99	109446.7	9.	5.99	164313.2
4.3	4.11	112738.3	9.3	6.07	166501.6
4.6	4.23	116029.9	9.6	6.16	168970.3
4.9	4.34	119047.3	9.9	6.24	171164.7
5.	4.46	122338.9	10.	6.32	173359.1

4 INCH DRAIN PIPE.

Fall in 100 feet. ft. in.	Velocity per second in feet.	Discharge in gallons in 24 hours.	Fall in 100 feet. ft. in.	Velocity per second in feet.	Discharge in gallons in 24 hours.
.3	1.08	43697.6	5.3	4.86	196639.4
.6	1.50	60691.2	5.6	4.97	201090.1
.9	1.83	74043.2	5.9	5.09	205945.3
1.	2.13	86181.4	6.	5.20	210396.
1.3	2.38	96296.6	6.3	5.30	214442.1
1.6	2.61	105602.6	6.6	5.41	218892.8
1.9	2.81	113694.8	6.9	5.51	222938.8
2.	3.	121382.3	7.	5.61	226984.9
2.3	3.19	129089.9	7.3	5.71	231031.
2.6	3.36	135948.2	7.6	5.81	235077.1
2.9	3.53	142826.5	7.9	5.91	239123.2
3.	3.68	148895.7	8.	6.01	243169.2
3.3	3.82	154560.2	8.3	6.10	246810.7
3.6	3.96	160224.7	8.6	6.19	250452.2
3.9	4.10	165889.2	8.9	6.28	253193.7
4.	4.24	171553.7	9.	6.37	257735.2
4.3	4.37	176813.6	9.3	6.45	260971.9
4.6	4.50	182073.5	9.6	6.54	264603.1
4.9	4.62	186928.3	9.9	6.63	268254.9
5.	4.75	192188.7	10.	6.71	271491.8

5 INCH DRAIN PIPE.

Fall in 100 feet. ft. in.	Velocity per second in feet.	Discharge in gallons in 24 hours.	Fall in 100 feet. ft. in.	Velocity per second in feet.	Discharge in gallons in 24 hours.
.3	1.13	95841.2	5.3	5.02	442401.3
.6	1.57	138362.[illegible]	5.6	5.14	452976.6
.9	1.90	167442.[illegible]	5.9	5.25	462670.6
1.	2.20	193881.	6.	5.37	473246.
1.3	2.45	215912.9	6.3	5.49	483820.4
1.6	2.70	237944.9	6.6	5.60	493514.6
1.9	2.90	255569.5	6.9	5.70	502327.4
2.	3.10	273195.9	7.	5.80	511140.2
2.3	3.29	289940.1	7.3	5.90	520052.
2.6	3.46	304921.9	7.6	6.	528766.5
2.9	3.64	320784.9	7.9	6.10	537578.7
3.	3.80	334885.4	8.	6.20	546391.5
3.3	3.96	348974.8	8.3	6.30	555204.5
3.6	4.11	362204.9	8.6	6.40	564017.
3.9	4.26	375424.1	8.9	6.49	571948.
4.	4.40	387762.1	9.	6.58	579880.
4.3	4.52	398337.5	9.3	6.66	586930.2
4.6	4.66	410675.3	9.6	6.75	594861.4
4.9	4.78	421250.6	9.9	6.84	602793.2
5.	4.90	430825.0	10.	6.93	610723.8

8 INCH DRAIN PIPE.

Area: 50.2640 inches.

Fall in 100 feet. ft. in.	Velocity per second in feet.	Discharge in gallons in 24 hours.	Fall in 100 feet. ft. in.	Velocity per second in feet.	Discharge in gallons in 24 hours.
.3	1.23	277487.7	5.3	5.35	1206959.3
.6	1.65	372239.7	5.6	5.47	1234031.3
.9	2.01	453455.7	5.9	5.59	1261103.3
1.	2.33	525647.7	6.	5.71	1288175.3
1.3	2.60	586559.7	6.3	5.83	1315247.3
1.6	2.85	642959.6	6.6	5.95	1343838.9
1.9	3.08	694847.6	6.9	6.07	1369391.3
2.	3.30	744479.7	7.	6.17	1391951.2
2.3	3.50	789599.6	7.3	6.27	1414531.1
2.6	3.70	844719.7	7.6	6.39	1441583.2
2.9	3.89	877583.5	7.9	6.50	1466399.3
3.	4.05	913679.5	8.	6.60	1488959.2
3.3	4.21	949775.6	8.3	6.70	1511539.1
3.6	4.37	971658.7	8.6	6.80	1534099.0
3.9	4.53	920447.4	8.9	6.90	1556658.9
4.	4.67	1055551.4	9.	7.	1579199.3
4.3	4.81	1086135.4	9.3	7.10	1601759.2
4.6	4.95	1116718.7	9.6	7.20	1624319.1
4.9	5.08	1146047.4	9.9	7.29	1644622.1
5.	5.22	1177631.3	10.	7.38	1664927.1

CHAPTER III.

DEPTH OF DRAINS.

THE proper depth of drains, depends on various conditions, but especially on the amount of outfall, and the nature of the soil. In some fields, it may be possible to obtain ready outfall for drains thirty inches in depth, where great expense would be incurred, if the drains were laid at depths of three or four feet; in such a case, it is better to use comparatively shallow drains and compensate by placing them at shorter distances. On the other hand, there are situations where the outfall is sufficient, and where a soil of porous materials lies on clay at a depth of four feet; it is then better, and in the end cheaper, to put the drain down upon the clay, and at proportionately greater distances apart. The cost will also be considered, in settling questions of depth and distance; deep drains are disproportionately expensive for the cost of taking out the lower foot of a four feet drain, and is almost equal to that of removing the upper three feet; while shallow drains, which require to be nearer together, involve a greater outlay for tiles. There has been an immense amount of controversy between the advocates of shallow drains, and the advocates of the deep, the one part preferring a depth of two feet six inches, the other a depth of four feet. Practical and experienced men have come to regard these two opinions, as indicating the possible range of useful drainage, either of which may be adopted under special circumstances, while in almost all cases the most useful and economical depth lies between these extremes, or about three feet.

The depth of the drain being one of the most important considerations, we will give at some length the opinions of those who have had ample experience, added to very extensive observations. The first authority we shall refer to, is Mr. Gisborne:

"Many experiments have shown that, in retentive soils, the temperature at 2 or 3 feet below the surface of the water table is, at no period of the year, higher than from 46° to 48°, *i. e.* in agricultural Britain. This temperature is little affected by summer heats, for the following short reasons. Water, in a quiescent state is one of the worse conductors of heat with which we are acquainted. Water warmed at the surface transmits little or no heat downward. The small portion warmed expands, becomes lighter than that below, consequently retains its position on the surface, and carries no heat downward.[1] To ascertain the mean heat of the air at the surface of the earth over any extended space, and for a period of eight or nine months, is no simple operation. More elements enter into such a calculation than we have space or ability to enumerate; but we know certainly that, for seven months in the year, air, at the surface of the ground, is seldom lower than 48°, never much lower, and only for short periods: whereas, at four feet from the surface, in the shade, from 70° to 80° is not an unusual temperature, and in a southern exposure, in hot sunshine, double that temperature is not unfrequently obtained on the surface. Now let us consider the effect of drains placed from 2 to 3 feet below the water table, and acting during the seven months of which we have spoken. They draw out water of the temperature of 48°. Every particle of water which they withdraw at this temperature is replaced by an equal bulk of air at a higher, and frequently at a much higher temperature. The warmth of the air is carried down into the earth. The temperature of the soil, to the depth to which the water is removed, is in a course of constant assimilation to the temperature of the air at the surface. From this

---

1 When water is heated from below, the portion first subjected to the heat rises to the surface, and every portion is successively subjected to the heat and rises, and each, having lost some of its heat at the surface, is in turn displaced. Constant motion is kept up, and a constant approximation to an equal temperature in the whole body. The application of superficial heat has no tendency to disturb the quiescence of water.

it follows necessarily, that during that period of the year when the temperature of air at the surface of the earth is generally below 48°, retentive soils which have been drained are colder than those which have not. Perhaps this is no disadvantage. In still more artificial cultivation than the usual run of agriculture, gardeners are not insensible to the advantage of a total suspension of vegetation for a short period. In Britain, we suffer, not from an excess of cold in winter, but from a deficiency of warmth in summer. Grapes and maize, to which our somber skies deny maturity, come to full perfection in many regions whose winters are longer and more severe than ours. However, we state the facts, without asking to put a large amount therefrom to the credit of our drainage.

"Mr. Parkes gives temperatures on a Lancashire flat moss, but they only commence at 7 inches below the surface, and do not extend to midsummer. At that period of the year the temperature at 7 inches never exceeded 66°, and was generally from 10° to 15° below the temperature of air in the shade, at 4 feet above the earth. At the depth of 13 inches the soil was generally from 5° to 8° cooler than at 7 inches. Mr. Parkes' experiments were made simultaneously on a drained and on an undrained portion of the moss; and the result was, that, on a mean of 35 observations, the drained soil at 7 inches in depth was 10° warmer than the undrained at the same depth. The undrained soil never exceeded 47°, whereas after a thunder-storm the drained reached 66° at 7 inches, and 48° at 31 inches. Such were the effects at an early period of the year on a black bog. They suggest some idea of what they are, when in July or August thunder-rain at 60° or 70° falls on a surface heated to 130°, and carries down with it into the greedy fissures of the earth its augmented temperature. These advantages porous soils possess by nature, and retentive soils only acquire them by drainage.[1]

---

1 The only temperature of thunder rain given in Mr. Parkes' Tables is 78. This, we imagine, must be an extreme heat. We have heard, with much satisfaction, that Mr. Parkes is, by means of his numerous staff stationed at the works which he is carrying on in many parts of Great Britain, Ireland, and (we believe) France, conducting a series of experiments on the temperatures of water of drainage, which tend to show an increase in some proportion to the length of time for which the drainage has been executed. We know no experiments connected with agriculture to the result of which we look with more hopeful expectation. Any agriculturist may, by means of a delicate thermometer, conduct and record such observations on his own farm. Probably the water of drainage from firm land may be

"In all soils the existence of the water table nearer than 4 feet from the surface of the land is prejudicial to vegetation.' Here open upon us the yelpings of the whole shallow pack. Four feet! The same depth for all soils! Here's quackery! We think Mr. Parkes must have stood in very unnecessary awe of this pack, when he penned the following half-apologetic sentence, which is quite at variance with the wise decision with which, in other passages of his works, he insists on depths of four feet and upward in all soils: 'In respect of the depth at which drains may, with a certainty of action, be placed in a soil, I pretend to assign no rule; for there can not, in my opinion, be a more crude or mistaken idea than that one rule of depth is applicable with equal efficiency to soils of all kinds.'[1] Those words—equal efficiency—are a sort of saving clause; for we do not believe that when Mr. Parkes wrote them, he entertained 'the crude or mistaken idea' of ever putting in an agricultural drain less than 4 feet deep, if he could help it. We will supply the deficiency in Mr. Parkes' explanation, and will show that the idea of a *minimum* depth of four feet is neither crude nor mistaken. And as to 'quackery'—which occurs *passim* in the writings and speeches of the shallow drainers—there is no quackery in assigning a minimum. Every drainer does it, and must do it. The shallowest man must put his drains out of the way of the plow and of the feet of cattle. That is his minimum. The man who means to subsoil must be out of the way of his agricultural implement. These two minima are fixed on mechanical grounds. We will fix a minimum founded on ascertained facts and on the principles of vegetation.

"Every gentleman who, at his matutinal or ante-prandial toilet, will take his well-dried sponge, and dip the tip of it into water, will find that the sponge will become wet above the point of contact between the sponge and the water, and this wetness will ascend up the sponge, in a diminishing ratio, to the point where the forces of attraction and of gravity are equal. This illustration is for gentlemen of the Clubs, of London drawing-rooms, of the Inns of Court, and for others of similar habits. For gentlemen who are floriculturists, we have an illustration

---

expected to be higher in temperature than that from the quoted bog in summer, and lower in winter.

1 Smith of Deanston may perhaps be open to some observation, for we believe that he did unadvisedly recommend, in thorough draining, an equal depth and equal distance for parallel drains in all soils.

much more apposite to the point which we are discussing. Take a flower-pot a foot deep, filled with dry soil. Place it in a saucer containing three inches of water. The first effect will be, that the water will rise through the hole in the bottom of the pot till the water which fills the interstices between the soil is on a level with the water in the saucer. This effect is by gravity. The upper surface of this water is our water table. From it water will ascend by attraction through the whole body of soil till moisture is apparent at the surface. Put in your soil at 60°, a reasonable summer heat for nine inches in depth, your water at 47°, the seven inches' temperature of Mr. Parkes' undrained bog; the attracted water will ascend at 47°, and will diligently occupy itself in attempting to reduce the 60° soil to its own temperature. Moreover, no sooner will the soil hold water of attraction, than evaporation will begin to carry it off, and will produce the cold consequent thereon. This evaporated water will be replaced by water of attraction at 47°, and this double cooling process will go on till all the water in the water table is exhausted. Supply water to the saucer as fast as it disappears, and then the process will be perpetual. The system of saucer-watering is reprobated by every intelligent gardener; it is found by experience to chill vegetation; besides which, scarcely any cultivated plant can dip its roots into stagnant water with impunity. Exactly the process which we have described in the flower-pot is constantly in operation on undrained retentive soil: the water table may not be within nine inches of the surface, but in very many instances it is within a foot or eighteen inches, at which level the cold surplus oozes into some ditch or other superficial outlet. At 18 inches, attraction will, on the average of soils, act with considerable power. Here, then, you have two obnoxious principles at work, both producing cold, and the one administering to the other. The obvious remedy is, to destroy their *united* action; to break through their line of communication. Remove your water of attraction to such a depth that evaporation can not act upon it, or but feebly. What is that depth? In ascertaining this point we are not altogether without data. No doubt depth diminishes the power of evaporation rapidly. Still, as water taken from a 30 inch drain is almost invariably two or three degrees colder than water taken from 4 feet, and as this latter is generally one or two degrees colder than water from a contiguous well several feet below, we can hardly avoid drawing the conclusion that the cold of evaporation has considerable influence at 30 inches, a much diminished influence at 4 feet, and little or none below that depth. If the water table is removed to the depth of 4 feet, when we have allowed 18 inches of attraction, we shall still have 30 inches of defense against evaporation; and we are inclined to believe that any prejudicial combined action of attraction and evaporation is thereby well guarded against. The facts stated seem to prove that less will not suffice.

"A farmer manures a field of four or five inches of free soil reposing on a retentive clay, and sows it with wheat. It comes up, and between the kernel and the manure it looks well for a time, but anon it sickens. An Irish child looks well for five or six years, but after that time potatoe feeding, and filth, and hardship, begin to tell. You ask what is amiss with the wheat, and you are told that when its roots reach the clay they are poisoned. This field is then thorough drained, deep, at least four feet. It receives again from the cultivator the previous treatment; the wheat comes up well, maintains a healthy aspect, and gives a good return. What has become of the poison? We have been told that rain water filtered through the soil has taken it into solution or suspension, and has carried it off through the drains, and men who assume to be of authority put forward this as one of the advantages of draining. If we believed it we could not advocate draining. We really should not have the face to tell our readers that water, passing through soils containing elements prejudicial to vegetation, would carry them off, but would leave those which are beneficial behind.[1] We can not make our water so discriminating; the general merit of water of deep drainage is, that it contains very little. Its perfection would be, that it should contain nothing. We understand that experiments are in progress which have ascertained that water, charged with matters which are known to stimulate vegetation, when filtered through four feet of retentive soil, comes out pure.[2] But to return to our wheat. In the first case, it shrinks before the cold of evaporation, and the cold of water of attraction, and it sickens because its feet are never dry; it suffers the usual maladies of cold and wet. In the second case, the excess of cold by evaporation is withdrawn; the cold water of attraction is removed out of its way; the warm air from the surface,

---

[1] We do not deny that some subsoils contain matter prejudicial to vegetation, but generally they are not worse than a *caput mortuum*; seldom quite so bad.

[2] Since this Essay was first printed, a portion of these experiments has been communicated to the public by Professor Way.

rushing in to supply the place of the water which the drains remove, and the warm summer rains, bearing down with them the temperature which they have acquired from the upper soil, carry a genial heat to its lowest roots. Health, vigorous growth, and early maturity are the natural consequences.

"Water can only get into drains by gravity, which only acts by descent—technically, by fall; the fall must be proportioned to the friction which the water encounters on its passage. Suppose drains four feet deep to be placed twelve yards apart on level land, it is plain that water at that depth, lying at the intermediate point between the two drains, will not get into either of them. A fall of some inches will be required to enable it to overcome the friction of six yards of retentive soil. In order, therefore, to lower the water table to four feet at all points, the drains must be some inches deeper than four feet. If the land lies on a slope (say four inches to the yard), drains of four feet, if driven on the line of steepest descent, will effect the object; because, though water at four feet, lying at the intermediate point between two drains, in a line at right angles to them, can not for want of fall get into either of them by traveling six yards, it will find a fall of four inches at less than seven, and of eight inches at less than eight, yards. If we must speak quite correctly, this intermediate water will never get into the drain till there is a fresh supply; it will descend perpendicularly, pushing out that which lies below it, and will be itself displaced by a fresh arrival from the heavens. In order that the whole soil, if homogeneous, or nearly so, may be drained evenly, it is manifest that the drains must be parallel. Extra friction in the soil must be met either by making the drains deeper, or by placing them nearer. On this point, which is one of practice rather than of principle, each case must be left to the sagacity of the operator. We doubt whether in any natural soil the friction is so great as to resist a fall of one inch in a yard. If we are right in this point, we should always attain the object of lowering the water table to four feet by 4-feet 6-inch drains, parallel, and twelve yards apart. We have already stated one advantage which results on a slope from driving the parallel drains in the line of steepest descent: to-wit, that when they are so driven, all water which lies at the same depth from the surface at the bottoms of the drains, can find a fall into one or the other by traveling a little more than half the distance between them; whereas, if the drains are driven across the slope, half the water so situated as to depth can only find a fall into the lower drain, and in order to reach it must travel distances varying from one half to the full interval between the two.

"We shall dismiss, with a very few words, two classes of writers on the subject of draining: 1. Those who limit the advantages of a drain to the water which is passed into it from its own surface, and who, therefore, enjoin that it should be filled with porous material, and that it should be shallow. 2. Those who will not drain 4 or 5 feet deep because it makes the ground too dry for the roots of plants. This idea must have come from some garret, having been conceived by an ingenious recluse brooding over his ignorance, and reasoning as follows: What makes vegetation burn up? The absence of water from its roots. What takes away the water? Deep drains. Ergo, deep drains are the cause of burning. We will supply a formula: Why does vegetation burn? Because its roots are very superficial. Why superficial? Because they won't face the cold of stagnant water. What removes the cold and the water? Deep drains. And the facts exactly coincide with our logic. Deep drained lands never do burn. Nothing burns sooner than a few inches of soil on a very retentive clay. No land is less subject to burn than the same soil when, by 4 or 5 feet draining, a range of 3 or 4 feet has been given to the previously superficial roots.

"Having dismissed these two small matters, we must treat more respectfully a lingering skepticism as to the efficacy of deep drains in very retentive soils; and instead of wondering at this skepticism, we wonder rather that deep thorough draining has so rapidly made converts. Representations are made of soils which consist of some inches of a moderately porous material reposing on a subsoil which is said to be impervious; and we are told that it is of no use to make the drain deeper into the impervious matter than will suffice for laying the conduit. If the subsoil is impervious, as glass or even as cast iron or caoutchouc are impervious, we at once admit the soundness of the argument. We only want to ask one question: Is your subsoil moister after the rains of midwinter than it is after the drought of midsummer? If it is, it will drain. Mr. Mechi asks, shrewdly enough: 'If your soil is impervious, how did you get it wet?' This imperviousness is always predicated of strong clays—plastic clays they are sometimes called. We really thought that no one was so ignorant as not to be aware that clay lands always shrink and crack with drought, and the stiffer the clay the greater the shrinking, as brickmakers well know. In the great drought thirty-six years ago, we saw, in a very retentive soil in the Vale of Belvoir, cracks which it was not very pleasant to ride among. This very summer, on land which, with reference to this very subject, the owner stated to be impervious, we put a walking-stick three feet into a sun crack without finding a bottom, and the whole surface was what Mr. Parkes not inappropriately calls a net-work of cracks. When heavy rain comes upon the soil in this state, of course, the cracks fill, the clay imbibes the water, expands, and the cracks are abolished. But if there are 4 or 5 feet parallel drains in the land, the water passes at once into them, and is carried off. In fact, when heavy rain falls upon clay lands in this cracked state, it passes off too quickly, without adequate filtration. Into the fissures of the undrained soil, the roots only penetrate to be perished by the cold and wet of the succeeding winter; but in the drained soil the roots follow the threads of vegetable mold which have been washed into the cracks, and get an abiding tenure. Earth worms follow either the roots or the mold. Permanent schisms are established in the clay, and its whole character is changed. An old farmer in a midland county began with 20 inch drains across the hill, and, without ever reading a word, or, we believe, conversing with any one on the subject, poked his way, step by step, to 4 or 5 feet drains in the line of steepest descent. Showing us his drains this spring, he said: 'They do better year by year; the water gets a habit of coming to them.' A very correct statement of the fact, though not a very philosophical explanation. Year by year the average dryness of the soil increases, the cracks are further extended, and seldomer obliterated. A man may drain retentive soils deep and well, but he will be disappointed if he expects what is unreasonable. No intelligent and honest operator will say more than that money judiciously expended in draining them will pay good, and generally very good, interest. If you eat off turnips with sheep, if you plow the land, or cart on it, or in any way puddle it when it is wet, of course the water will lie on the surface, and will not go to your drains. A 4-feet drain may go very near a pit or a watercourse without attracting water from either, because watercourses almost invariably puddle their beds, and the same effect is produced in pits by the treading of cattle, and even by the motion of the water produced by wind. A very thin film of puddle always wet on one side is impervious, because it can not crack.

"No system of draining can relieve soils of water of attraction. That can only be exhausted by evaporation. Retentive soils hold it in excess; its reduction by evaporation produces cold; and, therefore, retentive soils never can be so warm as porous. Expect reasonable things only of your drained retentive soils, and you will not be disappointed. Shallow drainers start with the idea of a drop of water falling on the top of the soil, and working its solitary way through narrow and tortuous passages to a drain; and they say that it would be lost in the labyrinth, which we think very likely. They have no idea that the water operated upon by the drain is that which lies at the level of its own bottom, which runs off, and is replaced by that which was immediately above it. And on account of this operation, which we have before explained, it is necessary in retentive soils, in which friction is greater than in porous, to have the drains deeper, in order to lower the water table to the same extent. A column of six inches may suffice to push water from the intermediate point between two drains in a porous soil, and it may require a 12-inch column in a retentive. In that case the drain in the retentive soil must be six inches deeper than in the porous. Ignorance says: Drain shallower because the soil is retentive. Experience and reason say: Drain deeper. We may here notice, that in clay lands the portion within one to two feet of the surface is almost always more retentive than that which lies below; simply, we apprehend, because its particles have been comminuted and packed close by the alternate influences of wet and dry, heat and cold. When dried below by drains, and above by evaporation, it is certain to crack and become permeable; and this operation may, if necessary, be assisted by subsoiling or other artificial means.

"Smith, of Deanston, first called prominent attention to the fertilizing effects of rain filtered through land, and to evils produced by allowing it to flow off the surface. Any one will see how much more effectually this benefit will be attained, and this evil avoided, by a 4-feet than by a 2-feet drainage. The latter can only prepare two feet of soil for the reception and retention of rain, which two feet, being saturated, will reject more, and the surplus must run off the surface, carrying whatever it can find with it. A 4-feet drainage will be constantly tending to have four feet of soil ready for the reception of rain, and it will take much more rain to saturate four feet than two. Moreover, as a gimlet hole bored four feet from the surface of a barrel filled with water will discharge much more in a given time than a similar hole bored at the depth of two feet, so will a 4-feet drain discharge in a given time much more water than a drain of two feet. One is acted on by a 4-feet, and the other by a 2-feet pressure."

The controversy between deep drains and shallow drains induced a great many experiments to be made. Among these experimenters was Lord Wharncliffe, who adopted a kind of compromise system, combining four feet and two feet drains.

Lord Wharncliffe states his principles as follows, and calls his method the combined system of deep and shallow drainage:

"In order to secure the full effect of thorough drainage in clays, it is necessary that there should be not only well-laid conduits for the water which reaches them, but also subsidiary passages opened through the substance of the close subsoil, by means of atmospheric heat, and the contraction which ensues from it. The cracks and fissures which result from this action, are reckoned upon as a certain and essential part of the process.

"To give efficiency, therefore, to a system of deep drains beneath a stiff clay, these natural channels are required. To produce them, there must be a continued action of heat and evaporation. If we draw off effectually and constantly the bottom water from beneath the clay and from its substance, as far as it admits of percolation, and by some other means provide a vent for the upper water, which needs no more than this facility to run freely, there seems good reason to suppose that the object may be completely attained, and that we shall remove the moisture from both portions as effectually as its quantity and the substance will permit. Acting upon this view, then, after due consideration, I determined to combine with the fundamental four feet drains a system of auxiliary ones of much less depth, which should do their work above, and contribute their share to the wholesome discharge, while the under-current from their more subterranean neighbors should be steadily performing their more difficult duty.

"I accomplished this, by placing my four feet drains at a distance of from eighteen to twenty yards apart, and then leading others into them, sunk only to about two feet beneath the surface (which appeared, upon consideration, to be sufficiently below any conceivable depth of cultivation), and laying these at a distance from each other of eight yards. These latter are laid at an acute angle with the main drains, and at their mouths are either gradually sloped downward to the lower level, or have a few loose stones placed in the same intervals between the two, sufficient to insure the perpendicular descent of the upper stream through that space, which can never exceed, or, indeed, strictly equal, the two additional two feet."

Speaking of the Wharncliffe system, Gisborne remarks:

"Were I to adopt his lordship's system, I must abandon, 1st., the principle of depth; and 2d., the principle of direction; and if I abandoned those two principles, I had much better put this treatise into the fire than send it to Mr. Murray for publication."

Alderman Mechi, speaking of deep drainage, says:

"Ask nineteen farmers out of twenty, who hold strong clay land, and they will tell you it is of no use placing deep four foot drains in such soils—the water can not get in; a horse's foot-hole (without an opening under it) will hold water like a basin; and so on. Well, five minutes after, you tell the same farmers you propose digging a cellar, well bricked, six or eight feet deep; what is their remark? 'Oh! it's of no use your making an underground cellar in our soil, you *can't keep the water* OUT!' Was there ever such an illustration of prejudice as this? What is a drain pipe but a small cellar full of air? Then, again, common sense tells us, you can't keep a light fluid under a heavy one. You might as well try to keep a cork under water, as to try and keep air under water. 'Oh! but then our soil is n't porous.' If not, how can it hold water so readily? I am led to these observations by the strong controversy I am having with some Essex folks, who protest that I am mad, or foolish, for placing 1-inch pipes, at four feet depth, in strong clays. It is in vain I refer to the numerous proofs of my soundness, brought forward by Mr. Parkes, engineer to the Royal Agricultural Society, and confirmed by Mr. Pusey. They still dispute it. It is in vain I tell them *I can not keep the rainwater out of* socketed pipes, twelve feet deep, that convey a spring to my farm-yard. Let us try and convince this large class of doubters; for it is of *national* importance. Four feet of good porous clay would afford a far better meal to some strong bean, or other tap roots, than the usual six inches; and a saving of $4 to $5 per acre, in drainage, is no trifle.

"The shallow, or non-drainers, assume that tenacious subsoils are impervious or non-absorbent. This is entirely an erroneous assumption. If soils were impervious, how could they get wet?

"I assert, and pledge my agricultural reputation for the fact, that there are no earths or clays in this kingdom, be they ever so tenacious, that will not readily receive, filter, and transmit rain water to drains placed 5 or more feet deep.

"A neighbor of mine drained 20 inches deep in strong clay; the ground cracked widely; the contraction destroyed the tiles, and the rains washed the surface soil into the cracks and choked the drains. He has since abandoned shallow draining.

"When I first began draining, I allowed myself to be overruled by my obstinate man, Pearson, who insisted that, for top water, 2 feet was a sufficient depth in a veiny soil. I allowed him to try the experiment on two small fields; the result was, that nothing prospered; and I am re-draining those fields at *one half* the cost, 5 and 6 feet deep, at intervals of 70 and 80 feet.

"I found iron-sand rocks, strong clay, silt, iron, etc., and an enormous quantity of water, all *below* the 2-feet drains. This accounted at once for the sudden check the crops always met with in May, when they wanted to send their roots down, but could not, without going into stagnant water."

Good results are always obtained from three feet drains, and there can be no doubt that the results would be more permanent with four feet drains. Where drains at three feet deep will accomplish all practical purposes for a period of twenty-five or thirty years, it will require very strong arguments indeed to induce the farmers to drain to the depth of four feet.

In this country, every farmer, as a general thing, owns the land he cultivates, and in a majority of instances has earned, with his own hands, every dollar that he paid for the farm and its improvements. In an improvement so permanent as underdraining proposes to be, the farmer "counts the cost" very closely and very *frequently*, before commencing it, and if he is fully satisfied that good results will attend his efforts, when draining at a depth of three feet, at an expense equal to three fourths of the amount that draining four feet would cost, scarcely any argument would induce him to drain at a depth of the additional foot. And the farmer is fully justified in this course, in the Middle and Western states. Landed estates change hands very rapidly in this country. Suppose a farmer incurs a debt of $1000, for any improvement over and above his immediate means. He can not mortgage his "crops in the ground," to secure this amount, because drought, hail, insects, or other adversities beyond his control may destroy them; he can not mortgage his sheep, because dogs may kill them, nor his cattle, because there is a possibility that they may die of pleuro-pneumonia, trembles, murrain, or a dozen other diseases; so that no other resource is left than to mortgage the farm itself. Should the mortgage mature, and crops be short, or prices unremunerative, the creditor can foreclose and the farmer be sold by the sheriff, in one hundred days or less. This is no fancy sketch; Ohio farmers have so frequently witnessed the fate of neighboring farmers, in this respect, that they have become exceedingly cautious so far as involving themselves in indebtedness is concerned.

In England or Germany, where lands seldom pass out of the hands of the family, even if the proprietor is absolutely bankrupt, larger amounts can be hazarded in improvements, without incurring the risk of losing the farm. In those countries they may insist on draining at four feet as a minimum depth. For reasons already given, we believe the minimum will be determined by each man for himself, without regard to system or theory, on the following basis, viz.: to lay the tile at such a depth that neither the plow nor subsoil plow will interfere with it, and that it will be beyond the range of frost. We think that these two points, namely, beyond the range of the frost, and out of reach of the subsoil plow, will determine the depth of drains in more instances, in this country, than all the illustrations that English or German draining engineers can adduce from experience in favor of very deep draining.

It is not an uncommon phenomenon to find the earth, in cultivated fields, consisting of loamy soils, frozen to the depth of 14 to 16 inches. In a cemetery we once saw a loamy clay frozen to the depth of 22 inches. Underdrained soils always freeze considerably deeper than undrained ones. If then the soil in a field freezes to the depth of 16 inches, it is safe to infer that if the field is well underdrained, the frost will find its way down fully two feet. We would not advise any one to underdrain at a depth less than 30 inches, and where the fall, and pecuniary means will warrant, we would insist that 3 feet should be considered the minimum, in all soils requiring underdraining.

The day is not far distant when subsoiling will be much more generally practiced. Improvements seldom terminate with the initiatory step, and the man who is sufficiently convinced of the importance of underdraining, and puts it in practice on his farm, will not hesitate to use the subsoil plow, and tile laid at a depth less than 30 inches, will not probably be beyond the reach of this plow.

The difference in cost between a three and a four feet drain is considerably more than one would at first suppose. A good English ditcher, in ordinary clay soil, will make eight rods of three feet drain per day, but will not make more than five rods of four feet in the same time—in fact, he seldom will make over four. To sink a three feet drain one foot lower, will cost nearly as much for the last foot as for the preceding three—for reasons that every practical man will at once understand. Drains for tile are narrowed from the top to the bottom—they are generally 14 to 18 inches wide at the top, and four inches only, or just wide enough to admit the tile, at the bottom. Now, although the last foot in a four feet drain contains no more earth to be removed than the last foot in a three feet drain, yet it is a foot *lower*, and must consequently be thrown a foot *higher* up, without taking into account the pile of earth already excavated on which or over which this last foot must be thrown.

When thorough drainage was first introduced into Scotland, it is said that 10,000 miles of drains were laid, at a depth of two feet, when it was discovered that this depth was not sufficient. In England large tracts were laid with tile at 12 to 18 inches deep. Of course the experimenters were gratified with the success which crowned their efforts—the land was in a tillable condition early in the seasons, and the surplus waters removed. But now the opposite extreme is advocated, and Alderman Mechi has gone so far as to make some drains 14 feet deep!

So far as draining the surface water, or the water falling in the shape of rain or snow is concerned, the Alderman says:

"After all that has been said and written on the subject, I have arrived at the following conclusions:

"1. That Mr. Parkes' statement is a convincing proof that one-inch pipes (*without* stones, straw or brushes) placed four feet deep, at intervals of thirty feet, will effectually and permanently drain the heaviest soils of the utmost quantity of surface water that can possibly fall, at a cost of from £2 to £3 per acre. That in mixed soils, the one-inch pipes, four feet deep and fifty feet apart, will perfectly drain such soils, at a cost of about 45s. per acre.

"2. That although those drains do not, the first year after being made, act so effectually as stones with pipes on my plan, which carry off the water at once; still the immense difference in cost, and greater depth, render Mr. Parkes' plan by far most desirable.

"3. There can be no doubt that it is the *depth* of the drain which regulates the escape of the surface water in a given time; regard being had, as respects extreme distances, to the nature of the soil, and a due capacity of the pipe. *The deeper the drain, even in the strongest soils, the quicker the water escapes.* This is an astounding but certain fact.

"4. That deep and distant drains, where a sufficient fall can be obtained, are by far the most profitable, by affording to the roots of plants a greater range for food.

"5. That had I to redrain my heavy land, I should do so, at least four feet deep, with inch-pipes at intervals of thirty feet, carrying each pipe with the fall of the land direct to an open ditch of ample capacity. I should thus economize several open ditches on my farm, which are at present a waste of ground. Each drain would thus be its own leader.

"I should place the pipes in the drains without stones, or other matter, merely covering them with the clay itself, leaving the drains open as long as possible, as practiced by Mr. Hammond. I should thus save £7 per acre on the cost of my draining, and have a greater depth of soil. The loss would be the difference between a perfect and imperfect drainage the first two years.

"In conclusion, I consider the balance of evidence, when stones and pipes are used, is in favor of the pipe being placed at the bottom."

CHAPTER IV.

DISTANCE BETWEEN DRAINS.

A RULE formerly adopted in England, that "the distance between parallel drains may be increased proportionably with their depth; and that drains may be laid as many perches apart as they are feet deep"—is no longer regarded as being correct. There are so many different kinds of soil that no general rule can be given which will be alike applicable to all; but each kind must be dealt with according to its inherent qualities. As a general thing, a depth of four feet for clay soils, appears to be uniformly adopted in England and Germany; but this is, perhaps, deeper than those who desire to drain in this country can afford to go, on account of the increased cost of the last foot in depth.

The *distance* between the minor drains, in thorough draining, depends on various circumstances, such as their depth, and the nature of the soil and subsoil. The greater the depth, the greater may be the distance; the more clayey and tenacious the soil, the nearer should the minor drains be placed. On stiff clay soils, the distance should be less than a rod and a half; on loose soils, resting on clay, a drain every two rods will be sufficient.

A system was at one time advocated of digging experimental or "*trial holes*," and regulating the depth according to the degree of moisture, but this produced such very variable results, both with regard to depth of drains and distance between them, as to afford no reliable data for a general system. These experimental holes gave rise, in the hands of Joshua Trimmer (celebrated as the author of a very comprehensive treatise on "Practical Geology and Mineralogy"), to the noted Keythorpe system of underdraining. In one of his pamphlets defending the system, Mr. Trimmer says:

"The peculiarities of the Keythorpe system of draining consist in this—that the parallel drains are not equidistant, and that they cross the line of the greatest descent. The usual depth is three and a half feet, but some are as deep as five and six feet. The depth and width of interval are determined by digging trial holes, in order to ascertain not only the depth at which the bottom water is reached, but the hight to which the water rises in the holes, and the distance at which a drain will lay the hole dry. In sinking these holes, clay banks are found with hollows or furrows between them, which are filled with a more porous soil.

"The next object is to connect these furrows by drains laid across them. The result is, that as the furrows and ridges here run along the fall of the ground, which I have observed to be the case generally elsewhere, the submains follow the fall, and the parallel drains cross it obliquely.

"The intervals between the parallel drains are irregular, varying, in the same field, from 14 to 21, 31, and 59 feet. The distances are determined by opening the diagonal drains at the greatest distance from the trial holes at which experience has taught the practicability of its draining the hole. If it does not succeed in accomplishing the object, another drain is opened in the interval. It has been found, in many cases, that a drain crossing the clay banks and furrows takes the water from holes lying lower down the hill; that is to say, it intercepts the water flowing to them through these subterranean channels. The parallel drains, however, are not invariably laid across the fall. The exceptions are on ground where the fall is very slight, in which case they are laid along the line of greatest descent. On such grounds there are few or no clay banks and furrows."

Another English doctrine was, that parallel drains may be laid as many *rods* distant from each other as they are *feet* deep. Thus, if the drains are four feet deep, they may be laid four rods apart. This doctrine is now rejected as being incorrect—at least for clay soils. It is not probable that any criterion other than such as experience may establish, can be given to determine the distance which minor drains should be apart. Those who assert that the distance between drains depends entirely upon the depth, take as a basis the direction formed by the water to the drains, for they say, "The deeper the pipes, the greater may the distance be between them, and yet afford the water the same angle of outlet."

This appears somewhat probable, but when we recur to hydrostatic laws, the assertion will perhaps lose its entire value.

The proper distance between the drains depends upon the space between the particles of the soil entirely; that is, upon the porosity of the earth. The reasons which induce this opinion are as follows:

The water existing in the earth forms a mutually-adhering mass of its particles, which, in a surface which may be desiccated by a system of drains, commonly stands on a level, with, perhaps, slight differences determined by local conditions of the soil.

The function of drains is to remove that part of this mass of water which lies so near the surface as to be injurious to the cultivated crop.

Now, if two drains of like depth are placed parallel in the earth, to intercept a portion of the water contained, the following conditions relatively arise:

The water below the drains, which can not be withdrawn by them, forms a resisting substratum which prevents the further sinking of the water above them. But to this the drains afford conduits in which it is compelled to find its way between the particles of the soil.

The specific gravity of the water which impels it to sink toward the center of the earth, determines the sinking of the whole mass into the conduits, until a level is reached, but this can not take place with the water above the pipes.

The stratum of water beneath the pipes is, at the same time, of great importance, because it forms in the drained surface the foundation, so to say, upon which the water to be drained rests, and gravity being exerted, causes the flow in a lateral direction, having no other impediment to overcome than friction among the earth particles. The original angle of the water line with regard to the pipes is a matter of less importance.

Accordingly, the known hydrostatic law, "every connected mass of water stands at a uniform level," could not exist were it not for the friction which the water must overcome in its passage to the drains.

Again, as the greater or less friction is dependent upon the greater or less proximity of the particles of the soil, this alone is a measure of the proper distance of the drains from each other; that is, they must be placed at such distances from each other that the friction can not neutralize the motive power (in this case, specific gravity).

We base this assertion upon the doctrine of physics, that adhesion exerts its influence as soon as it is stronger than the gravity which carries the adhering body downward.

Therefore, if the porosity of the soil affords drainage capacity which may be represented by a given triangle, the length of the sides of which represent the depth of drain and requisite distance apart, it would be a needless expenditure to place the drains nearer together than the base of the triangle of indication, or to lay them deeper than such base requires to drain a given space.

The objection which may be urged to this principle, that a greater angle contains a greater mass of water, and is thus calculated to remove it more rapidly and certainly, as gravity is the motive power, is not tenable, because, first, the lesser amount of water in a shallow triangle requires less time to flow off; and on the other hand, there is much less friction to overcome than in a deeper triangle of the same breadth or base. This, too, is a matter of importance, as even when the water between the particles of soil is conjoined to form one mass, the friction is a very powerful hindrance to the efflux of the water, as may be learned from the experiments referred to below. But if a simple, single infraction of the watercourse operate so powerfully upon its efflux, how much more must this be the case in the millions of curvatures determined in its passage of efflux by the relations of the particles of the soil?

Here it still is to be borne in mind that we speak only of sinkages which must first take place in the soil before the water reaches the pipes. The soil is more readily permeable than the subsoil, and if that be one fourth foot deep, the water will have traversed one third of its course when it reaches the latter, in case the drains are three and a half feet deep; only one fourth of the course will have been passed over when they are five feet deep.

Now, if it be objected that we have said that gravity is the motive power, and we must place the drains so far apart as to afford the water force (specific gravity, in this case), to overcome the friction, this position will be in opposition to the foregoing assertion that, in the supposed triangle, where the distance between the drains is the same, but the depth unequal, the gravity of the water is much greater in the deeper triangle, we need only state that the formation of the angle of efflux under consideration can be made by sinkage only, and that the surface lying between the extreme points of the triangle is the same whatever the depth of the drain, and consequently the sinkage, gravity and pressure must be the same; but aside from this, suppose drains at equal distances, one pair of which are five and another three feet deep, if all the spaces between the particles of earth in each instance were filled with water, the deeper interspace would contain two fifths more water, and there would be two fifths more friction to overcome in its efflux to the drains, and, as already stated, the porous soil in the shallow interspace affords one third, and in the deeper only one fifth of the friction in the mass of earth to be overcome. But as the entire pressure of the water is exerted downward, the gravity of the upper two strata of water being equal for equal surfaces, and in its efflux through the particles of the soil, it can only overcome the friction by equal pressure, and consequently, in similarly constituted soils, will sink with equal rapidity; and from this it follows that the water of one stratum, drained at five feet of depth, will sink as rapidly as the corresponding stratum of three feet, provided the subsoil above the five feet drain be equally permeable.

But if the subsoil below three feet be less permeable, the superficial strata will sink much more slowly than to the three feet drain, because the friction is much increased.

It is our opinion that this latter circumstance is greatly in favor of the shallower drainage, it being understood to refer only to temporary humidity.

In reference to the foregoing, we say to all who are enthusiastic in favor of deep draining, that their deep drain theories amount to nothing in all those cases where the drainage of temporary water is intended. The mass of water falling upon a given superficies is the same whatever be the depth of the drain, and hence the distance between drains must be the same to carry it off, and it must be longer in sinking to the deep drain than the shallow one. What evidence have deep drainers for their assertions?

Various drains have been made where the fall of the earth was such that the outlet pipes would be placed no deeper than two and three fourths feet, while the heading of the principal drain was placed at five feet of depth. Nevertheless, the effects upon all parts of the drained surface were equal—there was nowhere a difference in the condition of the crop cultivated or its produce.

This fact would be sufficient to confirm the assertion above made, but we will call the attention of every one who has drained large surfaces to one circumstance which most clearly verifies this statement.

Deep drainers themselves can not avoid, on account of the natural inequalities of the ground, placing the pipes at a less depth in some places than in the remaining portions of the drainings, while the distance between the drains remains the same. Nevertheless, the unprejudiced will certainly find no difference in the growth or produce of the crops.

It is self-evident that the depth of the drains must be such that the pipes shall be protected, from every external influence, and for this 3 feet of depth are quite sufficient.

Besides, whoever has done much drainage will certainly not dispute the fact, that whatever system may be adopted, in many portions of the county it will be found impracticable to place the drains at a depth of five feet, or even four feet, for want of sufficient fall, to secure a prompt outlet into the water of the lakes, rivers, etc., which receive the drained water, and that this mutual relation becomes more and more unfavorable for deep drainers at every foot of increased depth; so that in all drain systems only about 25 per cent. can have an average of five feet depth, 10 per cent., 5½ feet, and only 5 per cent. 6 feet depth, on account of the outlet. We repeat that the angle of descent, formed by water in its efflux, depends entirely upon the distance between the drains, and this again upon the porosity of the soil. (See Fig. 5, page 99.)

In very porous soil water sinks nearly horizontally. In all compact soils, as those consisting more or less of clay or loam, it naturally sinks in a perpendicular line over the pipes, more rapidly; first, because earth once dug up never regains its original compactness, and the water has less friction to overcome; and secondly, because the natural law of gravity is in a vertical or perpendicular direction.

If, now the stratum of water lying between the drains sinks equally rapidly at first, it has, nevertheless, greater friction to overcome, which will be greater in proportion to the distance from the drain.

The earth immediately surrounding the drain continues to yield its humidity, as this is removed by the drain, and consequently sloping lines of efflux will be found.[1] We are not aware, however, that these can be determined by deeper or shallower drains, but it appears to us that the greater or less pitch of these lines is to be measured alone by the distance between the drains. This has reference, however, only to those masses of water which fall immediately upon the surfaces to be drained.

When water has been conducted from a higher point, it tends, on account of the pressure from above, to raise

---

[1] The older the drain is the less perceptible will be this appearance, because the earth directly over the drain necessarily cast out, in the process of construction, becomes gradually more and more compact, even though it never regains its original solidity, yet its porosity constantly diminishes with time, while the soil of the interspaces becomes more mellow, and the surface of the water will be drawn off toward the drains much more horizontally than in newly completed drains.

itself again, and that in a greater degree as it meets with less obstruction. But in a drained surface, obstruction ceases when the rising water reaches the level of the drain pipe, as it then finds a free efflux, and to this point all the force of pressure tends; but when the water has once been forced into the drain, the law of gravity in the given case again comes into action, and the water flows immediately toward the discharge of the pipe.

It is, therefore, very important to determine the proper distance between the drains.

This problem has always appeared to me as one of the most important in the science of draining, as most, or we might with propriety say, all the success of a drain depends upon its solution.

This much is certain, if the distance be too great the drain either does not operate at all, or its effect is so tardy that it does no good to the crop; or if the distance be too small the work will be disproportionately expensive. To obviate these difficulties, it appeared to us necessary to examine closely, and find if there were any experiments, by means of which a rule for distances, according to differences of soil, could be determined. We have found some such in H. Wauer's work on drainage, from which we translate the following:

"For this purpose I instituted experimental drains in similar soil, at unequal distances, and observed their effect.

"After I had determined the distance of perfect drainage for the given soil, I took this as a basis for further observation and experiments, and proceeded as follows:

"I first ascertained the amount of clay contained in the soil, then desiccated a portion of this in an oven. I then filled a glass tube 18 inches long, two thirds full of this soil, and covered the lower end with a piece of thin linen, to permit water to flow off readily. I then added a certain quantity of water, and marked the time exactly when it had all escaped at the lower extremity of the tube, minus what was retained by the force of cohesion.

"This experiment I repeated with different grades of soil, and noted carefully the difference of time at each new experiment. I thus found that loamy earth, containing 35 per cent. of clay, permitted the passage of water in half the time required by clay soil containing 70 per cent. of clay; that loamy sand, with 15 per cent. of clay, yielded the water three times more rapidly than clay soil of 70 per cent., etc.; and upon this I based the calculation of the distance proper for the distance apart of the drains, as given in the following table:

1	In clay soil,	70	per cent. clay in	2	rods,	
2	"	65	"	2	"	3 feet.
3	"	60	"	2	"	6 "
4	"	55	"	3	"	9 "
5	"	50	"	3	"	—
6	loamy soil	45	"	3	"	4 "
7	"	40	"	3	"	8 "
8	"	35	"	4	"	—
9	"	30	"	4	"	6 "
10	"	25	"	5	"	—
11	loamy sand	20	"	5	"	6 "
12	"	15	"	6	"	—
13	"	10	"	6	"	6 "
14	sand	5	"	7	"	—
15	in sand at	0		7	"	7 "
16	in granular sand			8	"	6 "

"In turf and moor soils, devoid of clay or muck, the calculation resulted the same as in No. 12, and where many vegetable remains existed, as in No. 14.

"In calcareous soils, I found the percentage like that in clay soils.

"In the experiments afterward instituted, in the construction of drains, these calculations were verified exceedingly well, and they have been the basis of all my plans since then.

"That departures from this are required, for draining springs and ponds, is naturally to be supposed. Such cases require the technical knowledge of the drainer, and do not permit the application of fixed rules."

In draining meadows, which are designed to be used as

such, still a rule of distance is difficult to give, because stagnant water in them is due to such various causes, that it would be difficult to meet each case by general rules. If the soil is uniform, and if the water be temporarily collected, the calculation may be made according to the rules given by Wauer, for fields, but make the distances about one third greater.

In most cases, it will be sufficient to traverse the so-called swales with single drains. We recently read an essay upon draining which stated that meadows should be drained in the same manner as fields, but that the pipes should be smaller, so that the water might run off more slowly.

That would be sure to convert a meadow into a swamp. The pipes being filled, the remaining water would be forced into the ground, there to remain.

We must repeat, that too great a distance between drains, takes the water off too slowly, and no good is accomplished, because the interspace shows no effect of drainage, and nothing is left but the expensive method of putting down intermediate drains.

According to John,[1] the following distances, in the various kinds of soil named, are in accordance with the latest German experience, in this respect, where the drains are four feet deep.

I.	Heavy clay soil, - - - - - -	20 to 24 feet.
II.	Clay soil, - - - - - - -	24 to 30 "
III.	Loamy soil, - - - -	30 to 36 "
IV.	Light loamy soil, - - - -	48 "
V.	Sandy loamy soil, - - - - -	60 "
VI.	Very light soil, - - - - - -	80 to 120 "

Schönermark, in draining in Brunswick, established the

---

[1] Jahrbuch, 1854, I, 74.

following distances and depths for the various kinds of soil:

Soils.	Drain 4 feet deep.	Drain 3 feet 6 in. deep.	Drain 3 feet deep.
	feet apart.	feet apart.	feet apart.
I. Lime soil (soils containing from 30 to 60 per cent. of lime, and a mixture of sand, clay and humus). Clay soil containing 50 per cent. and upward of clay, - - - - - -	24 to 32	21 to 28	18 to 24
II. Marl soil (containing 10 to 30 per cent. of lime, and is mixed with clay, sand and humus), - -	32 to 40	28 to 35	24 to 30
III. Loamy soil, containing sand and clay, - - - - - -	40 to 48	35 to 42	30 to 36
IV. Sandy loam, containing from 10 to 30 per cent. of sand, - -	48 to 56	42 to 49	36 to 42
V. Loamy sand, containing 30 to 50 per cent. of sand, and humus soil, containing 30 to 50 per cent. of humus,	56 to 64	49 to 56	42 to 48
VI. Sandy soil, containing 50 to 70 per cent. of sand, and mixed with clay, humus, marl, etc., - - -	64 to 72	56 to 63	48 to 84

The table on next page, copied from *Henneberg's Jahrbuch Landwirthschaft*, shows the *greatest* length of drain admissible, according to the fall and distance between the drains. For example, the drains are three rods apart, and the fall is one inch per rod, how long may the drain be made with one inch tile? Find the table for one inch tile; then under the figure 3 in the first horizontal column (the figures in this column indicate the width between the drains in rods), find the number (1) corresponding with the fall, in the first left hand vertical column (which in this case will be 28); the number thus indicated will be the greatest possible length that the drain can be made to be effective. Should the fall be 3 inches per rod, the drain may be increased to 50 rods; or if 1½ inch pipe is used, and the fall is one inch, the drain may be 79 rods in length.

From this table it will be seen that 1½ inch pipe will

answer for almost all the minor drains that will be probably made in this country.

Fall, in inches, in a perch or rod.	One inch pipe tile.							One and one fourth inch pipe tile.						
	1½	2	2½	3	3½	4	4½	1½	2	2½	3	3½	4	4½
1-16	11	8	6	5	5	4	4	23	18	14	12	10	9	8
⅛	17	13	10	9	7	7	6	39	29	23	19	17	15	13
¼	27	20	16	13	11	10	9	46	35	28	23	20	18	16
½	39	29	24	19	17	15	13	67	50	40	33	29	25	23
¾	49	36	29	24	21	18	16	86	64	52	43	37	32	29
1	55	41	33	28	24	21	18	99	74	59	49	42	37	33
1½	70	53	42	35	30	26	23	125	93	75	62	53	47	42
2	83	62	50	41	37	31	27	145	108	87	72	62	54	48
2½	93	69	56	46	40	35	31	163	122	98	82	70	61	54
3	100	75	60	50	43	38	33	179	134	107	89	77	67	59
3½	109	82	66	55	47	41	37	195	146	117	97	83	73	65
4	117	88	70	59	50	44	39	209	157	126	105	89	78	69
4½	126	94	76	63	54	47	42	222	166	133	111	95	83	74
5	133	100	80	67	57	50	44	235	176	141	118	101	88	78
6	155	116	93	77	66	58	52	259	194	155	129	111	97	86

Fall, in inches, to a perch or rod.	One and one half inch pipe tile.							One and three fourths inch pipe tile.						
	1½	2	2½	3	3½	4	4½	1½	2	2½	3	3½	4	4½
1-16	32	24	19	16	14	12	10	45	34	27	23	19	17	15
⅛	52	39	31	26	22	19	17	69	51	41	34	29	26	23
¼	80	60	48	40	37	30	27	107	80	64	54	46	40	36
½	109	82	66	55	47	41	39	161	120	96	80	69	60	54
¾	140	105	84	70	60	52	47	203	152	121	101	87	76	68
1	157	118	94	79	67	59	52	233	175	140	117	100	87	78
1½	200	145	120	100	86	75	67	291	218	175	146	125	109	97
2	221	165	132	110	95	83	73	341	256	205	137	146	128	114
2½	257	192	154	128	110	96	86	390	292	234	195	167	146	130
3	286	214	172	143	123	107	95	418	313	251	209	179	157	139
3½	311	233	186	155	133	117	104	450	337	270	225	193	169	150
4	351	248	199	166	142	124	110	487	365	292	244	209	184	162
4½	352	265	212	177	151	132	118	517	388	310	259	222	194	172
5	371	278	223	186	159	139	124	545	408	327	272	233	204	184
6	399	299	239	199	171	149	133	595	446	357	297	255	223	198

As the cost of tile for draining may, perhaps, determine the distance between drains with some persons, in absence of any experimental or scientific reason, we have deemed it proper in this place to say a few words about the cost of tile, and to give a table from which the number of tiles necessary to drain an acre at several distances between the drains.

The cost of tile draining is made up of three items—the digging, the price of tiles at the kiln, and the expense of hauling them. It will readily be seen that each of these may vary considerably, and the total cost of the improvement be influenced accordingly.

If tiles are made on the farm, or in the immediate neighborhood, the cost of hauling is reduced to its lowest figure. Where they must be drawn several miles, the trouble and expense are great; five hundred of the smallest size being all that can readily and safely be put in a common two-horse wagon. Taking this item into account, the desirableness of concert of action among farmers is apparent; if several can agree to enter upon such improvements at the same time, they may manufacture in company, or, what is better, give their contracts to the nearest and best brick-maker, and get the tiles made at the most convenient point. Every farmer should consider it his interest to sustain any tile maker who has enterprise enough to commence the manufacture in his vicinity. There ought to be one or more good tile yards established immediately in every township in the state.

The price of tiles must vary in different localities, the cost of manufacture depending on the nature of the clay, the price of fuel and of labor. But these matters relating to the manufacture of tiles may be deferred to another time. Tiles are at present sold in Ohio at prices ranging from $8 to $12 per 1000 for the smallest size, or two inches in bore. Four inch tiles are about double the cost of the two inch; and six inch tiles are about double the cost of the four inch. A thousand tiles of ordinary length will lay sixty rods; thus, at the lowest figure stated above, the cost of tiles is a trifle over a shilling a rod.

The cost of digging, where men accustomed to the work, and proper tools can be obtained, will not exceed a shilling a rod for a three feet drain. The cost is proportionally greater for deep drains than for shallow ones; so that if the depth is diminished one third, the price should be lessened one half; or, if the depth is increased a third, about half the original price should be added. It will doubtless appear to some that such prices are low, compared with what they have been used to pay for ditching; this difference arises from the fact that not more than a third of the earth is removed in making a drain, that must needs be lifted in making an open ditch of the same depth.

The cost of thorough draining will depend, of course, on the frequency of the drains. At two rods asunder, there will be eighty rods to the acre; and this, at the prices already stated, or two shillings a rod, will amount to $20. To this it will sometimes be necessary to add ten per cent. for main drains. In general, about one tenth of all the drains in a field are main drains, and made at nearly double the cost of the minor drains. The profit or loss of underdraining, at such prices, will next be considered.

Table No. 1 presents, in the first column, the distance between the drains in feet; the second column shows the number of rods of drain in an acre—always supposing that the field to be drained is a square one—that is, a rectangle, having the opposite sides equal. The remaining columns show the number of tile of the different lengths required to lay the drains in an acre.

Suppose an acre is to be drained, with the distance of 20 feet between the drains; find the number 20 in the first column; and in the next column, opposite 20, will be found the number of rods of drain, 20 feet apart, in an acre; and in the fourth column will be found the number (2,011) of tile, 13 inches long—the usual length—required for the drains. Now, at $8 per 1000, the tile will cost $16 08; but, if the drains are placed 30 feet apart, the cost of tile will be $10 72 only.

TABLE NO. 1.

Width between drains. Feet.	Length of drains per acre, in statute rods of 5½ yards. Rods.	Yds.	No. of Tiles per acre. Length 12 in.	Length 13 in.	Length 14 in.	Length 15 in.	Length 16 in.	Length 18 in.
9	293	1¾	4840	4468	4149	3872	3630	3227
10	264		4356	4021	3734	3485	3267	2904
11	240		3960	3656	3395	3168	2970	2640
12	220		3630	3351	3112	2904	2723	2420
13	203	¼	3351	3094	2873	2681	2514	2234
14	188	3¼	3112	2873	2667	2490	2334	2075
15	176		3904	2681	2490	2324	2178	1936
16	165		2723	2514	2334	2178	2042	1815
17	155	1½	2563	2366	2197	2050	1922	1709
18	146	3¾	2420	2234	2075	1936	1815	1614
19	138	5¼	2293	2117	1966	1835	1720	1529
20	132		2178	2011	1867	1743	1634	1452
21	125	4	2075	1915	1778	1660	1556	1383
22	120		1980	1828	1698	1584	1485	1320
23	114	4½	1894	1749	1624	1516	1421	1263
24	110		1815	1676	1556	1452	1362	1210
25	105	3¾	1743	1609	1494	1394	1307	1162
26	101	3	1676	1547	1137	1341	1257	1117
27	97	4¼	1614	1490	1383	1291	1210	1076
28	94	1½	1556	1437	1334	1245	1167	1038
29	91	¼	1503	1387	1288	1202	1127	1002
30	88		1452	1341	1245	1162	1089	968
31	85	¾	1406	1298	1205	1125	1054	937
32	82	2¾	1362	1257	1167	1089	1021	908
33	80		1320	1219	1132	1056	990	880
34	77	3½	1282	1183	1099	1025	961	855
35	75	2¼	1245	1149	1067	996	934	830
36	73	1¾	1210	1117	1038	968	908	807
37	71	2	1178	1087	1010	942	883	785
38	69	2½	1147	1059	983	918	860	765
39	67	3¾	1117	1032	958	894	838	745
40	66		1089	1006	934	872	817	726
41	64	2¼	1063	981	911	850	797	709
42	62	4¾	1038	958	889	830	778	692
43	61	2¼	1014	936	869	811	760	676
44	60		990	914	849	793	743	660
45	58	3¾	968	894	830	775	726	646
46	57	2¼	947	875	812	758	711	632
47	55	1	927	857	795	742	696	618
48	55		908	838	778	727	681	605
49	53	4¾	889	821	762	712	667	593

TABLE No. 1—*Continued.*

Width between drains. Feet.	Length of drains per acre, in statute rods, of 5½ yards. Rods.	Yds.	No. of Tiles per acre. Length 12 in.	Length 13 in.	Length 14 in.	Length 15 in.	Length 16 in.	Length 18 in.
50	52	4½	872	805	747	697	654	581
51	51	4¼	855	787	733	684	641	570
52	50	4¼	838	774	719	671	629	559
53	49	4½	822	759	705	658	617	548
54	48	5	807	745	692	646	605	538
55	48		792	732	679	634	594	528
56	47	¾	778	719	667	623	584	519
57	46	1¾	765	706	656	612	574	510
58	45	2¾	752	694	644	601	564	501
59	44	4	739	682	633	591	554	492
60	44		726	671	623	581	545	484
61	43	1½	715	660	613	572	536	477
62	42	3¼	703	649	603	563	527	469
63	41	5	692	639	593	554	519	461

Table No. 2 shows the number of tiles of the length of 12, 13, or 14 inches, requisite to lay any number of rods of drain, from one rod up to 10,000 rods. Suppose it is required to know what number of 13 inch tile will be required to lay 8,765 rods of drain. From the table we find that

5,000 rods require	- -	76154 tiles.
3,000 "	- - -	45693 "
700 "	- -	10662 "
65 "	- - -	990 "
8,765 "	- -	133499 "

TABLE NO. 2.

Tiles required for Drains.				Tiles required for Drains.			
Length of drains in rods.	Length of tile 12 inches.	Length of tile 13 inches.	Length of tile 14 inches.	Length of drains in rods.	Length of tile 12 inches..	Length of tile 13 inches.	Length of tile 14 inches.
1	17	16	15	41	677	625	580
2	33	31	29	42	693	640	594
3	50	46	43	43	710	655	609
4	66	61	57	44	726	671	623
5	83	77	71	45	743	686	637
6	99	92	85	46	759	701	651
7	116	107	99	47	776	716	665
8	132	122	114	48	792	732	679
9	149	138	128	49	809	747	693
10	165	153	142	50	825	762	708
11	182	168	156	51	842	777	722
12	198	183	170	52	858	792	736
13	215	198	184	53	875	808	750
14	231	214	198	54	891	823	764
15	248	229	213	55	908	838	778
16	264	244	227	56	924	853	792
17	281	259	241	57	941	869	807
18	297	275	255	58	957	884	821
19	314	290	269	59	974	899	835
20	330	305	283	60	990	914	849
21	347	320	297	61	1007	930	863
22	363	336	312	62	1023	945	877
23	380	351	326	63	1040	960	891
24	396	366	340	64	1056	975	906
25	413	381	354	65	1073	990	920
26	429	396	368	70	1115	1067	990
27	446	412	382	80	1320	1219	1132
28	462	427	396	90	1485	1371	1273
29	479	442	411	100	1650	1524	1415
30	495	457	425	200	3300	3047	2829
31	512	473	439	300	4950	4570	4243
32	528	488	453	400	6600	6093	5658
33	545	503	466	500	8250	7616	7072
34	561	518	481	600	9900	9139	8486
35	578	534	495	700	11550	10662	9900
36	594	549	510	800	13200	12185	11315
37	611	564	524	900	14850	13708	12729
38	627	579	538	1000	16500	15231	14143
39	644	594	552	3000	49500	45693	42429
40	660	610	566	5000	82500	76154	70715

Table No. 3 shows the number of rods in drains at distances of 15 to 42 feet apart, in tracts of land from one fourth of an acre to 100 acres; and, in fact, to any number of acres exceeding 100, by multiplication or addition. Suppose the number of rods of drains 18 feet apart in a tract of 285 acres be required. From the table:

14666	rods,	3¾	yards	in	100	acres.
2						
29332	"	7½	"	"	200	"
11733	"	1¾	"	"	80	"
733	"	1¾	"	"	5	"
41798	"		"	"	285	"

If the number of 13 inch tile be required, by referring to Table No. 2, it will be seen that

5000	rods	require	76154	tiles.
8			8	
40000	"	"	609232	"
1000	"	"	15231	"
700	"	"	10662	"
90	"	"	1371	"
8	"	"	122	"
41798	"	"	636618	"

TABLE NO. 3.—LENGTH OF DRAINS.

Distance between Drains 15 feet.

Acres.	Length in rods. Rods.	Yards.	Acres.	Length in rods. Rods.	Yards.	Acres.	Length in rods. Rods.	Yards.
¼	44		15	2640		32	5632	
½	88		16	2816		33	5808	
¾	132		17	2992		34	5984	
1	176		18	3168		35	6160	
2	352		19	3344		36	6336	
3	528		20	3520		37	6512	
4	704		21	3696		38	6678	
5	880		22	3872		39	6864	
6	1056		23	4048		40	7040	
7	1232		24	4224		50	8800	
8	1408		25	4400		60	10560	
9	1584		26	4576		70	12320	
10	1760		27	4752		80	14080	
11	1936		28	4928		90	15840	
12	2112		29	5104		100	17600	
13	2288		30	5280				
14	2464		31	5456				

TABLE NO. 3—*Continued.*

Distance between Drains 18 feet.			Distance between Drains 21 feet.			Distance between Drains 24 feet.		
Acres.	Length in rods. Rods.	Yards.	Acres.	Length in rods. Rods.	Yards.	Acres.	Length in rods. Rods.	Yards.
¼	36	3¾	¼	31	2¼	¼	27	2¾
½	73	1¾	½	62	4¾	½	55	
¾	110		¾	94	1½	¾	82	2¾
1	146	3¾	1	125	4	1	110	
2	293	1¾	2	251	2¼	2	220	
3	440		3	377	¾	3	330	
4	586	3¾	4	502	4¾	4	440	
5	733	1¾	5	628	3¼	5	550	
6	880		6	754	1½	6	660	
7	1026	3¾	7	880		7	770	
8	1173	1¾	8	1005	4	8	880	
9	1320		9	1131	2¼	9	990	
10	1466	3¾	10	1257	¾	10	1100	
11	1613	1¾	11	1382	4¾	11	1210	
12	1760		12	1508	3¼	12	1320	
13	1906	3¾	13	1634	1½	13	1430	
14	2053	1¾	14	1760		14	1540	
15	2200		15	1885	4	15	1650	
16	2346	3¾	16	2011	2¼	16	1760	
17	2493	1¾	17	2137	¾	17	1870	
18	2640		18	2262	4¾	18	1980	
19	2786	3¾	19	2388	3¼	19	2090	
20	2933	1¾	20	2514	1½	20	2200	
21	3080		21	2640		21	2310	
22	3226	3¾	22	2765	4	22	2420	
23	3373	1¾	23	2891	2¼	23	2530	
24	3520		24	3017	¾	24	2640	
25	3666	3¾	25	3142	4¾	25	2750	
26	3813	1¾	26	3268	3¼	26	2860	
27	3960		27	3394	1½	27	2970	
28	4106	3¾	28	3520		28	3080	
29	4253	1¾	29	3654	4	29	3190	
30	4400		30	3771	2¼	30	3300	
31	4546	3¾	31	3897	¾	31	3410	
32	4693	1¾	32	4022	4¾	32	3520	
33	4840		33	4148	3¼	33	3630	
34	4986	3¾	34	4274	1½	34	3740	
35	5133	1¾	34	4400		35	3850	
36	5280		36	4525	4	36	3960	
37	5426	3¾	37	4651	2¼	37	4070	
38	5573	1¾	38	4777	¾	38	4180	
39	5720		39	4902	4¾	39	4290	
40	5866	3¾	40	5028	3¼	40	4400	
50	7333	1¾	50	6285	4	50	5500	
60	8800		60	7542	4¾	60	6600	
70	10266	3¾	70	8800		70	7700	
80	11733	1¾	80	10057	¾	80	8800	
90	13200		90	11314	1½	90	9900	
100	14666	3¾	100	12571	2¼	100	11000	

TABLE NO. 3—*Continued.*

Distance between Drains 27 feet.			Distance between Drains 30 feet.		Distance between Drains 33 feet.	
Acres.	Length in rods. Rods.	Yards.	Acres.	Length in rods. Rods. Yards.	Acres.	Length in rods. Rods. Yards.
¼	24	2½	¼	22	¼	20
½	48	5	½	44	½	40
¾	73	1¾	¾	66	¾	60
1	97	4¼	1	88	1	80
2	195	3	2	176	2	160
3	293	1¾	3	264	3	240
4	391	½	4	352	4	320
5	488	5	5	440	5	400
6	586	3¾	6	528	6	480
7	684	2½	7	616	7	560
8	782	1¼	8	704	8	640
9	880		9	792	9	720
10	977	4¼	10	880	10	800
11	1075	3	11	968	11	880
12	1173	1¾	12	1056	12	960
13	1271	½	13	1144	13	1040
14	1368	5	14	1232	14	1120
15	1466	3¾	15	1320	15	1200
16	1564	2½	16	1408	16	1280
17	1662	1¼	17	1496	17	1360
18	1760		18	1584	18	1440
19	1857	4¼	19	1672	19	1520
20	1955	3	20	1760	20	1600
21	2053	1¾	21	1848	21	1680
22	2151	½	22	1936	22	1760
23	2248	5	23	2024	23	1840
24	2346	3¾	24	2112	24	1920
25	2444	2½	25	2200	25	2000
26	2542	1¼	26	2288	26	2080
27	2640		27	2376	27	2160
28	2737	4½	28	2464	28	2240
29	2835	3	29	2552	29	2320
30	2933	1¾	30	2640	30	2400
31	3031	½	31	2728	31	2480
32	3128	5	32	2816	32	2560
33	3226	3¾	33	2904	33	2640
34	3324	2½	34	2992	34	2720
35	3422	1¼	35	3080	35	2800
36	3520		36	3168	36	2880
37	3617	4¼	37	3256	37	2960
38	3715	3	38	3344	38	3040
39	3813	1¾	39	3432	39	3120
40	3911	½	40	3520	40	3200
50	4888	5	50	4400	50	4000
60	5866	3¾	60	5280	60	4800
70	6844	2½	70	6160	70	5600
80	7822	1¼	80	7040	80	6400
90	8800		90	7920	90	7200
100	9777	4¼	100	8800	100	8000

TABLE No. 3—*Continued.*

Distance between Drains 36 feet.			Distance between Drains 39 feet.			Distance between Drains 42 feet.		
Acres.	Length in rods. Rods.	Yards.	Acres.	Length in rods. Rods.	Yards.	Acres.	Length in rods. Rods.	Yards.
¼	18	1¾	¼	16	5	¼	15	4
½	36	3¾	½	33	4¾	½	31	2¼
¾	55		¾	50	4¼	¾	47	¾
1	73	1¾	1	67	3¾	1	62	4¾
2	146	3¾	2	135	2	2	125	4
3	220		3	203	½	3	188	3¼
4	293	1¾	4	270	4¼	4	251	2¼
5	267	3¾	5	338	2½	5	314	1½
6	440		6	406	¾	6	377	¾
7	513	1¾	7	473	4¾	7	440	
8	586	3¾	8	541	3	8	502	4¾
9	660		9	609	1¼	9	565	4
10	733	1¾	10	676	5	10	628	3¼
11	806	3¾	11	744	3½	11	691	2¼
12	880		12	812	1¾	12	754	1½
13	953	1¾	13	880		13	817	¾
14	1026	3¾	14	947	3¾	14	880	
15	1100		15	1015	2	15	942	4¾
16	1173	1¾	16	1083	½	16	1005	4
17	1246	3¾	17	1150	4¼	17	1068	3¼
18	1320		18	1218	2½	18	1131	2¼
19	1393	1¾	19	1286	¾	19	1194	1½
20	1466	3¾	20	1353	4¾	20	1257	¾
21	1540		21	1421	3	21	1320	
22	1613	1¾	22	1489	1¼	22	1382	4¾
23	1686	3¾	23	1556	5	23	1445	4
24	1760		24	1624	3½	24	1508	3¼
25	1833	1¾	25	1692	1¾	25	1571	2¼
26	1906	3¾	26	1760		26	1634	1½
27	1980		27	1827	3¾	27	1697	¾
28	2053	1¾	28	1895	2	28	1760	
29	2126	3¾	29	1963	½	29	1822	4¾
30	2200		30	2030	4¼	30	1855	4
31	2273	1¾	31	2098	2½	31	1948	3¼
32	2346	3¾	32	2166	¾	32	2011	2¼
33	2420		33	2233	4¾	33	2074	1½
34	2493	1¾	34	2301	3	34	2137	¾
35	2566	3¾	35	2369	1¼	35	2200	
36	2640		36	2436	5	36	2262	4¾
37	2713	1¾	37	2504	3½	37	2325	4
38	2786	3¾	38	2572	1¾	38	2388	3¼
39	2860		39	2640		39	2451	2¼
40	2933	1¾	40	2707	3¾	40	2514	1½
50	3666	3¾	50	3384	3½	50	3142	4¾
60	4400		60	4061	3	60	3771	2¼
70	5133	1¾	70	4738	2½	70	4400	
80	5866	3¾	80	5415	2	80	5028	3¼
90	6600		90	6092	1¾	90	5657	¾
100	7333	1¾	100	6769	1¼	100	6285	4

TABLE No. 3—*Continued.*

Distance between Drains 45 feet.

Acres.	Length in rods. Rods.	Yards.	Acres.	Length in rods. Rods.	Yards.	Acres.	Length in rods. Rods.	Yards.
¼	14	3¾	15	880		32	1877	1¾
½	29	1¾	16	938	3¾	33	1936	
¾	44		17	997	1¾	34	1994	3¾
1	58	3¾	18	1056		35	2053	1¾
2	117	1¾	19	1114	3¾	36	2112	
3	176		20	1173	1¾	37	2170	3¾
4	234	3¾	21	1232		38	2229	1¾
5	293	1¾	22	1290	3¾	39	2288	
6	352		23	1349	1¾	40	2346	3¾
7	410	3¾	24	1408		50	2933	1¾
8	469	1¾	25	1466	3¾	60	3520	
9	528		26	1525	1¾	70	4106	3¾
10	586	3¾	27	1584		80	4693	1¾
11	655	1¾	28	1642	3¾	90	5280	
12	704		29	1701	1¾	100	5866	3¾
13	762	3¾	30	1760				
14	821	1¾	31	1818	3¾			

CHAPTER V.

MANUFACTURE OF TILES

SELECTION OF MATERIAL.

FOR drain tile it is necessary to manufacture an article of genuine earthenware, of sufficient strength to bear transportation and easy management, as well as to resist the action of water for a considerable period of time. These conditions may be found in a kind of earth less porous, more impervious, and finer in grain than that of which we make common brick; it must be similar in every respect to the earth of which roof tiles are made. We may, therefore, adopt as a general principle that, earth fit to make tile, is equally suitable for drain pipes, and that its preparation must, in both cases, be similar. Nevertheless, it may be remarked that flat and concave tiles for roofs are almost always manufactured by hand, while drain pipes are made cheaper and faster by machines.

The mortar about to be used ought to possess a degree of ductility and firmness which is not required for roofing tiles, especially when they are made flat. Pipe tile ought to be manufactured not far from the place where they are to be employed, on account of the cost of transportation, so as to render drainage easy and cheap. The materials of the compound must then be such as to furnish, at any time, at any place, cheap, substantial pipes.

Like other kinds of earthenware, this requires an essential distinction between the materials to be used for the composition of the mortar, and the elements that will constitute the piece completed or baked.

In the composition of the mortar, some compound foreign bodies are mechanically but not chemically combined. These compound bodies are materials for fabrication, but can be separated by water. In the baked or burnt mortar new combinations have been formed, against which water is powerless, so far, at all events, as to reduce the finished mass to the primitive materials. These combinations are multiple silicates, that is to say, silicic acid, combined with several bases—generally aluminum or lime—both in large quantities; at other times, and in less proportions, magnesia, oxide of iron, potash, soda and oxide of manganese. Burning, or baking, is the only means we have to obtain those fixed combinations that are subject to the action of neither acid nor water, and that are the more unalterable in proportion as the silicates are more exactly formed with their constituent elements, without any foreign admixture.

The essential elements are silicic acid and aluminum; from these may be obtained an earthenware which is fire proof, that is to say, it will not melt in the strongest fires of either forge or blast furnace. Aluminum, may sometimes be replaced in part by magnesia. The proportions of these indispensable elements are as follows:

Silica, - - - -	55 to 75 per cent.
Aluminum, - -	25 to 35 " "

When magnesia is present, it is generally found to the amount of 1 to 5 per cent.; there might be found as much as 25 to 35 per cent.

The accessory substances are still more variable in their proportions than the above; they are

Lime, - - - - -	0 to 19 per cent.
Potash, - - - -	0 to 5 " "
Protoxyd of iron, -	0 to 19 " "

These accessory elements give fusibility to earthenware, and, therefore, allow its constituent substances to combine in such a manner as to form a resisting body; and this is performed with a temperature lower in proportion as the accessory elements are more abundant. In some baked mortars, there is carbonic acid (0 to 16 per cent.), when lime is present in sensible proportions. Water is almost always totally driven out of the mortar by the heat; and is present only in the paste in preparation; but here it performs an important office, by assisting to mix together the various materials which will bring into the paste the elements that we have described; it serves also to give them the required softness, to endow them with a certain adhesive force, and to promote their plastic qualities.

We term *plasticity* that quality which some soft matters have of assuming, under the hand of artists and mechanics, the forms that they wish to reproduce. We term those *long pastes* which are possessed of this quality in the highest degree, and *short pastes* those which have it in a slight degree only.

Plasticity is not absolutely indispensable for the shaping of ceramic pastes; we can mold them, by pressing the materials which are in the very state of dust; but a plastic substance yields better to the easiest and most usual mode of giving shape, and it is, therefore, much more desirable.[1]

While plasticity is a condition of the first importance, in order to facilitate the shaping of the mortar into the desired forms, it offers great inconvenience when brought to an excessive degree. A paste which is too plastic dries

[1] A gentleman exhibited at the Ohio State Fair, at Dayton, in 1860, a tile machine, which made tile from hydraulic cement, without the aid of water. The cement, of course, possessed no plasticity, but the tile or pipes were made by enormous pressure. If the tile thus made were placed in drains the moisture of the ground would cause them to harden, so as to be serviceable for many years.

up with difficulty, and great unevenness; articles manufactured from it are very likely, in drying, to lose their proper shape; they are very apt to crack, both during the period of desiccation, and in the bake oven or kiln. Excessive plasticity may be modified by other materials, which are either natural or artificial.

Sand is the natural correcting or tempering material. All sands are composed of silicic acid, or silicum, and of some foreign substances, from one to 9 per cent.; these foreign matters are aluminum, magnesia, lime, oxyd of iron, potash, etc.

The artificial tempering materials are: 1. Fragments of burnt brick or tile, reduced to powder. 2. Scoria, from the forges. 3. Sometimes sawdust.

As far as the drain pipe is concerned, it will not be necessary to discuss all the other materials which are used in the various productions of the ceramic art.

Any kind of sand may be employed for making drainage pipes, provided it be free from gravel, as it would interfere seriously with the molding.

As to plastic materials, although they may all be used, their qualities must be discriminated, in order to know how they shall be mixed together, and what proportion of tempering material, that is to say, sand, ought to be added to them. It may happen that some kind of earth may be found susceptible of being employed alone, and without any mixture. Let us see then what qualities each ought to possess:

1. The earth having received a sufficient quantity of water, must be malleable enough to assume all forms that may be wished; it must be firm enough to preserve those forms; it must be composed of particles sufficiently adherent, so that, when passing through the dye plate, this adherence is not impaired.

2. The earth ought not to contain any particle of pure chalk, even so small as the fiftieth part of an inch; baking it would produce lime; and lime, in contact with water, would slack and burst the pipe. There ought not to be any particle of either sulphuret of iron or of pyrites, as these would produce the same result.

3. It must dry readily, and with evenness.

4. The process of drying must be carried on, in such a manner as to evaporate the water which gave adherence to the particles, without producing cracks or deformities in the pipes.

We will now examine the various kinds of plastic materials which may be used in the fabrication of drainage pipes.

Natural plastic materials comprise clay, and clayey marl.

Clay is, in the potter's sense, an earth which forms a paste with water, working easily and hardening by fire.

Clay is *plastic* when it contains nothing but silicum and aluminum. This variety of clay which often bears the name of potter's clay, on account of its tenacity, does not readily admit water to penetrate, but when saturated it is very retentive of moisture.

Clay is *fuliginous* when it contains some lime, in the maximum proportion of 5 to 6 per cent., part of it as carbonate, and part may be in the state of silicate. This clay is still coherent, but less tenacious than the plastic above mentioned. It produces a slight effervescence with the acids, but this effervescence, caused by an emission of carbonic acid gas, soon ceases.

These two kinds of clay may be combined with an oxide of iron, and sometimes with particles of gypsum (sulphate of lime), or plaster.

Plastic clay, when not combined by these bodies, is altogether fire proof, that is, it will not melt at any temperature of our furnaces; should these refractory qualities be wanted for any purpose, that clay must be tempered by using sand formed of pure silicum or flint.

In regard to drain pipes, both kinds of clay must be tempered, but with common materials as above described.

In the special point of view, for which alone we are writing, we will say that, both plastic and fuliginous clay should be used only to give plasticity to other materials.

No person can learn any trade, be it ever so simple, by reading alone. No matter how carefully the books which treat of such trade may be written, practice is necessary to perfect the workman in it. Books are a great auxiliary to effect a perfection of knowledge, and to those who have already made some progress in a practical acquaintance with any handiwork, reading greatly enhances their ability to pursue their occupation profitably. Those who have been engaged in the fabrication of pottery, tiles, or brick, may, by means of the theoretical information to be derived from books, cursorily learn sufficient of the operations of making drain tiles, as to succeed in their fabrication very well.

Clay suitable for drain tile is such as is proper for the fabrication of roofing tile, or even fine brick. It should not be too poor, or meager of clay constituents, and should be free from pebbles and pieces of limestone, although it is an error to suppose that it should be entirely void of lime, as this, in small quantity, and very evenly commingled, assists very much in melting the mass when subjected to heat. If the amount of sand contained in the clay intended for drain pipe be too small, in proportion to the other constituents, they are .too easily curved in handling before dry, and are very liable to crack, both while drying and subjected to heat; and, therefore, when a bed of clay is worked for this manufacture, too pure for the purpose, a proper proportion of sand must be added, to give the due consistence to the tiles when molded. Too great care in preparing the materials, can not be taken, and the results of proper precautions in this particular, will be a diminution of the average cost of the product.

Clays of Ohio.—Fortunately for the farmers of Ohio, clay suitable for tile making may be found in nearly every part of the state; in fact, almost any clay that will make good bricks, may, with care, be made into tiles. The principal requisite is, that the tile maker thoroughly understand the character of the material he uses.

The blue clay which lies directly upon the shale, or limestome, in most of the northern and north-western counties of Ohio, has been found to be well adapted to the manufacture of tiles. This clay contains a large amount of carbonate of lime, in a state of minute division; it also contains water-worn fragments of limestome, shale and primitive rocks. It varies considerably in the degree of plasticity, and in the amount of stones in different localities. In one tile yard, near Cleveland, it is taken directly from the bed to the tile machine, without the need of any tempering whatever. In general, however, it requires much working, and the employment of the screen or rollers.

Above the blue clay, there is, in the same localities, a layer of yellow clay. It contains nearly the same rocky fragments; it has but little lime diffused through it, but contains much more oxide of iron. This clay is extensively used for brick making, and is found to make excellent tiles. The use of the screw or roller mill, is generally needed where the yellow clay is used. The pottery clays, peculiar to the coal regions, will, of course, make excellent tiles. It has, however, been supposed by some, that tiles should be porous, so as to permit the water to filter through their sides, and, therefore, that clay suitable for stoneware or earthenware, would not be proper for tiles. This opinion is founded on an error, for the water in drains does not filter through the tiles, but enters at the joints, and there is not the least objection to having tiles made of the hardest and most compact material. Indeed, the manufacturing of tiles from the better clays has this advantage, that they may be made much lighter, and therefore cost less for carriage.

The marly clays of south-western Ohio, are well adapted to tile making. The lime acts as a bind or flux, to effect the semi-fusion of the other constituents, and extremely hard and durable tiles are made of such material, the main point being to secure a thorough burning.

In almost every township of the state are small swamps, or basins, with clayey bottoms. The clay of these swamps, although identical in composition with that of the surrounding uplands, has a far greater value for tile making. From being kept constantly wet, it possesses a degree of solidity, uniformity of texture and plasticity, that can only be given to the clay of hillsides, after much working. In many places, these swamp clays require no labor to bring them to a proper condition for use.

The following letter from a very successful tile maker in Ohio, with respect to materials for tiles, may be of interest to those who intend manufacturing their own tiles:

"WOODSTOCK, Champaign county, O.

DEAR SIR—As I promised you when at Columbus I would send you a description of the different kinds of clay suitable for making drain tile, and where it is likely to be found, I now undertake to communicate to you what I know about it.

Clay, suitable for making tile for underground draining, may be generally found through that portion of Ohio that I am acquainted with, in the following described localities, viz: in small pond holes (such as contain water the greater part of the year), by some called cat swamps, usually found on white oak ridges. The clay found in these holes is rather on the blue order, from a bluish black to a blue gray color. This clay is of a depth of from two to ten feet. There is also a clay found on almost all white oak land that will answer to make tile of, but it is not of the best quality, and somewhat difficult to get it free from limestone gravel; this clay is of a reddish cast, and runs from one and a half to three feet in depth. There is also a clay found on low, wet prairies, where wild grass grows. The real blue clay in some places you will find near the surface, but frequently you will have to dig from two to four feet before coming to it; it runs to a great depth in the ground. It requires stronger clay to make good tile than it does to make brick. In making brick, if your clay is too strong, you have to add sand with the clay; but not so in making tile: the stronger the clay, the better the tile. My impression is, there is hardly a township in the state but that has clay in it suitable for making tile.

Respectfully yours, D. KENFIELD."

One important step in the preparation of the materials is:

Throwing up the clay before the commencement of winter, as one of the principal means of success in fabricating good tile. Every farmer knows that a field plowed in the fall into rough furrows, and left fallow, becomes much more mellow than it would be if left undisturbed by the plow until spring. This effect is likewise produced in clay dug up and exposed to the frost during winter, and is produced by the expansion of the water during the process of freezing, which separates the particles of clay from each other, and thus, by lessening to some degree its adhesiveness, fits it for more easy manipulation in the process of fabrication. In order to secure this object most thoroughly, the clay should be placed in heaps and frequently turned over, so as to expose all parts to the action of the frost, which does not readily act upon very stiff clay. A great saving of labor and also of time is thus secured, inasmuch as what is done in winter leaves so much the less to do after making begins. Where the clay requires grinding, this process may very well go on in connection with the digging; for, although the clay does not grind so easily when first dug, there is a very great advantage in giving it time to become settled together and thoroughly united before it is used. If the grinding is done at the time of the manufacture, it is necessary to pug the clay or tread it, because it comes from the rollers in too loose a state for the tile machine. A few months of exposure, after grinding, is worth as much as pugging, and will render that operation unnecessary. When the iron rollers are not required, a pug mill, similar to those used in the manufacture of bricks, will effect all the tempering that is needed.

The purification of the clay intended for the manufacture of drain tile is necessary, when the material contains pebbles or pieces of limestone, or is too meager of clayey elements, and can only be effected by mixing it thoroughly with water, so as to dissolve it, as it were, and then permit the heavier matters, as pebbles, limestones and coarse sand, to be deposited upon the bottom of the vat or pit in which the operation is performed, draining off the superfluous water and leaving the clay remain until evaporation shall have restored it to a proper degree of consistence.

To effect this object the clay is placed in a properly constructed vat or pit, of suitable size, provided with a sliding gate, by means of which to drain off the water which remains after settling. The clay is then mixed with water and stirred until the whole mass becomes fluid; it is then permitted to settle, when all the supernatant water can be drained off by the sliding gate; or, after being reduced to a lime condition, the clay may be passed through a wire sieve, of suitable strength and fineness, which will readily separate the coarse materials from it.

This mode is more particularly applicable when the clay is of the desired composition, except the existence in it of pebbles and insoluble lumps.

A very convenient form of a mixing pit is furnished by the common mortar box and bed of plasterers; the mortar box may be used to mix the clay with water, and the bed will answer for a fit receptacle in which a complete separation of the light and heavy materials can occur. As soon as this has taken place, the superabundant water can be removed by means of a draining gate, and the mass left to dry out sufficiently for working. The mixing can be effected, where but small quantities of clay are used, by means of hoes, such as are used by plasterers; but, when the manufacture is carried on upon a large scale, a suitable method is to make use of a *mixing machine*, the form and capacity of which may best be determined by the quantity of labor intended to be performed by it. The common pug mill used in brick yards may be so modified as to answer the purpose very well; taking care to adjust it so that the clay may be mixed with a sufficient quantity of water before it is permitted to flow off by the outlet gate.

Another convenient mixing machine may be constructed in the following manner: Take a large hollow log, of suitable length, say five or six feet; hew out the inequalities with an adze, and close up the ends with pieces of strong plank, into which bearings have been cut to support a revolving shaft. This shaft should be sufficiently thick to permit being transfixed with wooden pins long enough to reach within an inch or two of the sides of the log or trough, and they should be so beveled as to form in their aggregate shape an interrupted screw, having a direction toward that end of the box where the mixed clay is designed to pass out. In order to effect the mixing more thoroughly, these pins may be placed sufficiently far apart to permit the interior of the box to be armed with other pins extending toward the center, between which they can easily move. The whole is placed either horizontally or vertically, and supplied with clay and water in proper quantities, while the shaft is made to revolve by means of a sweep, with horse power, running water or steam, as the case may be. The clay is put into the end farthest from the outlet (if horizontal), and is carried forward to it and mixed by the motion, and mutual action and reaction of the pins in the shaft and sides of the box. Iron pins may, of course, be substituted for the wooden ones, and have the advantage of greater durability and of greater strength in proportion to their size, and the number may therefore be greater in a machine of any given length. The fluid mass of clay and water may be permitted to fall upon a sieve or riddle, of heavy wire, and afterward be received in a settling vat, of suitable size and construction to drain off the water and let the clay dry out sufficiently by subsequent evaporation. A machine of this construction may be made of such a size that it may be put in motion by hand, by means of a crank, and yet capable of mixing, if properly supplied, clay enough to mold 800 or 1000 pieces of drain pipe per day.

In Ohio, where clay of suitable character is accessible in so very many localities, this process of mixing and preparing by filtration and settling, need be instituted only in very few instances; and the question of comparative cost, between tiles purchased and transported from manufactories situated where natural facilities are greater, and the home manufacture of such tiles, under the disadvantages accruing from less suitable clay beds, must be determined by every one for himself. Where the home manufacture of drain pipe is found preferable, the preceding hints will be found of invaluable assistance in obviating difficulties otherwise almost insurmountable.

There are other methods by which this evil may be remedied; the simplest is by the use of a screen in the tile machine; the other method is to crush the clay between heavy iron rollers. The choice between these methods depends on the number and character of the stones to be disposed of. Where they are few in number, or anything else than small limestones, the screen will be sufficient. If, however, the stones are numerous, and among them are many fragments of limestone, of the size of peas, or smaller, the rollers will be better than the screen; for though such small limestones would not interfere with the molding, they will occasion the loss of tiles in burning. The presence of stones requiring the rollers, is no serious objection to the use of a clay, otherwise suitable. Grinding the clay will add to the cost of manufacturing about fifty cents a thousand. The purchaser can much better afford to pay this extra price, than to add two or three miles of carriage from some more distant locality. Five hundred two inch tiles will fill the box of a lumber wagon; the labor of extra carriage may therefore easily exceed the additional cost of manufacture.

When the clay is not free from the admixture with stones or pebbles, or when it contains too little sand, some manufacturers use a clay cutter; but such a machine can not supersede the mixing apparatus mentioned. A very useful machine for working clay, and one that will obviate much hand labor, may be constructed of two or more cast iron rollers, referred to above, and which we will soon explain more fully, so arranged as to rotate closely together. Between such rollers the clay may be passed, and by this means made to assume a proper consistence and plasticity for molding.

Moistening the clay to a proper degree is always necessary, whether it may have been necessary to purify it by washing and settling or not. For this purpose, pits are dug in the earth, and walled up or lined, so as to have five or six feet of length and breadth, and four feet in depth, clear. The clay is removed into these pits from its winter beds, and thoroughly mixed with water by means of any suitable instrument, as shovels, so as to be uniformly moist throughout. The degree of moisture should be about equal to that of potters' luting clay, and is necessary as a preliminary step to its further working upon the kneading board.

The process of preparation, carried still further upon the kneading board, in some foreign manufactories of excellent drain tile, is, in short, as follows:

The clay is taken from the moistening pit, or settling bed, as the case may be, after being reduced to the proper condition, and spread in thin layers upon the kneading board. If too rich, the proper proportion of sand is added, and the whole mass is thoroughly trodden by men. It is then piled up in a low heap, and well worked by means of a stirrer, shaped something like a saber, fixed in a handle three feet long, moistened, if necessary, and again thoroughly trodden and kneaded by the feet. By these means, any pebbles existing in it may be discovered and removed.

After being thus worked, it is piled up in the form of a large sugar-loaf, four or five feet high, and of about two feet base. The pile is begun by placing a layer of about six inches hight, and then beating this down with a large wooden maul, as firmly as possible. Upon this a second layer is superimposed, and beaten down in like manner, until the cone is sufficiently high. A workman then, with a scraper, having a handle at each end, proceeds to shave down the whole cone into thin shavings. By this means the smallest pebble is discovered and readily removed.

The clay shavings are then cast upon another plank table, where it is beaten into masses of suitable size and form, to fit the box of the molding press. This last hammering must be performed very carefully, to drive out any air which may yet be confined in the clay, or intervene between the clay blocks and the ridges of the molding machine, as the existence of air in the press would materially interfere with the perfection of the tile.

This method of preparation has several advantages over those in which machinery is employed, upon the durability of which the profits largely depend. A clay cutting machine is expensive, and yet the clay can not be best prepared, by its use, for the molding press. Some persons make use of sieves, or perforated plates, particularly when the clay has been prepared by machinery, through which the materials are forced to strain the pebbles remaining; but they are continually liable to become clogged, and hinder the progress of the work by the loss of time necessary to keep them clean, and demand a great deal of power to drive the clay through their small interstices. These difficulties are obviated by working the clay by hand, and a sufficient force can be set to work to produce the desired amount of prepared material with certainty; which can not be always accomplished by the best of machines, on account of the accidents to which they are liable.

If the prepared clay can not, when made into masses for the press, be worked up fast enough, the blocks may be kept moist by covering them with wet cloths, until such time as they may be needed for use.

It is a matter of the greatest importance that the clay to be worked should be properly tempered, and kept so, from the beginning of the process; and to this point particularly, the attention must be constantly directed, but especially when it is put into the press. If it be worked too moist, the sides of the pipes fall together, or collapse, as they come from the mold, or they shrink greatly and become curved and wry. If the clay is too stiff, the work is difficult, and the pipes are rough, cracked, and shelly, when burnt, and often fall to pieces. Beside, it must be observed, that the same degree of moisture is not suitable for the fabrication of pipe tile of different diameters. Large pipes are made of stiffer clay than small ones. If clay of the proper temper for making inch pipe were pressed through a four inch mold, not a single piece would be found perfect—every one would be flattened and distorted. Experience is the only guide, and by this alone can a proper acquaintance with the matter be obtained.

Whatever method may be adopted for removing small stones from the clay, it is very necessary that this clay should be pugged, except, indeed, in such very rare cases as the clay formation at Cleveland.

1. *The Pug Mill* of tile yards differs little from that commonly used in the manufacture of bricks; the only material difference being in the arms or pins by which the clay is tempered. The clay being used in a much stiffer state for tiles than bricks, iron knives are needed for cutting the clay, in the place of wooden pins. These knives are made strong and sharp, and when set in the upright shaft, the advancing edge is raised a little, so that the effect of their movement is to press the clay downward toward the opening. Instead of several iron knives, some use a smaller number of heavier ones, and these have riveted into them, at right angles, a number of short knives, which are so arranged as to pass each other some-

what closely, and serve to cut the lumps of clay to pieces thoroughly.

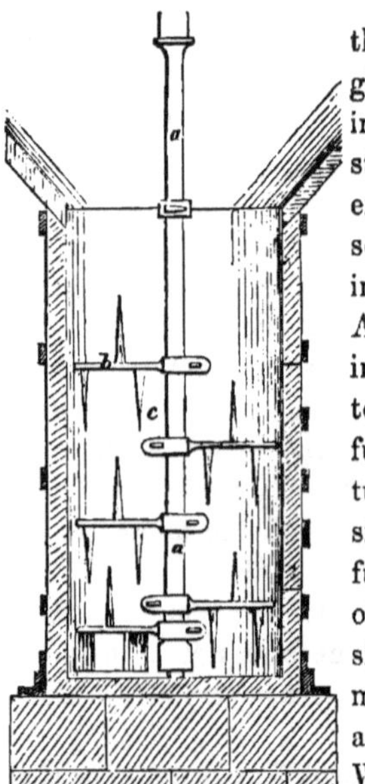

Fig. 43.—Section of a Pugging Mill.

The subjoined cut represents the section of an excellent pugging mill. It consists of a cylindrical body, well bounded by stout hoops—the upper portion expands outward from the body, so as to form a hopper or funnel, into which the clay is thrown. A stout iron bar, *a*, *a*, is placed in the center of the body, so as to revolve. This center bar is furnished with stout iron armatures, *b*, placed on alternate sides of the bar; the armatures furnished with three teeth, *c*, one of which is placed on the upper side of the armature, and just midway between the two which are placed on the lower side. Wherever this mill has been used, it has given the most ample satisfaction.

2. *The Roller Mill*, which is found necessary in some localities, consists of two iron rollers, each about fifteen inches in diameter. Some prefer to have one roller smaller than the other, so that if they revolve in equal times, the surface of one will move faster than that of the other, and combine a rubbing with the crushing movement. This, however, is probably of no real benefit, and is attended with the disadvantage of not feeding as well as a mill where the rollers are both large. The rollers are about 30 inches in length; they are thick, but not

solid, and should weigh about 400 pounds each. They are cast in an iron mold or chill. This secures a perfectly hard and smooth surface, and much more durability than when they are cast in the common sand molds. The shafts of the rollers work in boxes lined with babbit metal, which, by means of set screws, are made to slide upon the iron frame, and give the rollers any degree of closeness that may be desired. From one eighth to one sixth of an inch is a distance that will permit no stones to pass, large enough to do mischief, either in molding or burning. The gearing of the mill should be so adjusted to the power, as to give only about ten to fifteen revolutions of the rollers in a minute. A rapid movement not only requires a greater force, but greatly increases the danger of breakage of the machinery. A plank hopper is set over the rollers, large enough to hold a wheelbarrowful of clay. To the underside of the iron frame, on which the rollers rest, it is necessary to attach scrapers of iron or steel plate, to scrape the clay from the rollers, otherwise they will clog and become immovable. In setting the mill, it should, if possible, be elevated four or five feet above the level of the yard, and placed on horizontal timbers of some length, rather than upon posts set immediately under. The object of this is to secure space for the ground clay under the mill. The whole expense of such a clay mill, at the Cuyahoga Steam Furnace, in Cleveland, will be about one hundred dollars.

3. *The Horse Power* to drive the mill, whether the endless chain or lever power be used, should be arranged for two horses. A single horse, unless very strong, or the mill be geared for a slow movement, will, if there be many stones, or the clay be very lumpy, find the work rather severe. It is better, therefore, in the first instance, to obtain a power on which two horses may be used, if

necessary, or a single horse, if one is found to be sufficient. If the clay be dug up in the fall or winter, and thoroughly frozen, or if it be turned over and well wetted a few days before grinding, the work will be much easier; if taken fresh from a bank or hillside, and ground immediately, a good deal of additional power will be required.

Tile Machines.—It would be a useless task to describe all the different forms of tile presses in use. All possible forms of construction have been used, but those known as Clayton's or Whitehead's, are among the preferable kinds. These machines are strong, simple, and require comparatively little power to drive them, and are not apt to get out of repair.

A passing description of both these machines, may be introduced with propriety.

The clay box of the Clayton press consists of a perpendicular cylinder, terminating below in the mold box. The cover of the clay box is a kind of piston head, which is made to drive the clay downward, while working, by means of a cogged piston rod, in the cogs of which mash the cogs of a small wheel, which is driven by a larger cog-wheel, and this in turn by a smaller cog-wheel attached to the handle or working lever of the machine. The pipes are pressed out at the bottom, and hanging free, are received upon the prongs of a fork, which correspond in number to the number of pipes pressed out at one time. The tiles are cut off by a wire, which is attached to the machine. Two cylinders properly belong to this kind of machine, one of which is removed when emptied, and replaced by the other full. Latterly, this machine has been so modified, that the pipes are forced out horizontally, and received upon a truckle-bed, and are not cut off until this is full, when they are separated and borne away upon forks.

Whitehead's machine has a flat-lying quadrangular box, closed by a cover. In front is the mold, and by means of a cogged piston rod, the plunger, consisting of the entire posterior end of the box, is driven forward, which presses the clay through the mold. When empty, the action is reversed, the plunger again becomes the back wall of the box, the cover is raised, more clay is filled in, and the work proceeds again. The cogged piston or plunger rod is worked horizontally, by means of three cog-wheels meeting with each other, as do those of the Clayton machine. The pipes are received upon a truckle-bed as they are expelled.

Neither of these machines is without its advantages and defects, and yet six or eight thousand tiles can be molded daily, by some of the latter machines, while the former may be made to produce more, and is therefore better calculated, perhaps, for use in large manufactories. Proper machines can be manufactured after models, in almost any machine shop, but purchasers should always take care to secure good and warranted machines.

Mons. Barrall, in his excellent treatise on drainage, gives a detailed description, accompanied by engravings, for the most part, of *fifty-nine* different tile machines, used in England, France, and Germany. Many of these machines are very expensive, but at the same time, manufacture a large number of tiles daily. One machine which is there figured and described, would require *eight* active boys to carry away the tiles as fast as they are made—each boy taking six tiles at a time!

In this country, several gentlemen have invented machines for the manufacture of tiles. Of those in most general use, in this state, are the Mattice & Penfield machine and the Daine's machine. We present a cut and a short description of each.

Fig. 44.—Mattice & Penfield's Drain Tile Machine.

This machine not only grinds the clay, and molds the tile, but places them upon the drying boards. 4, represents the die; 3, the tile; and 2, 2, the drying boards, which are cut the length of three tiles; and placed upon the carriage, 1, the portion of which, under the machine, is covered with an endless belt, upon which these boards are placed, on the rear of the carriage, and are drawn under by the tiles as they issue from the die, and deposit themselves upon the boards. 7, 7, is a frame, held together by the handles, across which small wires are stretched, 8, 8, for the purpose of cutting the tiles. This frame is movable, for the purpose of cutting the tiles where the end of the board occurs. 6, is the shaft which passes through the machine, upon which iron knives are fastened to grind the clay. To the lower ends, eccentrics are fastened, that move the plunger in the clay box, to which the die, 4, is fastened. 5, is the lever by which the cut-off plate is driven over the clay box, after it is filled, to prevent the clay from pushing back up in the machine when the plunger pushes it out. 9, is the yoke upon which a slide is fastened, driven by an eccentric on the shaft that moves the lever, the plunger throwing it back when making the plunge, where it remains, leaving the cavity open again. A, is the sweep. The machine makes a plunge at every turn of the shaft. Less than one fourth of the time required to make a turn of the shaft, makes a plunge, which gives the man that cuts the tiles ample time to do so, and set them on the drying racks, which are placed upon the carriage for the purpose of moving them from the press, when dried, to the kiln.

The American Tile maker.—The Tile Maker is only eight feet in length, including aprons. It is mounted upon wheels, and is simple in construction, easily kept in order, and not liable to accident from any ordinary cause. It will make horseshoe or sole tile of any size, according to the nature of the die which may be used; the power applied to drive the clay through the dies is the screw, worked by a small balance wheel, as shown in the engraving. This machine is made of cast iron, and consists of a box set on feet, to which are attached small wheels, by which it can be moved from place to place. The iron box or frame is about five feet in length, and fourteen inches wide; at one end is fastened the die, which is easily taken off or put on by screws. The box into which the clay is put, and in which the square plunger compresses the clay through the die, to form the tile, is the main division of the frame, and occupies about two feet in length; one half of this division is covered with an iron plate, screwed down solid; the other consists of a lid, which lifts with a handle, and which, when the clay is filled in is shut and fastened by strong iron latches on each side, which swing into their place by weights. The other two feet of the frame is occupied by the iron tube, in which the screw of

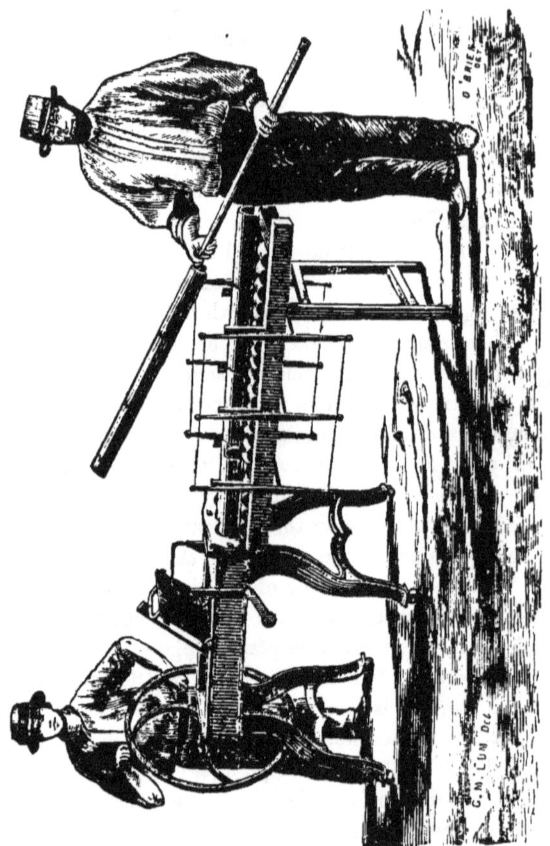

Fig. 45.—Daine's American Tile Machine.

the piston or plunger works, which is worked by a handle attached to a small balance wheel; attached to the end where the tiles are made, is a small wooden frame, supported on a level with the lower line of the die, by legs that fold up when it is taken off to be moved or packed away; it is about three and a half feet in length, and is made in three divisions, of twelve inches each; these divisions have each a series of small wooden rollers, on which a cloth apron moves when the clay is forced through the die. It comes out in three long parallel tubes of tile, moved and supported on these aprons, each of which is the length of a tile; when the table is full, the tiles are cut into exact lengths by wires which are passed down through gauges, which form a part of the wooden framework of the apron stand. The whole is easily worked in a space of eight by ten feet.

But the following cut illustrates the simplest and cheapest tile machine of which we have any knowledge. We propose to name it the "Buckeye" tile machine; it may be made by any ordinary mechanic, at a cost not exceeding $5.

FIG. 46.—THE "BUCKEYE" TILE MACHINE.

It consists of a stout box, A, whose sides are about eight inches high, twelve long, and the ends about eight wide. The back part of the box is occupied by a post, G, eight inches wide, and four thick, and from two feet to thirty inches high. In the top of this post is fastened a lever, H, and to this latter is fastened the plunger, F. The dies are represented at B. The box is filled with mortar, the plunger placed on the mortar, and by the lever is then pushed home; this operation forces the clay through the dies and forms the tiles. The carriage consists of twenty rollers, or five sets of four rollers each; over each set of rollers is an apron—the force and weight of the tile issuing from the dies, causes them to rotate so as to carry off the tile the entire length of the carriage. When the tiles are forced through the die, and cover the extent of the carriage, the frame, E E, is closed like a lid over the tile, and cuts them by means of the wires, D D D D D, into proper lengths. They are then removed from the apron to the dryer.

This machine can be operated, in all its departments, by a "man and a boy"—the man to fill the box, press the tile and cut them off, while the boy uses an implement shaped somewhat like the letter **Y**, or rather, like a two-pronked table fork—each prong about ten inches in length, and one inch in diameter. The prongs are inserted into the cavity of the tiles and thus borne away to the dryer.

Not much reliance can be placed upon statements, as to the amount of tiles which may be made in a day upon any of the machines—some days double, if not triple the amount can be made than on other days. Daine's machine claims to make 250 two inch tiles in an hour—this would amount to 2,500 in a day of ten hours. From 700 to 900 would be a fair day's operation on the "Buckeye."

Pressing the pipes is a very simple business. The blocks of clay are to be placed in the press-box, and hammered in the filling, to prevent the retention of any air, as this might occasion the bursting of the pipes, or the formation of air-cells in their walls, to such an extent as to render them useless. When the clay is properly packed in, the cover shut down and secured, and the press put in motion, the wince is turned until the truckle-bed is filled—the cutting apparatus is brought down, and one pressing of the rough pipes is completed.

The smaller kinds of pipe must be handled by means of properly made forks, with extreme care, and placed upon a drying rack. If great care be not taken, the sides of the pipes will either fall together, or the soft clay will be pressed out of shape, and the passage more or less obstructed. The larger kinds are taken off the truckle-bed by hand, and set up perpendicularly for drying.

As all kinds of clay and loam shrink more or less in drying, this change of volume must be regarded in the pressing; and because different qualities of clay have different shrinkage in drying and burning; and because of the different degrees of humidity at which the clay is worked, and the different length of time the working is continued, and that of drying and burning required—all have their influence upon this shrinkage—no rule can be given of general application, and every manufacturer must learn by experience to give a proper length and thickness to his drain tiles. As a general thing, if the *green* tile are 13 inch in length, they will scarcely be 12 inches long when burned; and tile measuring 2 inches from outside to outside, when green, will not measure over 1¾ when burned.

Drying tiles is a matter of great importance, and special attention must be directed to this part of the manufacturing process. The tile to be good must be dried in a shed; in fact, a good shed is indispensable to the manufacture of tiles. The clay must be tough to retain its shape after running through the dies of the tile machine;

and such clay will warp and crack in drying, unless the process is conducted in the shade. If the manufacture goes on under a shed, no time is lost on account of rainy days, and the tiles, while drying, are protected alike from rain and sun.

A convenient arrangement of the shed is of considerable importance; in form, it is long and narrow, and must be so set as to allow the kiln to be put directly at one end, while the clay bank or pit and pug mill are at the other. Where it is intended to make from one hundred to one hundred and fifty thousand tiles in a season, the shed will need to be sixty feet in length by eighteen in width. It is not necessary to put up an expensive frame, a lighter structure answering equally well. Four sills, either of timber or plank, may be laid upon the ground, and leveled to receive the feet of the posts. The sills are laid parallel to each other, and lengthwise of the shed the inner ones ten feet apart, and the outer ones, one on each side, and four feet from the inner. The posts made of scantling, four inches square, stand upon the sills, making two rows on either side of the central space. The outer posts may be six feet in length, and the inner eight feet six inches, the tops being halved to receive the rafters, of two-by-four scantling. It is convenient to have these posts and rafters of a uniform distance of six feet apart, through the whole length of the shed. The rafters may be tied together by three pieces of the same scantling, and these so placed as to give the best support to the roof boards, which lie lengthwise up and down the roof. The rafters and roof boards should be fourteen feet in length, so as to project about three feet beyond the outer posts; this is to prevent the rain from beating under and injuring the tiles. The supports for the shelves are narrow strips of board nailed to the scantling posts, the top of one being eight inches from the top of the next. The shelf boards should be twelve feet long; they will then have a support at both ends; and in the middle they should be made of narrow but straight and well seasoned oak stuff, and laid loose upon their supports, and at a distance of about an inch from each other. The tiles dry better on these than upon wide boards. In this way the shed will have shelving on each side of the central space in which the tiles are made; each shelf inside the posts, will be a little more than three feet wide, which is sufficient for three tiles endwise in the green state; there will be nine tier of shelves, one over the other. A shed of this size will dry about ten thousand tiles at a time.

Through the center of the shed a railway track should be laid. This may be made of two-by-four scantling set endwise, and tied together by cross pieces, and sunk nearly to the level of the floor. Upon this a little four-wheeled car runs, carrying the clay from the pug mill to the tile machine, and afterward the tiles from the shelves to the kiln.

A shed something like what is described above is needed where hand tile machines, similar in principle to Daines', are used. If it be intended to use Penfield's tile machine, which works by horse power, and has another arrangement for drying, scarcely any shedding is absolutely required. In this method, the tile machine being a fixture, drying carriages are constructed, and these are put on a track connecting the machine to the kiln, and are moved along as they are filled.

The internal arrangement of the shed should be such that the tile machine may be placed as near the center as possible. In the east and west ends of the shed "racks," as represented in the following cut, should be placed, on which to dry the tile. Tile should always *be dried in the*

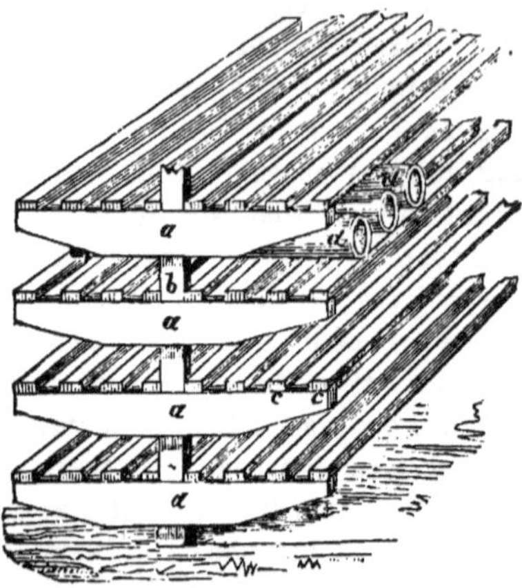

FIG. 47.—RACK FOR DRYING TILE.

shade—they dry more uniformly there than in the sun; beside, should inclement weather intervene, they are then protected. The rack is very cheaply and simply made; *b* is an upright, made of scantling, say, 3-by-4 inches, on which are fastened, with 4 or 5-inch spikes, the bats *a, a, a, a;* the slats or dryers, *c, c*, on which the tile are placed, may be of lath one-by-one and half inches. The uprights (*b*) should be no more than 6 feet apart—in fact, 4 feet is a good distance—in order to prevent the slats from warping, or "*sagging*," as the tile makers say.

The cut is intended to represent a rack to dry 16-inch tile; but it is best to make them wide enough, so that three tile may be laid on endwise. The vertical spaces between the slats *c*, *c*, or bats *a*, *a*, *a*, *a*, should vary with the size of the tile made—thus the distance from *a* to *a* should be greater for three than for 1½-inch tile. When the tile are molded by the machine they are carried away and placed upon the dryers, as represented at *d*, *d*.

Or the drying racks may be conveniently made as follows: Two upright posts of roofing lath should be fixed sufficiently far apart to permit one end of a drying board to go between them, and at the length of this board two more to receive the other end. When the tiles are cut off, they should be closely laid upon drying boards, the length of which should, for convenience, be about four or five feet, and the width equal to the length of the tiles. When the board is full it should be placed in the rack, upon pieces of scantling or other supports, and upon each end should be placed a similar piece of scantling, half an inch or so thicker than the tile, and upon these the next drying board filled is to be placed—the other supporting scantlings at the ends and drying boards upon them until the rack has been filled to the desired hight. The number of racks and their distance apart must be determined by the size of the dry house and the necessary movements between them.

Upon these racks the tiles remain to be dried, but they still require constant care and watching to effect drying properly. To keep the tiles straight, they should be placed close together, and if, in the process of drying, they become curved, the bow should be twined upward, so that they may assume their straight form again. The admission of air should be carefully regulated to dry the tiles uniformly; otherwise, they are liable to crack. In point of fact, the tile should be dried by the winds—not by hot southern winds, but cool north or northwest ones.

If the clay is well prepared, and proper attention paid to the pipes during the drying process, the remaining parts of the fabrication will go on well. The admission of air in proper quantity is always a matter of importance.

The larger kinds of pipe, which are placed upright while drying, should be reversed frequently, until hard enough to be laid down without injury, because the upper end always dries the most rapidly. When the pipes become somewhat dry, they may be laid in piles of several pipes in hight, according to their dryness; and, when dry enough to burn, five or six of even the largest size may be superimposed upon each other.

Some manufacturers dry their pipes upon hurdles or frames made of laths, in order to favor the admission of air; but this mode has scarcely any advantage over the simple drying board, and is far more expensive, and the pipes are more liable to be bent by being placed upon such racks.

Rolling and rimming the tiles is to be performed to secure a faultless product, and is done when they have lain long enough to be somewhat stiff, but still not hard enough to crack when handled and bent. This step in the progress of fabrication is too much neglected in this country, but in England is considered indispensable.

Rolling the pipes is thus performed: A round, smooth stick, one quarter or one third of an inch less in diameter than the clear capacity of the pipe, and long enough to reach through and afford a hand hold at each end, is passed through the opening, and the pipe gently rolled upon a smooth table two or three times to straighten it, and thus prevent any inequality which may have occurred during the progress of drying from becoming permanent. After rolling, the pipes are rimmed, by inserting into each end alternately and turning around the rimmer a wooden instrument, which is constructed of a round, smooth stick, just large enough to fill the end of the pipe, around the end of the shaft of which, and between it and the handle, is a collar, or square offset. This instrument, properly used, gives an exactly square end to the pipe, and insures their closely fitting together when laid down. The top or shaft of the instrument should be somewhat tapering, so as to favor its insertion and only exactly fill the opening at the shoulder or collar. This shape favors its insertion and prevents the clay from being pushed into ridges when it is inserted.

The operations of rolling and rimming are very important to secure good tiles, and the expense is so slight that it will be more than repaid in the better quality of the product.

For convenience, a light table, about fifteen inches broad (where the tiles are twelve inches long), should be used. The tiles can be lifted off and on the drying board by means of the rolling pin, and the table moved forward as the work progresses, and in this manner the process may go on very rapidly.

Tile burning.—This is performed when the tiles are perfectly dry, and can only be done well by a person acquainted with the business. No extended description can supply a want of practical knowledge, but a word of admonition, in regard to important moments in the process, may be of great utility. One indispensable matter is a proper burning kiln. Almost any kind of lime or potters' kiln may be made use of, but an oven especially adapted will be found of great advantage.

In establishing a tile yard, it is usual to make and burn a clamp of bricks, in the first instance; then to use the scoving, the soft and other waste bricks for building the kiln. If the intention is to make from one to two hundred thousand in a season, a kiln 11 feet by 13 in the inside, and 10 feet high, will be a suitable size. A kiln of such dimensions will hold about 15,000 tiles, the number varying, of course, according to their size, beside bricks enough to fill to the top of the arches. The kiln must be built at one end of the shed, and directly in a line with it, so that the doorway into the side of the kiln, through which the tiles are carried to be set, may be on a line with the car track of the shed. Directly opposite this doorway there should be a similar opening on the other side of the kiln, through which the burnt tiles may be carried. The fire holes will be through the narrowest sides of the kiln, or those which correspond with the sides of the shed. For a kiln of the size named, there will be four fire holes, each end of which will be open. The walls of the kiln should not be less than two feet six inches in thickness at the bottom, and three feet is still better. They are carried up perpendicularly on the inside, but gradually becoming thinner toward the top, by drawing in on the outside. They are better built of tempered clay, mixed with a considerable proportion of sand, than of lime mortar. Some 25,000 bricks will be required to build such a kiln as the one described.

Tile kilns, of the following construction, will be found very appropriate for the purpose. There are two principal forms of construction in vogue in Europe—one of which is the *high* kiln, and the other the *low* kiln. The high kilns are commonly 20 to 24 feet long, 10 to 12 feet wide, and the arch 10 to 12 feet high, measured in the clear. The walls are made four courses of brick thick, and are supported by buttresses in the longitudinal walls. Between the buttresses, on each side, there are 8 furnace holes, 15 inches wide and about twice as high, and provided with a grate and ash box. They permit the firing to be done by means of wood, coal or turf. The door, placed at one end, should be wide enough to admit of wheeling in the tiles, and must be walled up when the burning is begun. Inside, between each pair of furnaces, there is a small flue, 3 to 4 inches square, which, passing up the wall and along the arch, terminates in low chimneys formed conveniently of tile pipe of proper size.

There are beside 6 or 8 rows of small smoke stacks, 4 to 5 inches in clear diameter, and 2 feet high, which pass through the arch, that may be opened or closed on the outside at pleasure, and by means of which the heat may be regulated according to requirement during the process of burning. Such a kiln resembles very much a common tile kiln. 40,000 or 50,000 pieces, of different dimensions, may be burnt in such a kiln, if space be economized, by placing the smaller pipes inside of the larger ones. The furnaces must be covered with an arch of masonry, else the pipes placed immediately upon the top of the furnace would be over-burnt.

The *low* kiln resembles a common potter's oven, and is now greatly in vogue, as it is easily built, and yields a well burnt product. Such a kiln consists of a long arch, 8 to 10 feet high, 14 to 16 feet long, and 10 to 12 feet wide in the clear. The walls and arch may be built very thin, if supported by iron arch bands—6 or 8 inch walls being sufficient. But latterly the walls have been built thicker and supported by buttresses. At one end is placed the chimney, and at the other end the door for wheeling in the tiles. On each side of the door is built a furnace, of 18 to 20 inches breadth, and 10 to 15 inches hight, and a third furnace is fixed in the doorway when this is walled up. Immediately behind the doorway wall is placed the ash pit, 2½ feet broad and 6 to 10 inches deep. The hearth of the oven lies a little higher than the opening of the ash pit, and behind this again there is a depression 9 inches wide and 6 deep, in which originate four flues, which, leading through the walls, terminate in the chimney. The chimney is not placed, as in the potter's oven, upon the arch, but at the end, so that the fire may be forced to pass over all the pipes, which are thus uniformly burnt in all parts. The chimney is about 15 to 18 inches clear in diameter. In the walls and gable ends are vents, which during burning are walled up, but are opened when this is finished, so as to favor cooling.

Twenty to twenty-five thousand pipes of different sizes can be placed in such a kiln. When the tiles are wheeled in for arrangement, bricks are placed upright upon the hearth, and the pipes are set upon these perpendicularly, so that the fire may readily draw through the whole. When the kiln is filled, a sieve-like wall of a single course of brick, is built up to force the fire to spread equally through the entire oven, and at the same time to protect, in some measure, the first courses of tiles from the excessive action of the fire.

The following precautions must be observed in burning:

The tiles should not be placed in the oven before they are perfectly dry; but in case it is necessary to do so, they must be dried there, by being subjected, very gradually, to the heat of a slow fire, in order to dry them thoroughly, before heating them very much, as tiles burnt rapidly, in a damp condition, are nearly always bent and full of cracks.

The pipes are placed in the oven, perpendicularly upon the hearth and brick work which forms the furnace passages. Small pipes are put into larger ones, but not so nearly of a size, as to hinder a free play of the fire between them. Six inch pipe may be filled with three or four inch pipe, and these with inch pipe. This mode of placing is for the purpose of saving space. The upper tier of pipes may be placed horizontally, but the lower ones could not sustain the superincumbent pressure were they so placed.

It is very important to be provided with good fuel, and to keep the heat at an even temperature throughout the process. If the draught of the wind cause the heat to be excessive upon one side, this can be remedied in the *high oven*, by opening or closing the smoke stacks; and in the *low oven*, by varying the intensity of the fire upon one or the other side, as the case may require.

When the burning is completed, precaution is necessary to prevent too rapid cooling, otherwise the tiles will be found much cracked. It often happens that tiles are found imperfectly baked, and are denominated "pale" or "soft." These can not be used, as they crumble to pieces in the wet, and should be reburnt with the next kiln full. If at any particular part of the kiln the tiles are commonly imperfectly burned, the "soft" tiles of one burning may be placed in that part for the next burning, and they will thus become sufficiently baked.

Fuel.—In many localities coal is cheaper than wood, and fortunately it answers the purpose equally as well. Where wood is employed, the soft kinds are greatly preferred to the hard. Soft maple, basswood, whitewood and chestnut, are the best, making a steadier heat and more flame. For kiln use, wood must be thoroughly seasoned, split tolerably fine, of the length of the holes or shorter. A cord of good wood should burn about three thousand tiles.

Burning.—When holes are made on both sides of the kiln, as recommended, the burning is effected on one side at a time. By this method, more time is probably consumed, though not more wood, and there is less danger of an unequal or insufficient burn. In burning tiles, it should be borne in mind that the soft burnt are worthless, and only those that are thoroughly hard, and will ring when struck, are of any value. It requires from two days and a night, to four days and nights, to burn a kiln of tiles, the difference depending on the kind of fuel, the character of the clay, and the method pursued—they should be allowed from 48 to 60 hours to cool.

We translate the following from Barrell's work on drainage:

"*Burning.*—The operation of burning the pipes comprises three divisions: 1. Placing the tiles into the oven. 2. Conducting off the fire. 3. Cooling and removing from the oven.

"The burning of pipes is of great importance, for it affects both the quality and the price of the manufactured article. Therefore, the more perfect is the oven, or kiln, the better and cheaper will be the pipe. We can not enter into any description of the numerous improvements on the subject which have transpired, a whole book would not suffice; but we will give general outlines in the expressions of Mr. Brongniart, the most competent writer on the 'Ceramic Art:'

"'An oven contains four principal parts, viz: The fire place, the mouth, the laboratory and the chimney; in the fire place is thrown the fuel, whatever it may be; the mouth is an opening through which the air is introduced which is to sustain combustion; the laboratory is the place where the articles to be burned are placed; the chimney is a channel though which the gases escape after having produced their effect.

"'Some ovens have no special chimney—it is a part of the laboratory—into these the flame or gas is directed from the fire place through holes or openings named '*carneaux;*' when the flame is not admitted into the laboratory, and goes directly into the chimney, the heat is received by radiation.'

"'We will suppose an ordinary potter's kiln is employed, and proceed to the operation of placing the pipes into the laboratory: A layer of common brick is to be disposed in a vertical position, at a small distance from each other, on the floor of the oven; upon these, the pipes of the largest diameter are to be placed, one upon the other, so as to form layers up to the top; some manufacturers place the pipes upright in the same position as they were arranged to dry; this system is evidently favorable, because it results therefrom that each series of pipes placed on ends form as many chimneys, which favor the draft, and distribute more equally the heat. The fire is next kindled, and kept at first very slow; after a few days heat may gradually be increased up to the highest possible degree; during that time the utmost care and constant attention are necessary to avoid accidents.

"'When the burning is not complete, or otherwise defective, the pipes remain tender, earthy, with a dull color, either white or red; they are not sonorous, and break or shell off under the influence of the air; they facilitate the generation of saltpeter, crumble to pieces, are destroyed in a very short period or time, and finally ruin the drains.

"'Should, on the contrary, the fire be *untimely*, or excessive, the material will melt in part, the pipes become dark brown or black, out of shape, and stick to each other in cooling. From this may be seen how important it is to secure the right degree of heat, which produces tile between a dark and a very bright red; this may be easily watched, by keeping, within reach show pieces, which may be extracted through convenient holes; it is advisable not to hurry the operation of burning. When the fire has been brought to the proper degree of intensity, it must be gradually diminished, and suppressed altogether; then the mouth and chimney of the oven are to be closed so as to exclude carefully the cold air; all must be left in this state, during several days, to permit the pipes gradually to cool down, otherwise they would crack or burst to pieces.

"'A skillful burner will, at the proper time, remove the pipes from the oven, with hardly any breakage, or at most from two to five per cent.; whereas the loss may be considerable from want of skill or care.'"

The price of tile at tileries, throughout Ohio, is yet entirely too great to induce farmers, generally, to adopt tile draining, where they are obliged to rely upon the tileries for supplies. There is no good reason—other than the fact that tile making is yet a new business, and not thoroughly understood—why tile should cost any more than common brick. The amount of material used in a single brick will make from two to four or five tile, according to size; while the amount of heat required to burn one brick will burn more tile than can be made from the material in the brick. True, a little more care is necessary in arranging the tile in the kiln; a much smaller quantity can be burned at a time than of brick, and every defective or warped tile is worthless; these, of course, are drawbacks, but in course of time they will in a very great degree be obviated. Good tile can be obtained at Cleveland, Columbus, Cincinnati, Woodstock, Painesville, Springfield, Claridon, etc., in Ohio, at reasonable rates.

CHAPTER VI.

HOW WATER ENTERS THE PIPES.

THIS question is asked by all persons who, for the first time, direct their attention to the subject of drainage, and the solution of the problem involved in the inquiry, is rather a subject of scientific interest than a matter of practical moment; for the water does find ingress, as experiment proves. But nevertheless, there are some practical bearings in the question which demand investigation.

In the ordinary arrangement of strata of earth, there is a very permeable layer or soil and subsoil, and below, a less permeable stratum or "hard-pan." The water of rains descends to this stratum, and is there retained for a longer time than in the more permeable soils above; and it is a consequence of this retention, that the upper strata become submerged with water.

When drains are laid much above the level of this retentive stratum, they do not begin to carry off the surface water until this has completely saturated the whole depth of soil from the "hard-pan" up to the level of the drains, which thus obtains the water which enters it from below. It was at one time supposed to be disadvantageous to the object intended, if the surface water made its way immediately downward into the drains, as it was supposed to be not sufficiently filtered, and much of the soil enriching contents would be carried away into the drain, when it should have remained in the soil. To obviate the immediate descent of the water into the drain, it was recommended to cover the newly-laid pipe with a layer of sand, or other porous material, two or three inches, and then overlay it with a covering of stiff clay, which would cause the more even and natural descent of the surface water to the impermeable stratum below, and its subsequent ascent to the drain level, into the bottom of which it finds entrance. But recent experiments have shown the fallacy of this doctrine. We have shown, in the experiments of Liebig and others, that the soil at once absorbs all the nutritious properties borne down by the rains. The permeable strata will not yield their moisture to the drain until the point of saturation has been reached below.

The manner in which the water finds admission into the drain pipe, when it has once found its way to it, is very simple and easy of explanation. If the whole drain were one continued, unbroken pipe, submerged into a supersaturated soil, a portion of water would find its way by means of what may be termed soakage, through the somewhat porous walls of the pipes, as water makes its way slowly through bricks. This soaking or sweating process would go on more readily through soft, poorly burned pipes; but in tiles very thoroughly burned, it would go on very slowly; so slowly as to defeat the purpose for which such tiles are laid down. The proportion of water, however, which enters the jointed pipes (the only ones used) by soakage, is so inconsiderable, that we must look for some other mode of entrance, in answer to the question, "How does it get in?"

No jointed pipe can be made and laid down, in which the joints will fit sufficiently close to prevent the free access of water to the empty space within the tube. The facility for entrance, by this means, afforded by a pipe of any size, under four inches, 200 feet long, made of 13 inch sections, will exceed, by far, the capacity of the same pipe to discharge the stream which might thus find entrance. The water, then, enters at the joints, which can not be made close enough to prevent its ingress, and, when properly laid down, the water entering the drain has its course from below upward.

Kielman appears to doubt, that sufficient space would occur between the joints of twelve or thirteen inch pipe to carry off the water which would collect. But being satisfied that the joints were the only place at which water could enter, he manufactured tiles having a length of nine inches only, in order to facilitate the admission of water. This we consider very bad policy; because it makes not only more joints than are necessary, but because short joints are more subject to disturbances than long ones. In fact, sixteen or eighteen inch tiles afford sufficient joint apertures for all the water they can convey away. There have been many calculations with regard to the amount of space between the joints of pipes; and although we have quoted Messrs. Shedd and Edson, at page 282, as being correct in the main, we yet prefer, as a matter of mathematical precision, those made by Vincent. He says, in effect, that water requires no other means of entering the pipes than the spaces at the joints. The inner circumference of a one inch pipe, amounts to about three inches. If, then, the width between the joints is assumed to be one eighth of a line, or one ninety-sixth part of an inch, which, in all probability, is the least possible space which is likely to occur, under ordinary circumstances, it produces an entrance space equivalent to one thirty-second of a square inch. The section or opening of a one inch pipe would then have a capacity of nearly three fourths of a square inch. Then, twenty-four or twenty-five joints, each having an entrance capacity at the joints of one ninety-sixth of an inch, will have an aggregate joint entrance capacity equivalent to the caliber of the pipe itself. In less than two rods, we have upward of twenty-five joints, therefore the minimum capacity of admission at the joints more than equals the caliber of the pipe every two rods.

But as it is not at all likely that drainage water will fill the pipes every two rods, the joints might even be made closer than one ninety-sixth of an inch, and yet admit all the water that is likely to find its way into the drain. On the other hand, there are scarcely any tiles manufactured whose joints will fit closer than one half a line, or the one twenty-fourth of an inch; therefore the water would find its way into the pipes in sufficient quantities, even if the tiles were two feet, instead of one foot long.

CHAPTER VII.

DURABILITY OF TILE.

THIS question has not been tested fully in this, and perhaps in no other, country. The length of time since the first pipe tiles have been laid down here, has not been long enough to determine this question. All the information that can be gathered from direct experiment, and analogical reasoning, goes to show that drains of properly burned tiles, may be considered "*permanent*" improvements.

A few references to known cases of durability of tiles, and other objects of similar constitution, may aid in arriving at a proper estimate of the indestructibility of tile drains.

In Wigtonshire, England, the celebrated Marshal, Earl of Stair, had constructed some drains of brick, laid upon the clay subsoil, beneath the vegetable mold, one hundred years ago, which, when examined after the lapse of that time, were found to be uninjured, both as to materials and permeability. They were laid, in one instance, by setting two courses of bricks lengthwise, about four inches apart, and covering the space inclosed by laying other bricks endwise across. In another case, the drain was made by laying down bricks side by side, as a foundation, upon the edges of which other brick were set up sideways, and the whole covered with flat stones. In both cases the work was next inclosed with a packing of broken bricks, or "bats," and then earth superimposed.

In France, there are tile and brick drains laid down in the early part of the seventeenth century, still in good repair, and fit for the purpose intended, which proves sufficient durability to warrant the construction of drains (if properly performed), with the reasonable expectation that they will outlast the generation of those who perform the work. There are, indeed, in England, certain legal enactments and regulations made to promote and favor the construction of drains, which contemplate fifty years as the minimum period of durability, which may be assigned to this species of improvement, if properly made.

The almost indestructible nature of the materials, when properly protected, may be inferred from the fact, that at Ninevah and Babylon, bricks have been exhumed after having lain in the earth for more than thirty centuries, in a state of perfect preservation. In Italy and Greece, specimens of ancient pottery are found, the age of which is often not less than two thousand years. Even in Ohio the antiquary can point to the remains of a *very inferior kind of earthenware*, of an age coeval with the mound-builders, the cycle of whose life and labors is lost in the utter oblivion of forgetfulness, while their fragile potters-ware remains to tell us that "art is long, though life is short," and insure the duration of the work of our hands, until our name, and even nation, may pass away and be forgotten.

In the Great Basin of Utah Territory, may be found the volcano-burnt clays of a period so remote in the world's geologic history, that no number of years can satisfactorily designate the durability which this clay, like that of our tiles in composition, has already shown; and no guess as to when the common causes of its destruction will have disintegrated it again, can assign the limit of its future permanence.

The useful durability of our tile drains, depends upon

the following circumstances: 1. A properly constituted clay, suitable for making a "hard tile," that is, a semi-vitrified product. 2. The perfect burning of this into properly shaped hard pipes. 3. The laying of these so deeply in the earth, as to protect them from the frost, a most powerfully disturbing and destructive agent. 4. An observance of the proper rules of construction, so as to avoid curves, up and down, to such an extent as to favor the deposition of sand and rubbish, which may find their way into the tiles, through the crevices of the joints. Sand will be arrested at any depressed point in the course of a drain, and clog the conduit so as to prevent the flow of the water. And, last, the protection of the entrance and exit extremities of the pipes, from the admission of small animals, reptiles, and the like, or the treading of cattle. This object can best be attained by the use of tile plates, perforated with fine holes at each end, and inclosing the exit with a fence, or walling it up to prevent the cattle, attracted by the water flowing out, from treading the tiles to pieces. (See page 382.)

In regard to the kind of pipes which are most durable, it may be remarked that "pale," or "soft" tiles are readily softened and broken by the action of the water, while tile may be made perfectly indestructible, if sufficiently burned, by any means save violence, frost, or powerful chemical re-agents, against all of which means of destruction a proper mode of deposit will entirely protect it; and a drain thus constructed, can have no limit assigned to its useful durability. In common phrase, it will "last forever."

CHAPTER VIII.

LAYING OUT DRAINS.

In laying out drains, the first thing to be determined is the amount of fall. Therefore, the lowest spot on the field or fields to be drained must be selected as the starting point. The amount of fall which can be obtained at the lowest point necessarily determines the depth of the drains. After having determined the amount of fall, the next thing to be determined is, whence comes the water? Should it be ascertained that the water comes from an underground spring, then a drain on the Elkington plan may be advisable. If the water appears in concavity, on the side of a hill, it will, perhaps, be well to examine the soil immediately underneath, and, if an impervious bed underlies, which is in turn succeeded by a porous bed, it may be bored through at short distances, drawing the water into the lower and pervious stratum. Should the water make its appearance at the bottom of the hill, flowing over an impervious stratum, a drain might be dug parallel with the base of the hill, which will remove the water coming from above, and the spring will be cut off. Again, from the bottom of this drain auger holes might be bored through the impervious bed into the next below, should it be found pervious. (See illustration, Fig. 48.)

In this case the purpose is merely to collect and carry off springs that come to the surface—a knowledge of the character and arrangement of the earth a few feet below the surface, therefore, is very desirable. Where the water washes its way to the surface, in a layer of sand or gravel, lying upon a layer of clay or rock, as is usually the case, the work is very simple. A ditch or drain is made up to the foot of the hill or ridge, from some creek or other place, where sufficient outfall can be obtained; it is then carried along the foot of the hill or ridge, usually a little above where the water makes its appearance. The drain must be low enough at the mouth to allow of cutting entirely through the layer of sand or gravel that carries the water, or much will escape under the drain. It is of little use to run drains endwise into banks, for the purpose of drainage, though it is sometimes done successfully when the object is only to obtain a supply of stock water.

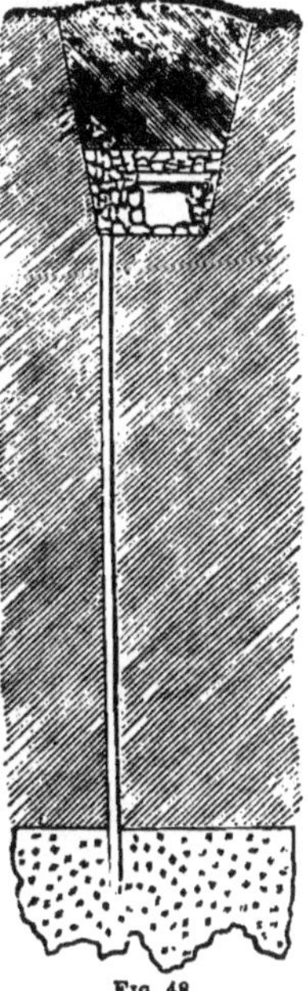

Fig. 48.

In the drainage of swamps, or small basin-like depressions, it is customary to cut a main drain through the center, at a depth sufficient effectually to drain the lowest point. In the direction, for example, from 4 to the top of the hill, 1. Then other drains, as at 6, 6, 6, 7, which empty into the first from both sides, commencing as near as may be to the edge of the swamp

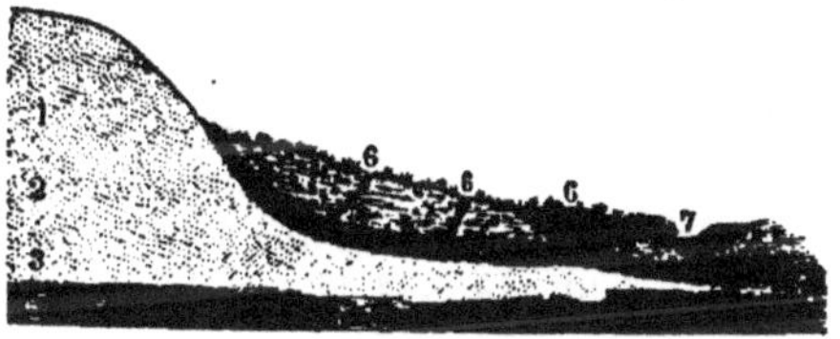

Fig. 49.

to catch the water in its descent from the higher lands. Without these side drains, or a drain encircling such depressions to a greater or lesser extent, they frequently continue wet and cold, notwithstanding the existence of a good central drain or ditch.

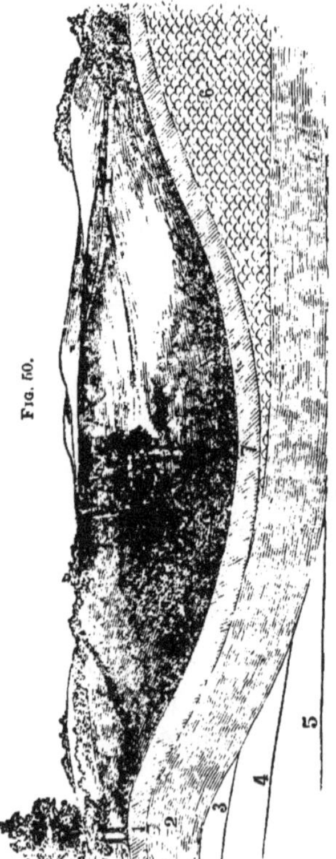

Fig. 50.

Where there is a basin-shaped field, as in the annexed cut (Fig. 50), in which 1 represents a clay soil, 2 a bed of hard-pan, 3, 4 and 5 different layers of rock and shales, 6 gravel, drains may be cut centering at 7, like those at G, G, G, G, in Fig. 51 (next page), at H, cut through the strata into a pit or well; and, if necessary, minor drains may be cut leading into those figured.

In thorough draining, sufficient fall having been obtained from the lowest point of the land to be drained, that becomes the proper starting point. If the field has a regular descent toward one of its sides, along that side the main drain is carried, and all the minor drains start from and run parallel one to another. If the lowest part of the land to be thoroughly drained be not along one of its sides, the main drain is carried along the lowest place, whether straight or otherwise, and the minor drains start from it on both sides. If the direction of the minor drains be at right angles to the main drain, it is better to curve the end of the minor drain for a few feet, where it enters the main, so that its current may not be across that of the main drain, but partly in the same direction.

The fewer main drains and general outlets to a field, the better. In the drainage of hillsides, it has been a question whether the parallel drains should be carried down the line of greatest descent, or obliquely to it; but longer experience has settled the question, where tiles are

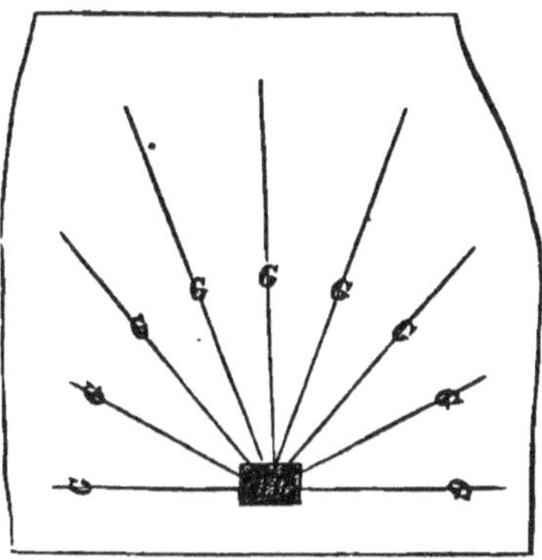

FIG. 51.

used, in favor of the line of greatest descent, or, in other words, running the minor drains straight down the slope.

One should think that a question apparently so self-evident would require no argument. But we find, in the works of the various writers on this subject, that a great diversity of opinion exists. One party insists that if a drain be cut across the foot of the hill, as at 1, in Fig. 52, it will completely drain not only the stratum 3, but also that indicated by 2, and all above it; and, therefore, object to making drains in the direction of the greatest descent. Another party would make a drain to carry off the water from each stratum which would crop out from the hillside. But, in order to drain land effectually, it is essentially necessary that we have a correct idea of the

FIG. 52.

sources from which the water is derived that is to be carried off; whether the water is directly from the clouds, or is derived from fields enjoying a greater elevation, and sloping toward it, so that the water comes down, like on a roof, from the other fields; or whether it comes up in springs, which find vent in particular spots, as indicated at 7, Fig. 49. If the water is not derived from adjoining fields but from the clouds direct, a different mode of draining is required than would be if the water came from higher fields. When lands are situated midway on an undrained slope, from which the water spreads over the surface of the land, such a system must be adopted as will not only drain the field in question, but also to cut off the supply of water from the higher fields.

One thing must be borne in mind, that water runs down hill, and does not spread so as to run laterally. From the fact that water always seeks the lowest level by force of gravitation, and drains are simply *lower levels* to conduct the surplus water away, in order to decide correctly what direction a drain should have, it is not only necessary to have a correct idea of the sources of water, and the superposition of strata, but a definite idea as to the special office the drain is to perform so as to carry off the surplus water and drain the land.

As before stated, drains should be dug up and down the slope, as from 1 to 2, Fig. 52. Suppose a man has a field lying on a slope, which he wishes to drain. If he lay out his drains thirty feet apart, and cut them up and down the line of greatest descent, it is very evident that the drains will then intersect all the strata, and bear away the water from all of them. But, if he lay out his drains the same distance apart across the line of greatest descent, the lower drain will receive the water from the thirty feet next above it; the next drain from the thirty feet next above that, and so on; thus compelling the water to traverse or percolate through thirty feet of soil before reaching a drain. But in the other case, the water will traverse a distance of fifteen feet only to find a conduit. The line of the greatest fall is the only line in which the drain is relatively lower than the land on either side of it. The water must be disposed of which rests upon the impervious strata, whether it has found its way there from fields or strata above, or whether it is water from the clouds, and has recently found its way there. But, in order to drain a field lying on a slope, with higher lands above it, it is, perhaps, as well to cut the upper drain across the line of greatest descent, and lead it, as a sub-main, down the line of greatest descent, at the side or center of the field, to the outlet. This answers the purpose, as these drains significantly have been termed, of mere catch-waters.

Now, looking at the operation of drains across the slope, and supposing that each drain is draining the breadth next above it, we will suppose the drain to be running full of water. What is there to prevent the water from passing out of that drain in its progress, at every point of the tiles, and so saturating the breadth below it? Drain pipes afford the same facility for water to soak out at the lower side, as to enter on the upper, and there is the same law of gravitation to operate in each case. Mr. Denton gives instances in which he has observed, where drains were carried across the slope, in Warwickshire, lines of moisture at a regular distance below the drains. He could ascertain, he says, the depth of the drain itself, by taking the difference of hight beween the line of the drain at the surface, and that of the line of moisture beneath it.[1] He says again:

"I recently had an opportunity, in Scotland, of gauging the quantity of water traveling along an important drain carried obliquely across the fall, when I ascertained with certainty, that, although the land through which it passed was comparatively full of water, the drain actually lost more than it gained in a passage of several chains through it."

So far as authority goes, there seems, with the exception of some advocates of the Keythorpe system, of which an account has been given, to be very little difference of opinion. Mr. Denton says:

"With respect to the direction of drains, I believe very little difference of opinion exists. All the most successful drainers concur in the line of the steepest descent, as essential to effective and economical drainage. Certain exceptions are recognized in the west of England; but I believe it will be found, as practice exends in that quarter, that the exceptions have been allowed in error."

In another place, he says:

"The very general concurrence in the adoption of the line of greatest descent, as the proper course for the minor drains in soils free from rock, would almost lead me to declare this as an incontrovertible principle."

We will suppose A, B, Fig. 53, to represent a portion of the higher field above. Then the catch-water or drain across the line of greatest descent will be represented by A, H, E, H, B; and when the nature of the

---

[1] French *on Drainage*.

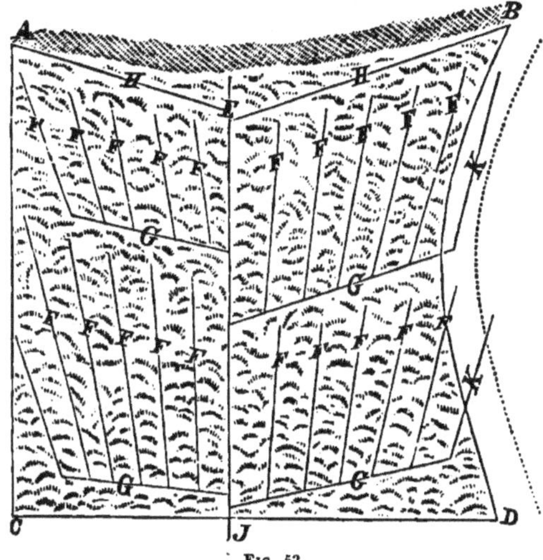

Fig. 53.

ground will admit, or should there be a depression toward the center of the field, the catch-water may be led from E to J, as a sub-main, being some distance below J, the main drain. The minor drains then should run parallel, or nearly so, to E, J.

Where the distance from E to J is considerable, it is always advisable to run the minor drains F, F, F, etc., into sub-mains, G, G, G, G. In draining a piece of land, situated like that represented in Fig. 52, which would involve the cutting of ditches to the depth of eight or ten feet between 1 and 2, so as to have the drains of a proper depth at 3, it will be found advisable to lead the minor drains into a sub-main from 4 to 3, and then commence a new series of drains between 2 and 1, and lead them into another sub-main at 1.

Some good drainers advise, that when works stop on a

slope, a drain called a *header* should connect the tops of the minor drains, thus preventing the water lying between the upper sub-main, A, E, B, of Fig. 53, and the minor drains F, F, F, F, etc., from passing down into the ground between the minor drains, and also relieving the minor drains from the pressure of the water above them, and by which they will the more easily become clogged than when protected. However, when the sub-main is dug above the minor drains, as in the figure, the necessity of headers is very slight, except when the quantity and pressure of water is sufficient to cause it to flow over the sub-main.

Even the sub-main will not drain the slope above it entirely. Capillary attraction, and the resistance offered to the descent of the water will prevent the sub-main from bringing about a complete drainage. The cuttings of our railways and high banks of rivers show that no depth of ditch can remove the moisture from a very considerable distance. This part of the subject has been more fully discussed in the Chapter on Distance of Drains.

The sub-main draining the highest portion of the slope should be independent of all minor drains and branches, for being directly in contact with the head of water from above, it will necessarily carry down more mud and silt, and have a tendency, if allowed, to choke up the minor drains.

It is sometimes found advantageous to construct a tank, sink, or silt-basin, in both surface and covered drains. This is more especially the case where an open enters into a covered drain. From this sink the water flows off comparatively clear. This arrangement will not be found to answer its purpose, when the amount of water flowing through the drains is very great, for then the motion of the stream passing through the tank will prevent the mud from depositing. It will also be necessary to have the tank frequently cleaned from its deposit, for when filled with mud it is only an obstruction.

We have now described the proper method of cutting off the supply of water from an underground spring, as well as draining the underground water from an adjoining slope, and it yet remains to say a few words upon conveying away the amount discharged by the clouds. This is a subject upon which much has been written, and is even yet an exceedingly controverted point. It is, in fact, the egg of Columbus for drainers, as it involves not only a calculation of the distance between drains, the depth of drains, fall, and size of tile, but also evaporation and filtration. All of these points have been discussed in the preceding pages. We may assume that the meteorological precipitations for Ohio, will average 43 inches per annum (see page 77). The precipitations then will be 10·34 inches during the spring months; 13·40 during the summer; 9·60 during autumn, and 9·66 during winter. Assuming, then, in the absence of positive experiments, that evaporation is the same, pro rata, as in Continental European countries, it will amount to 15 inches per annum in Ohio, leaving 28 inches to be filtrated, and to flow off the surface. Of this, about one half, or 14 inches, finds its way into the soil, and the remainder into brooks, creeks, etc. Now, if these assumptions are correct, then underdrained soils inaugurate a vast change in these proportions; because where observations have been correctly registered, it was found that eleven twentieths of the summer precipitations were discharged by the drains, and often more than three fifths of the autumn and spring precipitations, while the discharges from the drains averaged more than three fourths of the winter precipitations. Hence, the assumption, that one third of the precipitations are absorbed by filtration, is no criterion for the drainer. He must assume that at least one half of the meteorological precipitations are to be carried off by the drains. Now, the summer and autumn precipitations must not be taken as a basis, upon which to predicate either the distance between the drains or the capacity of the tiles, because the soil is then in a condition to dispose of the precipitation without any obstruction. But the winter and spring precipitations will constitute a more reliable basis. Freezing during the winter months, arrests the operation of the drains, and when the genial weather in spring time sets in, the water of the two seasons have both to be drained at once. Now, if we take the amount of the precipitation of the three winter months, and add to it that of two spring months, this will give us the largest mass of water to be drained in the shortest period of time, so as to relieve the growing crops from sustaining any injury. The period in which this water should be drained away, should never exceed fourteen days.

Having given tables in the preceding pages, of fall, width between drains, and capacity of tiles, each one may make his own calculations for the piece of ground intended to be drained.

CHAPTER IX.

MAIN DRAINS.

THE main drain should be located on the lowest portion of the farm. It should be an open ditch, at least four feet deep, but when circumstances will permit, six feet. The side should have a slope of a foot and a half to each foot of depth. If then the drain be four feet deep, and eighteen inches wide at the bottom, the width at the top will be thirteen and a half feet. The ground excavated, if thrown up on the sides, will form a capital fence. In fact, the *ha-ha* fences of England are built in this manner, for the reason that they occupy less space, and are equally as preventive as hedges are against the irruptions of unruly cattle. The main should invariably be made before the minor drains, for very obvious reasons, prominent among which is the determination of the amount of fall and depth of the minor drains. The main drain should invariably be a foot or eighteen inches lower than the outlet of the minor drains, if they discharge immediately into the main drain; but where submain drains are employed, the main should be at least eight inches below the outlets of the sub-mains, while the sub-mains should be at least 6 inches lower than the minor drains. Of course where these proportions are not practicable, less fall between the minor drains and sub-mains, and between the sub-mains and mains, must be admissible. But where these proportions can be attained, greater security will be given to the drains, against disturbances by frogs, lizards, or other amphibious animals. Where a sub-main or minor drains empty into the main drain, the exit pipe should be secured by a system of masonry, similar to that represented in Fig. 54. This effectually prevents the entrance of frogs, craw-fish and other "*var-mints.*"

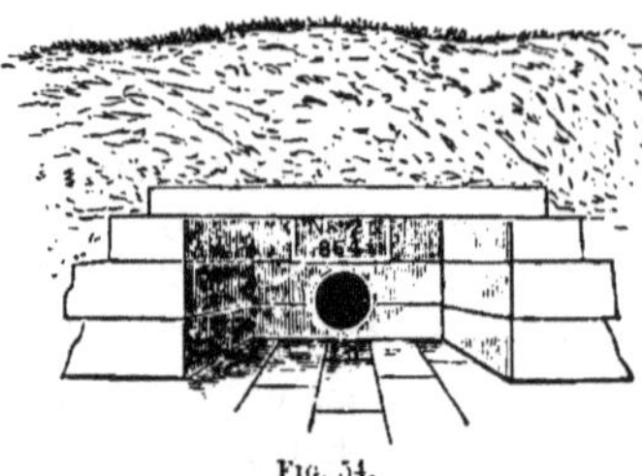

FIG. 54.

We have mentioned minor drains emptying into the main drain, or open ditch, thus making a separate outlet for each minor drain. We do not wish to be understood as recommending this method, by any means, because these outlets are not only liable to be frozen up in winter time, but are exposed to cattle and to mischievous boys, and to become obstructed by deposits which are discharged by the drains themselves. A much better plan is to have the minor drains empty into a sub-main, as G G, emptying into J, in the lower portion of Fig. 53. The smaller the number of outlets, in any system of draining, the better.

Some may object to our plan of one outlet, on the ground that, should any obstruction occur in the minor drains, it will be more difficult of inspection. This is true in a certain sense; but we think that surface indications will show when and where any serious obstruction takes place, with as much certainty as the open end of the drain. And surely, the additional security of having a few openings, well protected, is a much greater advantage than a drain left open for the purpose of investigation. However, to obviate any difficulty which might arise from either of the above methods, some good drainers recommend that "peep-holes" should be placed at regular distances, by which, should any derangement occur, its locality and extent could be easily determined. The construction of these "peep-holes" may be varied to suit the taste or means of the proprietor. A very easy method of making them will be to sink a stout barrel or hogshead over the drain. This, however, will be but a temporary concern. Another form, more in place with the whole system, may be constructed after the annexed cut, Fig. 55, either of earthenware or cast iron. It should be well

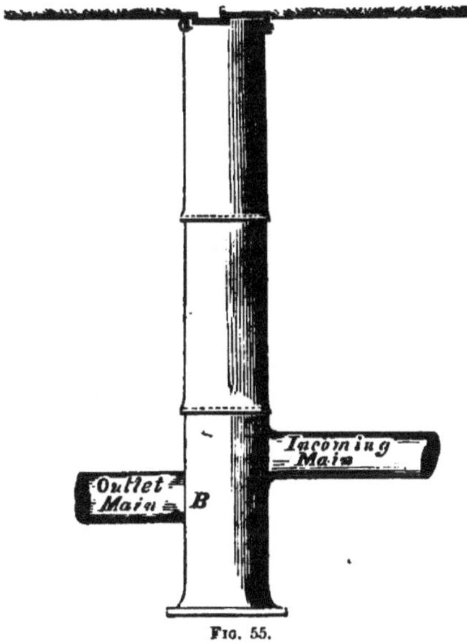

FIG. 55.

protected at the surface of the ground, against cattle, etc., by a strong cover, as represented. This arrangement will furnish ample means for investigating drains, convincing the incredulous, and also, of making observations on the working of the system in different portions of the work.

We have before spoken of sinks or silt-basins. These should not be confounded with "peep-holes." The accompanying cut gives a very good idea of their construction. They should be built of solid masonry, large enough to admit of being cleaned out without inconvenience. A relief-pipe, as shown in the figure, will not always be ne-

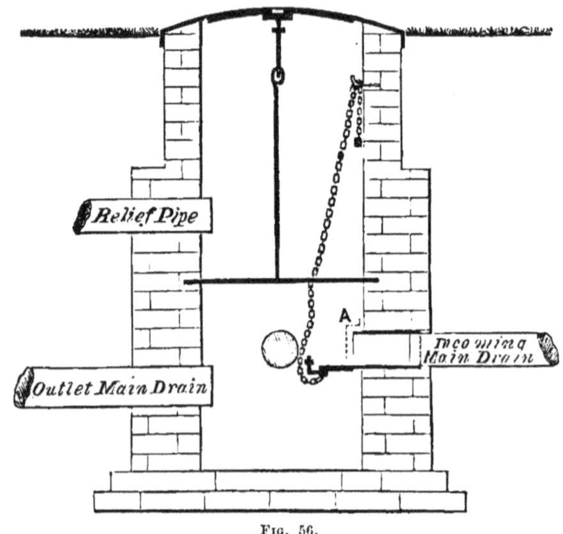

Fig. 56.

cessary, and may give rise to some inconvenience. The chain, which is attached to the flap covering the incoming drain, is operated from above. The object of this flap or valve is to prevent the water from flowing through the drain for any desirable length of time. The pent-up water, when released, rushes down with force, sufficient to carry down the sand and other impediments from the tiles above, also effecting a partial cleansing of the basin

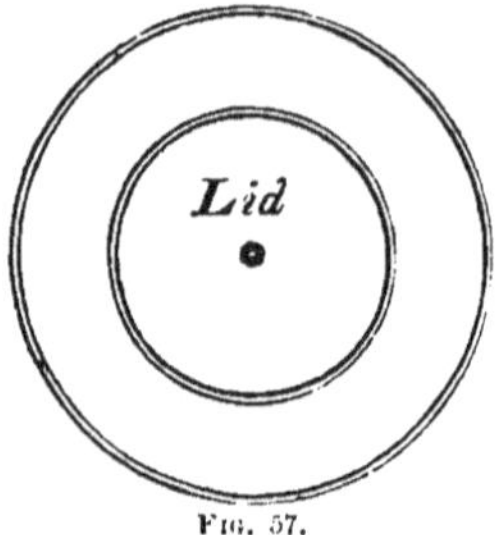

Fig. 57.

itself. The *lid*, Fig. 57, should be made of cast iron, and firmly fixed, to prevent displacement and accidents.

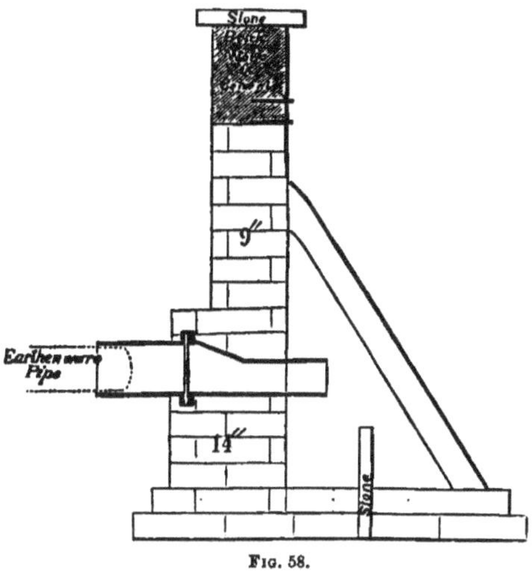

FIG. 58.

Large Outlet.—No portion of the whole drain requires to be more substantially constructed than the large outlet; and none is more likely to be neglected. The drains we expect to last a lifetime, and certainly the outlet, which is the foundation and abutment of the whole, should be built with the same expectation. We have before spoken of the outlets of the minor drains, where they are emptied into the open or main ditch. We have now to speak of a preferable plan, namely, where the minor drains are united, forming a sub-main, and of the outlet which this sub-main should have. On this subject Mr. Denton says:

"Too many outlets are objectionable, on account of the labor of their maintenance; too few are objectionable, because they can only exist where there are mains of excessive length. A limit of twenty acres to an outlet, resulting in an average of, perhaps, fourteen acres, will appear, by the practices of the best drainers, to be about the proper thing. If a shilling an acre is reserved for fixing the outlets, which should be *iron pipes, with swing gratings*, in masonry, very substantial work may be done."

We present, in Fig. 58, preceding page, a section of such outlet as has been found to answer its purpose effectually. It is composed of solid masonry, strongly braced. The exit pipe is of cast iron, projecting a few inches from the work. The exit should be some inches, or even a foot and a half, if that distance can be had, from the bottom of the main drain, both that the water may flow off readily, and that it may be protected from any backwaters ascending the main drain from the stream or pond in which it flows. It would be still better, if a fall could be given to the main drain before discharging its water into the creek or pond, thus preventing any backwater whatever.

CHAPTER X.

DRAINING TOOLS, INSTRUMENTS, ETC.

The instruments used in the construction of drains are simple and few in number. They consist, mainly, of shovels, such as are used for ordinary purposes, spades, scoops, and picks. In addition to these, a pipe-layer will be necessary, for narrow drains, and a drain gauge and level are very convenient, if not necessary.

Some of these tools are not made in this country, at present; they must either be imported, or some substitute obtained of an ingenious blacksmith.

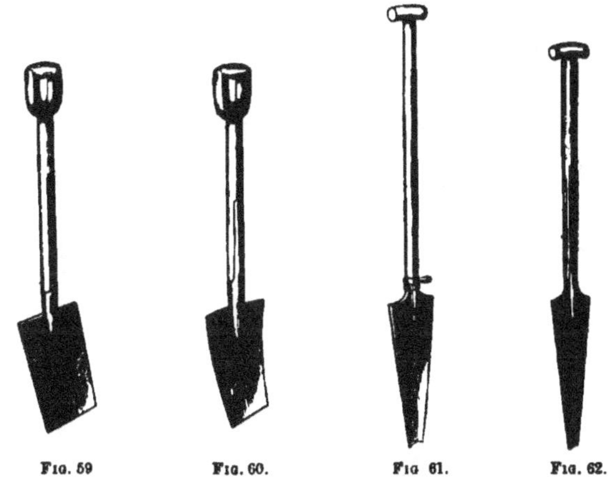

Fig. 59 Fig. 60. Fig 61. Fig. 62.

Shovels.—Ordinary shovels will be very useful in removing the earth, when the ditch is not less than one foot in width. They should be made of the best material and strongly braced by two slips of iron, extending some distance up the handle from the socket. The long-handled, pointed, scoop shovel, in common use on our railroads, will be found very useful in removing light soil or gravel, after being turned up with the pick.

Spades. — Three spades are all that are necessary. These should be of different sizes, gradually diminishing in width, to suit different depths. When the ground contains stones, or other impediments, they should be made perfectly flat, as in Figs. 59, 60, and 61 (preceding page). When the soil is free from all impediments, a curved form, represented in Fig. 63, will be found advantageous.

Morton, in the *Cyclopædia of Agriculture*, gives the spades, Figs. 61, 62, and 63, as those most in general use, for digging the last, or lowest portions of the drain.

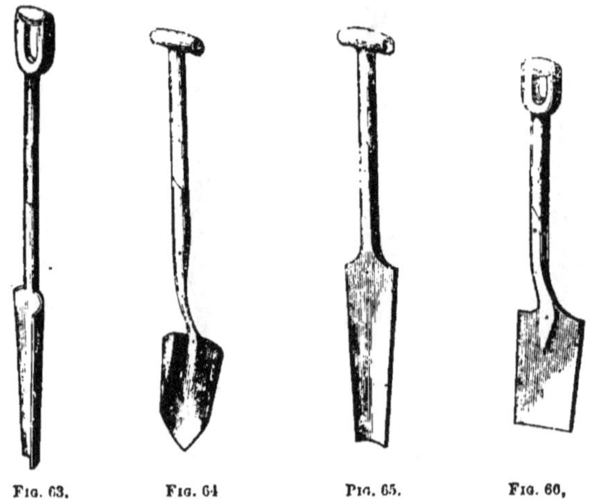

Fig. 63. Fig. 64 Fig. 65. Fig. 66.

Fig. 64 represents a broad and curved shovel, somewhat triangular in shape, with a bent handle. This is used for removing dirt from large drains.

Scoops.—For removing the soil from the bottom, and shaping out the ditch for the reception of the tile, scoops are necessary. For small and narrow ditches, different forms are used, as shown in Figs. 67 and 70. These are to be used standing on the surface of the ground. The instrument shown in Fig. 70, is especially adapted to fitting the bottom for round tiles or pipes.

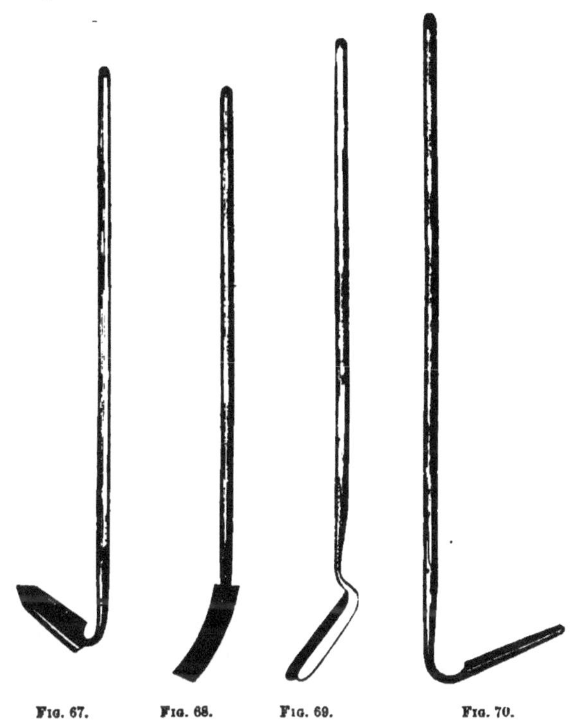

FIG. 67. FIG. 68. FIG. 69. FIG. 70.

When the ditches are made with flat bottoms, such a tool as represented by Fig. 68 is used for scooping it out. Where the bottom is soft, or the crumbs mixed with water, a similar tool, with the sides turned up, as represented by Fig. 69, is used to clean out the ditch.

Picks.—Where the subsoil is stony, or hard-pan, a pick will be necessary to loosen it. The dirt is then removed with the long scoop shovel. The common pick (Figs. 71

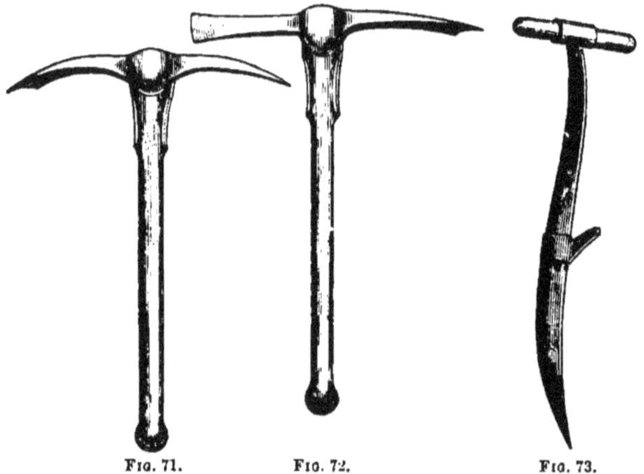

Fig. 71. Fig. 72. Fig. 73.

and 72) is all that is necessary for this purpose, though, in some cases, a foot pick (Fig. 73) may be advantageously used.

Pickaxes may be made either heavy or light, as suits the workman. They should be strongly made, and the usual form, with a pick at one end and chisel at the other, is best.

Pipe layer is a convenient tool; the handle is long and light, like that of a rake; from the end of this passes a stout piece of iron wire or rod, a foot in length, and having a direction almost at right angles with the handle. This is for the purpose of laying the tiles or pipes into the drain; and, if the drains are made as narrow as they ought to be, it will then be not only convenient but highly useful. It is better understood by reference to the cut (Fig. 74) than from description.

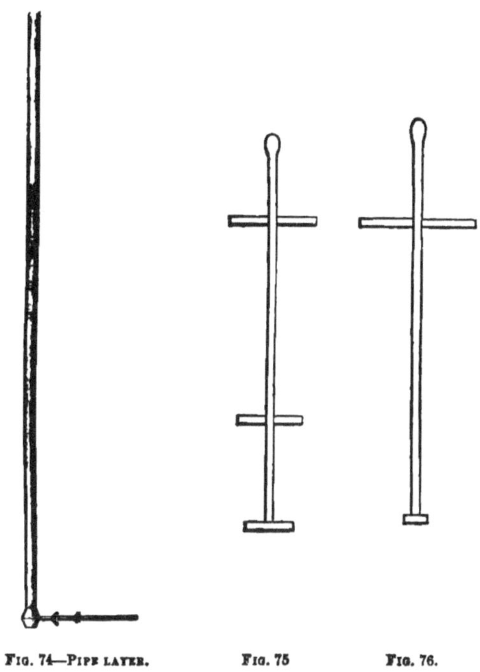

Fig. 74—Pipe layer. Fig. 75 Fig. 76.

Drain gauge.—This necessary though simple instrument is shown in Figs. 75 and 76. It should be strongly made, not liable to be altered, either by accident or design on the part of the workman. It may be constructed according to either figure, and shows both the depth of the drain and its width at top and bottom. If stones are used, it may be made to show the depth of filling.

A water level is the first instrument of which one who

has lands to drain should possess himself. This need not be an expensive article, for one of very simple construction will answer every purpose. Take a piece of lead pipe, two or three feet in length, and about half an inch in bore, bend up an inch or two at each end to a right angle; then take a small glass phial that will slip into the tube, break off the bottom, which may easily be done by making a crease round on the corner of a grindstone; then secure the phial in the tube with sealing wax; both ends are to be fixed alike. The level should be fastened to a small piece of wood, to give it stiffness and security. A nail, or screw, or peg, is put through the middle of the wood, just on one side of the lead pipe, to serve as a pivot in directing the instrument. For a tripod, three notches may be made in a little block of wood, and each leg secured by a nail, so as to make a movable joint; then bore a hole in the top of the block, to receive the pivot of the level. When about to be used, the level is filled with colored water, about half way up both phials, which are then corked, so that it may be carried about. When the level is put on the tripod, and as near right as can be guessed, the corks are removed, and the fluid in the phials stands at a water level. There is then no difficulty in obtaining accurate levels in any direction. Instead of the lead pipe, a glass tube may be substituted, and the ends bent up, after heating in a spirit lamp. Descriptions and plates of this water level are given in "*Thomas on Farm Implements*," "*Munn's Practical Drainer*," and in the "*Register of Rural Affairs*." It is much better to use a level in laying out all draining work than to depend on the best estimates otherwise obtained. It is not only desirable to know the lowest points of the field to be drained, and the highest, but also to know the exact difference in inches, in order to have the fall regular and uniform.

A span level is the best instrument for determining the exact fall in a drain that is being dug when no water is present. Three narrow strips of board are required, each about six feet in length; these are nailed together in the form of the letter A, the span or stretch being exactly half a rod. (See Fig. 77). From a nail or pin at the top a plummet is suspended. It is then placed, for the purpose of marking, upon a floor or piece of timber, which is perfectly level, and the place where the plumb line touches the cross bar marked; one foot is then raised one fourth of an inch, and the place where the line crosses the bar again marked, and will show a rise or fall one half inch to the rod. The foot is then raised to half an inch and the bar marked, indicating one inch to the rod. These markings can be made to any extent desired, and the instrument, by dropping it into the drain occasionally, will show that the drain is dug with uniform fall, and precisely that determined on at the outset.

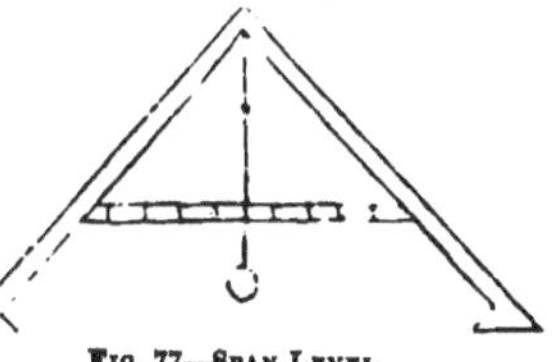

FIG. 77—SPAN LEVEL.

We have not aimed at prescribing a set of tools which are absolutely necessary, being too well acquainted with the genius of the western people, and knowing too well that they will make almost any kind of tool answer the purpose; but we deemed it necessary to give a general description of the tools employed by expert drainers.

CHAPTER XI.

DIGGING UNDERDRAINS.

AFTER proper levels have been taken, and the rate of fall ascertained, the digging may commence, the workman being kept straight by a line, as represented in Fig. 78.

FIG. 78.*

The dotted line represents the bottom of the drain; the dotted lines forming a triangle, or wedge-shape, represents a section of the ditch, as seen from the body of the ditch. Every three or four rods, two narrow boards, having a slit sawed in from the upper end, should be placed on a line with the center of the ditch. A line is then placed in the slit of the board, at the end of the ditch, and continued to the other board, supported by frames or braces resembling on iron square—these latter are placed at the side of the ditch, and the line suspended over the projecting arm, to keep it *taut*, or to prevent it from "*sagging.*" If the line is properly placed it will always enable the workman to ascertain whether the drain is of the proper depth, because the distance from the line to the bottom of the

* This cut is from French's work—but the plan has been adopted by ditchers in Ohio during the past twenty-five years.

drain must always be precisely the same, whether the surface of the ground is level or full of undulations.

Without some care, a ditch will not be dug straight even where a line is used, for in passing over swells or elevations, if the surface of the top is not removed enough wider to allow for the regular slope of the sides, the bottom will not be straight, or the sides will be too perdicular. To correct this latter difficulty, a draining gauge, Fig. 79 or 80, is employed. These gauges consist of an upright wooden strip, say, four feet in length, with a foot at the bottom, the precise width of the tile to be laid; and near the top a cross piece, the length of which is the exact width of the drain. Where great precision in the slope of the sides is required a central cross piece, as in Fig. 79, having for its length the exact width of the drain at that point, or rather a mean between the foot piece and upper cross piece.

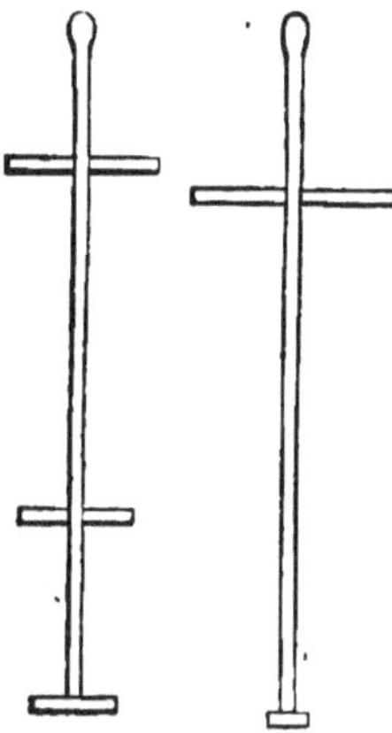

Fig. 79. Fig. 80.

The first *spit*, or spade depth of turf, or surface soil, is usually removed by a common spade; a stronger one being required than would be chosen for gardening purposes. The width of the drain, on the top, must always depend on the depth required; skillful drainers dig a much narrower drain than the unskilled. The narrowness of the drain is an advantage, there being less earth to throw out, and of course less to return. For a depth of three feet, one foot on top is abundantly wide, and many drains would not require so much. The crumbs are all shoveled out with a common shovel. It is usual, at this stage of the work, to bring the bottom of the drain to its true level, at least so far as to correct any noticeable unevenness of the surface. The span level before described must be used occasionally, unless water be present. Sometimes a turf of only a few inches is taken off before the first full spit is dug.

The second spit is dug with the narrower spade or proper draining tool, and the crumbs are removed by a draw scoop; or a long handled shovel, with the sides turned up, will answer very well. The removal of the second spit brings the drain to two feet in depth, and seven inches in width on the bottom, unless greater width and depth are required. The third and last spit of a three feet drain is cut with the same narrow spade as the second, or one still narrower. The bottom is made of the exact width of the tiles to be put in, and when these are less than four inches across outside, the tool must be narrower; or if it be required to cut a channel three inches wide on the bottom, with a tool four inches in width, this is readily done, where the tool is a little curved, by holding it obliquely, instead of transversely across the drain. The crumbs are removed, and the bottom fitted for the tiles with the draw scoop. The drainer never sets his foot on the bottom of a narrow drain; in fact, he could not get it there. Whatever the size of the tile used, that must be the width of the bottom of the drain; there should be just room to admit the tile, but not the least possibility of its getting out of place.

New beginners in digging drains, as a general thing, remove double the quantity of earth necessary to make the drain. This is an error, however, which generally corrects itself by practice. Some drainers prefer making the ditch, say 18 inches wide at the top, and give the sides *c*, Fig. 81, a gentle slope, until a depth of two feet is attained—leaving the bottom of the ditch, *b*, *b*, fourteen or fifteen inches wide. This part of the ditch may be made with the ordinary spade, Figs. 59, or 60. Then the narrow spade, Figs. 61, 62 or 65, is used to excavate the remaining foot of earth, *a;* this leaves the bottom, 2, 3, or 4 inches wide—according to the tool used—and just the size for the tile. When this style of ditching is adopted, the tools, Figs. 67 and 70 are used to clear the bottom of all pieces of ground which may have fallen in, as well as to remove any inequalities in the bottom. The tile is then taken up with the short arm of the pipe layer, Fig. 74, laid in the bottom of the ditch and properly adjusted.

FIG. 79.

Alderman Mechi, says:

" *On Digging a Drain.*—Before I proceed to describe my mode of digging, I will remark that a very great mistake is made by most drainers in removing more earth than is necessary. My men, for a 5-feet drain, only open the surface 18 inches wide, and at 4 feet they can do it in 12 to 14 inches; at 6 feet deep they allow themselves 22 inches; this is when the land is tolerably dry; when very wet and adhesive, they sometimes allow themselves an inch or two more, to prevent the earth touching their clothes. As they are paid by the piece, they are very particular not to remove a bit more earth than is absolutely necessary. In stony and hard soils, requiring the frequent use of the pickaxe, the workmen require rather a wider opening; but even so deep as 6 feet deep, it is seldom necessary to open 2 feet wide. It must always be borne in mind that the pipes can not be placed by the hand in such narrow drains, the bottom not being 2 inches wide. The drainers have a stick with a piece of iron like a long cock's spur, on which they place the pipe, and standing astride on the top of the opening, place the pipes abutting against each other in a continuous line, giving them a tap or two to set them firm in their places. Great care is required to scoop out all the crumbs, leaving the bottom of the drain smooth, with a sufficient fall. The bottom of the drain should not be wider, if possible, than the outside diameter of the pipe; it is thus kept firmly in its place. A common carpenter's level answers very well; but the workmen are generally sure to give fall enough to spare their labor in going too deep. We never plow out for the laborers. They stretch a garden line, so as to open their work straight and true. The ordinary spades are not at all calculated or proper for draining in tenacious soils. We use the patent grafting tools, made by Mr. Lyndon, of Birmingham; they are thin, well plated with steel, and ring like a bell, and will go easily into hard clays, when the common spades could not be used at all. They may be had of Mr. Lyndon direct, or ordered through the iron-mongers. The middle spits are removed by a narrow three quarter spade, with a projecting iron for the foot; and the lowest spit is taken out by a long 14-inch dagger-like spade, with two cutting edges, a sharp point, and an iron rest for the foot; this is worked edgewise first, and then removes a considerable thin, but broad deep mass. The scoop follows for the crumbs. All these tools may be had of Mr. Lyndon."

As digging ditches for drains is frequently done by contract, "*by the job*," or by the rod, we have deemed it proper to insert the following table, giving the number of cubic yards of earth to be removed in digging ditches:

CUBIC YARDS OF DIGGING IN DRAINS.

Depth 2 feet 6 inches.

Length.	Width 9 inches.		Width 10 inches.		Width 11 inches.		Width 1 foot.		Width 1 foot 6 inches.		Width 2 feet.		Width 2 feet 6 inches.		Width 3 feet.	
Yards.	Yds.	ft.	Yds.	ft.	Yds.	ft.	Yds.	ft.	Yds.	ft.	Yds.	ft.	Yds.	ft.	Yds.	ft.
½		3		3		3		4		6		7		9		11
1		6		6		7		7		11		15		19		22
2		11		12		14		15		22	1	3	1	10	1	18
3		17		19		21		22	1	7	1	18	2	2	2	13
4		22		25	1		1	3	1	18	2	6	2	21	3	9
5	1	1	1	4	1	7	1	10	2	2	2	21	3	13	4	4
6	1	7	1	10	1	14	1	18	2	13	3	9	4	4	5	
7	1	12	1	17	1	21	1	25	2	25	3	24	4	23	5	22
8	1	18	1	23	2	1	2	6	3	9	4	12	5	15	6	18
9	1	24	2	2	2	8	2	13	3	20	5		6	7	7	13
10	2	2	2	8	2	15	2	21	4	4	5	15	6	25	8	9
11	2	8	2	15	2	22	3	1	4	16	6	3	7	17	9	4
12	2	13	2	21	3	1	3	9	5		6	18	8	9	10	
13	2	19	3		3	8	3	16	5	11	7	6	9	1	10	22
14	2	25	3	6	3	15	3	24	5	22	7	21	9	19	11	18
15	3	3	3	13	3	22	4	4	6	7	8	9	10	11	12	13
25	5	6	5	21	6	10	6	25	10	11	13	24	17	10	20	22
40	8	9	9	7	10	5	11	3	16	18	22	6	27	21	33	9
55	11	12	12	20	14		15	7	22	25	30	15	38	5	45	22
70	14	16	16	5	17	22	19	12	29	4	38	24	48	16	58	9
85	17	19	19	18	21	17	23	16	35	11	47	6	59	1	70	22
100	20	22	23	4	25	12	27	21	41	18	55	15	69	12	83	9
200	41	18	46	8	50	25	55	15	83	9	111	3	138	24	166	18
500	104	4	115	20	127	8	138	24	208	9	277	21	347	6	416	18
1000	208	9	321	13	354	17	277	21	416	18	555	15	694	12	833	9

Depth 2 feet 9 inches.

Length.	Width 9 inches.		Width 10 inches.		Width 11 inches.		Width 1 foot.		Width 1 foot 6 inches.		Width 2 feet.		Width 2 feet 6 inches.		Width 3 feet.	
Yards.	Yds.	ft.	Yds.	ft.	Yds.	ft.	Yds.	ft.	Yds.	ft.	Yds.	ft.	Yds.	ft.	Yds.	ft.
½		3		3		4		4		6		8		10		12
1		6		7		8		8		12		16		21		25
2		12		14		15		16		25	1	6	1	14	1	25
3		19		21		23		25	1	10	1	22	2	8	2	20
4		25	1		1	3	1	6	1	22	2	12	3	1	3	18
5	1	4	1	7	1	11	1	14	2	8	3	1	3	22	4	16
6	1	10	1	14	1	18	1	22	2	20	3	18	4	16	5	13
7	1	16	1	21	1	26	2	4	3	6	4	7	5	9	6	11
8	1	22	2	1	2	6	2	12	3	18	4	24	6	3	7	9
9	2	2	2	8	2	14	2	20	4	3	5	13	6	24	8	7
10	2	8	2	15	2	22	3	1	4	16	6	3	7	17	9	4
11	2	14	2	22	3	2	3	10	5	1	6	19	8	11	10	2
12	2	20	3	1	3	10	3	18	5	13	7	9	9	4	11	
13	2	26	3	8	3	17	3	26	5	26	7	25	9	25	11	25
14	3	6	3	15	3	25	4	7	6	11	8	15	10	19	12	22
15	3	12	3	22	4	5	4	16	6	24	9	4	11	12	13	20
25	5	20	6	10	7		7	17	11	12	15	7	19	3	22	25
40	9	4	10	5	11	5	12	6	18	9	24	12	30	15	36	18
55	12	16	14		15	11	16	22	25	6	33	16	42		50	11
70	16	1	17	22	19	16	21	10	32	2	42	21	53	13	64	4
85	19	13	21	17	23	22	25	26	38	26	51	25	64	25	77	25
100	22	25	25	12	28		30	15	45	22	61	3	76	10	91	18
200	45	22	50	25	56		61	3	91	18	122	6	152	21	183	9
500	114	16	127	8	140	1	152	21	229	4	305	15	381	25	458	9
1000	229	4	254	17	280	2	305	15	458	9	611	3	763	24	916	18

CUBIC YARDS OF DIGGING IN DRAINS.

Depth 3 feet.

Length.	Width 9 inches.		Width 10 inches.		Width 11 inches.		Width 1 foot.		Width 1 foot 6 inches.		Width 2 feet.		Width 2 feet 6 inches.		Width 3 feet.	
Yards.	Yds.	ft.	Yds.	ft.	Yds.	ft.	Yds.	ft.	Yds.	ft.	Yds.	ft.	Yds.	ft.	Yds.	ft.
½		3		4		4		4		7		9		11		13
1		7		7		8		9		13		18		22	1	
2		13		15		16		18	1		1	9	1	18	2	
3		20		22		25	1		1	13	2		2	13	3	
4	1		1	3	1	6	1	9	2		2	18	3	9	4	
5	1	7	1	10	1	14	1	18	2	13	3	9	4	4	5	
6	1	13	1	18	1	22	2		3		4		5		6	
7	1	20	1	26	2	4	2	9	3	13	4	18	5	22	7	
8	2		2	6	2	12	2	18	4		5	9	6	18	8	
9	2	7	2	13	2	20	3		4	13	6		7	13	9	
10	2	13	2	21	3	1	3	9	5		6	18	8	9	10	
11	2	20	3	1	3	10	3	18	5	13	7	9	9	4	11	
12	3		3	9	3	18	4		6		8		10		12	
13	3	7	3	16	3	26	4	9	6	13	8	18	10	22	13	
14	3	13	3	24	4	7	4	18	7		9	9	11	18	14	
15	3	20	4	4	4	16	5		7	13	10		13	13	15	
25	6	7	6	25	7	18	8	9	12	13	16	18	20	22	25	
40	10		11	3	12	6	13	9	20		26	18	33	9	40	
55	13	20	15	7	16	22	18	9	27	13	36	18	42	22	55	
70	17	13	19	12	21	10	23	9	35		46	18	58	9	70	
85	21	7	23	16	25	26	29	9	42	13	56	18	70	22	85	
100	25		27	21	30	15	33	9	50		66	18	83	9	100	
200	50		55	15	61	3	66	18	100		133	9	166	18	200	
500	125		138	24	152	21	166	18	250		333	9	416	18	500	
1000	250		277	21	305	15	333	9	500		666	18	833	9	1000	

Depth 3 feet 3 inches.

Length.	Width 9 inches.		Width 10 inches.		Width 11 inches.		Width 1 foot.		Width 1 foot 6 inches.		Width 2 feet.		Width 2 feet 6 inches.		Width 3 feet.	
Yards.	Yds.	ft.	Yds.	ft.	Yds.	ft.	Yds.	ft.	Yds.	ft.	Yds.	ft.	Yds.	ft.	Yds.	ft.
½		4		4		4		5		7		10		12		15
1		7		8		9		10		15		19		24	1	2
2		15		16		18		19	1	2	1	12	1	22	2	4
3		22		24	1		1	2	1	17	2	4	2	19	3	7
4	1	2	1	5	1	9	1	12	2	4	2	24	3	16	4	9
5	1	10	1	14	1	18	1	22	2	19	3	16	4	14	5	11
6	1	17	1	22	2		2	4	3	7	4	9	5	11	6	13
7	1	24	2	3	2	9	2	14	3	21	5	1	6	9	7	16
8	2	4	2	11	2	17	2	24	4	9	5	21	7	6	8	18
9	2	12	2	19	2	26	3	7	4	24	6	13	8	3	9	20
10	2	19	3		3	8	3	16	5	11	7	6	9	1	10	22
11	2	26	3	8	3	17	3	26	5	26	7	25	9	25	11	25
12	3	7	3	16	3	26	4	9	6	13	8	18	10	22	13	
13	3	14	3	25	4	8	4	19	7	1	9	10	11	20	14	2
14	3	21	4	6	4	17	5	1	7	16	10	3	12	17	15	4
15	4	2	4	14	4	26	5	11	8	3	10	22	13	15	16	7
25	6	21	7	14	8	7	9	1	13	15	18	1	22	15	27	2
40	10	22	12	1	13	6	14	12	21	18	28	24	36	3	43	9
55	14	24	16	15	18	6	19	23	29	21	39	19	49	18	59	16
70	18	26	21	2	23	5	25	7	37	25	51	15	63	5	75	22
85	23	1	25	16	28	4	30	19	46	1	61	10	76	20	92	2
100	27	2	30	2	33	3	36	3	54	4	72	6	90	7	108	9
200	54	4	60	5	66	5	72	6	108	9	144	12	180	15	216	18
500	135	11	150	12	165	14	180	15	270	22	361	3	451	10	541	18
1000	270	22	300	25	331		361	3	541	18	722	6	902	21	1083	9

CUBIC YARDS OF DIGGING IN DRAINS.

Depth 3 feet 6 inches.

Length.	Width 9 inches.		Width 10 inches.		Width 11 inches.		Width 1 foot.		Width 1 foot 6 inches.		Width 2 feet.		Width 2 feet 6 inches.		Width 3 feet.	
Yards.	Yds.	ft.	Yds.	ft.	Yds.	ft.	Yds.	ft.	Yds.	ft.	Yds.	ft.	Yds.	ft.	Yds.	ft.
½		4		4		5		5		8		10		13		16
1		8		9		10		10		16		21		26	1	4
2		16		17		19		21	1	4	1	15	1	25	2	9
3		24		26	1	2	1	4	1	20	2	9	2	25	3	13
4	1	4	1	8	1	11	1	15	2	9	3	3	3	24	4	18
5	1	12	1	17	1	21	1	25	2	25	3	24	4	23	5	22
6	1	20	1	25	2	4	2	9	3	13	4	18	5	22	7	
7	2	1	2	7	2	13	2	19	4	2	5	12	6	22	8	4
8	2	9	2	16	2	23	3	3	4	18	6	6	7	21	9	9
9	2	17	2	25	3	6	3	13	5	7	7		8	20	10	13
10	2	25	3	6	3	15	3	24	5	22	7	21	9	19	11	18
11	3	6	3	15	3	25	4	7	6	11	8	15	10	19	12	22
12	3	13	3	24	4	7	4	18	7		9	9	11	18	14	
13	3	21	4	6	4	17	5	1	7	16	10	3	12	17	15	4
14	4	2	4	14	5		5	12	8	4	10	24	13	16	16	9
15	4	10	4	23	5	9	5	22	8	20	11	18	14	16	17	13
25	7	8	8	3	8	25	9	19	14	16	19	12	24	8	29	4
40	11	18	12	26	14	7	15	15	23	9	31	3	38	24	46	18
55	16	1	17	22	19	16	21	10	32	2	42	21	53	13	64	4
70	20	11	22	18	24	26	27	6	40	22	54	12	68	1	81	18
85	24	21	27	15	30	8	33	1	49	16	66	3	82	17	99	4
100	29	4	32	11	35	17	38	24	58	9	77	21	97	6	116	18
200	58	9	64	22	71	8	77	21	116	18	155	15	194	12	233	0
500	145	22	162	1	178	6	194	12	291	18	388	24	486	3	583	0
1000	291	18	324	2	356	13	388	24	583	0	777	21	972	6	1166	18

Depth 3 feet 9 inches.

Length.	Width 9 inches.		Width 10 inches.		Width 11 inches.		Width 1 foot.		Width 1 foot 6 inches.		Width 2 feet.		Width 2 feet 6 inches.		Width 3 feet.	
Yards.	Yds.	ft.	Yds.	ft.	Yds.	ft.	Yds.	ft.	Yds.	ft.	Yds.	ft.	Yds.	ft.	Yds.	ft.
½		4		5		5		6		8		11		14		17
1		8		9		10		11		17		22	1	1	1	7
2		17		19		21		22	1	7	1	18	2	2	2	13
3		25	1	1	1	4	1	7	1	24	2	13	3	3	3	20
4	1	7	1	10	1	14	1	18	2	13	3	9	4	4	5	
5	1	15	1	20	1	25	2	2	3	3	4	4	5	6	6	7
6	1	24	2	2	2	8	2	13	3	20	5		6	7	7	13
7	2	5	2	12	2	18	2	25	4	10	5	22	7	8	8	20
8	2	13	2	21	3	1	3	9	5		6	18	8	9	10	
9	2	22	3	3	3	12	3	20	5	17	7	13	9	10	11	7
10	3	3	3	13	3	22	4	4	6	7	8	9	10	11	12	13
11	3	12	3	22	4	5	4	16	6	24	9	4	11	12	13	20
12	3	20	4	4	4	16	5		7	13	10		12	13	15	
13	4	2	4	14	4	26	5	11	8	3	10	22	13	15	16	7
14	4	10	4	23	5	9	5	22	8	20	11	18	14	16	17	13
15	4	19	5	6	5	20	6	7	9	10	12	13	15	17	18	20
25	7	22	8	18	9	15	10	11	15	17	20	22	26	1	31	7
40	12	13	13	24	15	7	16	18	25		33	9	41	18	50	
55	17	5	19	3	21		22	25	34	10	45	22	57	8	68	20
70	21	24	24	8	26	20	29	4	43	20	58	9	72	25	87	13
85	26	15	29	14	32	13	35	11	53	3	70	22	88	15	106	7
100	31	7	34	19	38	5	41	18	62	13	83	9	104	4	125	
200	62	13	69	12	76	10	83	9	125		166	18	208	9	250	
500	156	7	173	16	190	26	208	9	312	13	416	18	520	22	625	
1000	313	31	347	6	381	25	416	18	625		833	9	1041	18	1250	

CUBIC YARDS OF DIGGING IN DRAINS.

Depth 4 feet.

Length.	Width 9 inches.		Width 10 inches.		Width 11 inches.		Width 1 foot.		Width 1 foot 6 inches.		Width 2 feet.		Width 2 feet 6 inches.		Width 3 feet.	
Yards.	Yds.	ft.	Yds.	ft.	Yds.	ft.	Yds.	ft.	Yds.	ft.	Yds.	ft.	Yds.	ft.	Yds.	ft.
½		4		5		5		6		9		12		15		18
1		9		10		11		12		18		24	1	3	1	9
2		18		20		22		24	1	9	1	21	2	6	2	18
3	1		1	3	1	6	1	9	2		2	18	3	9	4	
4	1	9	1	13	1	17	1	21	2	18	3	15	4	12	5	9
5	1	18	1	23	2	1	2	6	3	9	4	12	5	15	6	18
6	2		2	6	2	12	2	18	4		5	9	6	18	8	
7	2	9	2	16	2	23	3	3	4	18	6	6	7	21	9	9
8	2	18	2	26	3	7	3	15	5	9	7	3	8	24	10	18
9	3		3	9	3	18	4		6		8		10		12	
10	3	9	3	19	4	2	4	12	6	18	8	24	11	3	13	9
11	3	18	4	2	4	13	4	24	7	9	9	21	12	6	14	18
12	4		4	12	4	24	5	9	8		10	18	13	9	16	
13	4	9	4	22	5	8	5	21	8	18	11	15	14	12	17	9
14	4	18	5	5	5	19	6	6	9	9	12	12	15	15	18	18
15	5		5	15	6	3	6	18	10		13	9	16	18	20	
25	8	9	9	7	10	5	11	3	16	18	22	6	27	21	33	9
40	13	9	14	22	16	8	17	21	26	18	35	15	44	12	53	9
55	18	9	20	10	22	11	24	12	36	18	48	24	61	3	73	9
70	23	9	25	25	28	14	31	3	46	18	62	6	77	21	93	9
85	28	9	31	13	34	17	37	21	56	18	75	15	94	12	113	9
100	33	9	37	1	40	20	44	12	66	18	88	24	111	3	133	9
200	66	18	74	2	81	13	88	24	133	9	177	21	222	6	266	18
500	166	18	185	5	203	19	222	6	333	9	444	12	555	15	666	18
1000	333	9	370	10	407	11	444	12	666	18	888	24	1111	3	1333	9

Depth 4 feet 6 inches.

Length.	Width 9 inches.		Width 10 inches.		Width 11 inches.		Width 1 foot.		Width 1 foot 6 inches.		Width 2 feet.		Width 2 feet 6 inches.		Width 3 feet.	
Yards.	Yds.	ft.	Yds.	ft.	Yds.	ft.	Yds.	ft.	Yds.	ft.	Yds.	ft.	Yds.	ft.	Yds.	ft.
½		5		6		6		7		10		13		17		20
1		10		11		12		13		20	1		1	7	1	13
2		20		22		25	1		1	13	2		2	13	3	
3	1	3	1	7	1	10	1	13	2	7	3		3	20	4	13
4	1	13	1	18	1	22	2		3		4		5		6	
5	1	24	2	2	2	8	2	13	3	20	5		6	7	7	13
6	2	7	2	13	2	20	3		4	13	6		7	13	9	
7	2	17	2	25	3	6	3	13	5	7	7		8	20	10	13
8	3		3	9	3	18	4		6		8		10		12	
9	3	10	3	20	4	3	4	13	6	20	9		11	7	13	13
10	3	20	4	4	4	16	5		7	13	10		12	13	15	
11	4	3	4	16	5	1	5	13	8	7	11		13	20	16	13
12	4	13	5		5	13	6		9		12		15		18	
13	4	24	5	11	5	26	6	13	9	20	13		16	7	19	13
14	5	7	5	22	6	11	7		10	13	14		17	13	21	
15	5	17	6	7	6	24	7	13	11	7	15		18	20	22	13
25	9	10	10	11	11	12	12	13	18	20	25		31	7	37	13
40	15		16	18	18	9	20		30		40		50		60	
55	20	17	22	25	25	6	27	13	41	7	55		68	20	82	13
70	26	7	29	4	32	2	35		52	13	70		87	13	105	
85	31	24	35	11	38	26	42	13	63	20	85		106	7	127	13
100	37	13	41	18	45	22	50		75		100		125		150	
200	75		83	9	91	18	100		150		200		250		300	
500	187	13	208	9	229	4	250		375		500		626		750	
1000	375		416	18	458	9	500		750		1000		1250		1500	

CUBIC YARDS OF DIGGING IN DRAINS.

Depth 5 feet.

Length.	Width 9 inches.		Width 10 inches.		Width 11 inches.		Width 1 foot.		Width 1 foot 6 inches.		Width 2 feet.		Width 2 feet 6 inches.		Width 3 feet.	
Yards.	Yds.	ft.	Yds.	ft.	Yds.	ft.	Yds.	ft.	Yds.	ft.	Yds.	ft.	Yds.	ft.	Yds.	ft.
½		6		6		7		7		11		15		19		22
1		11		12		14		15		22	1	3	1	10	1	18
2		22		25	1		1	3	1	18	2	6	2	21	3	9
3	1	7	1	10	1	14	1	18	2	13	3	9	4	4	5	
4	1	18	1	23	2	1	2	6	3	9	4	12	5	15	6	18
5	2	2	2	8	2	15	2	21	4	4	5	15	6	25	8	9
6	2	13	2	21	3	1	3	9	5		6	18	8	9	10	
7	2	25	3	6	3	15	3	24	5	22	7	21	9	19	11	18
8	3	9	3	19	4	2	4	12	6	18	8	24	11	3	13	9
9	3	20	4	4	4	16	5		7	13	10		12	13	15	
10	4	4	4	17	5	2	5	15	8	9	11	3	13	24	16	18
11	4	16	5	2	5	16	6	3	9	4	12	6	15	7	18	9
12	5		5	15	6	3	6	18	10		13	9	16	18	20	
13	5	11	6		6	17	7	6	10	22	14	12	18	1	21	18
14	5	22	6	13	7	3	7	21	11	18	15	15	19	12	23	9
15	6	7	6	25	7	17	8	9	12	13	16	18	20	22	25	
25	10	11	11	15	12	20	13	24	20	22	27	21	34	19	41	18
40	16	18	18	14	20	10	22	6	33	9	44	12	55	15	66	18
55	22	25	25	12	28		30	15	45	22	61	3	76	10	91	18
70	29	4	32	11	35	17	38	24	58	9	77	21	97	6	116	18
85	35	11	39	9	43	8	47	6	70	22	94	12	118	1	141	18
100	41	18	46	8	50	25	55	15	83	9	111	3	138	24	166	18
200	83	9	92	16	101	23	111	3	166	18	222	6	277	21	333	9
500	208	9	231	13	254	17	277	21	416	18	555	15	694	12	833	9
1000	416	18	462	26	509	7	555	15	833	9	1111	3	1388	24	1666	18

Depth 5 feet 6 inches.

Length.	Width 9 inches.		Width 10 inches.		Width 11 inches.		Width 1 foot.		Width 1 foot 6 inches.		Width 2 feet.		Width 2 feet 6 inches.		Width 3 feet.	
Yards.	Yds.	ft.	Yds.	ft.	Yds.	ft.	Yds.	ft.	Yds.	ft.	Yds.	ft.	Yds.	ft.	Yds.	ft.
½		6		7		8		8		12		16		21		25
1		12		14		15		16		25	1	6	1	14	1	22
2		25	1		1	3	1	6	1	22	2	12	3	1	3	18
3	1	10	1	14	1	18	1	22	2	20	3	18	4	16	5	13
4	1	22	2	1	2	6	2	12	3	18	4	24	6	3	7	9
5	2	8	2	15	2	22	3	1	4	16	6	3	7	17	9	4
6	2	20	3	1	3	10	3	18	5	13	7	9	9	4	11	
7	3	6	3	15	3	25	4	7	6	11	8	15	10	19	12	22
8	3	18	4	2	4	13	4	24	7	9	9	21	12	6	14	18
9	4	3	4	16	5	1	5	13	8	7	11		13	20	16	13
10	4	16	5	2	5	16	6	3	9	4	12	6	15	7	18	9
11	5	1	5	16	6	4	6	19	10	2	13	12	16	22	20	4
12	5	13	6	3	6	19	7	9	11		14	18	18	9	22	
13	5	26	6	17	7	8	7	25	11	25	15	24	19	23	23	22
14	6	11	7	3	7	23	8	15	12	22	17	3	21	10	25	18
15	6	24	7	17	8	11	9	4	13	20	18	9	22	25	27	13
25	11	12	12	20	14		15	7	22	25	30	15	38	5	45	22
40	18	9	20	10	22	11	24	12	36	18	48	24	61	3	73	9
55	25	6	28		30	22	33	16	50	11	67	6	84	1	100	22
70	32	2	35	17	39	6	42	21	64	4	85	15	106	25	128	9
85	38	26	43	8	47	17	51	25	77	25	103	24	129	23	155	22
100	45	22	50	25	56		61	3	91	18	122	6	152	21	183	9
200	91	18	101	23	112	1	122	6	183	9	244	12	305	15	366	18
500	229	4	254	17	280	2	305	15	458	9	611	3	763	24	916	18
1000	458	9	509	7	560	5	611	3	916	18	1222	6	1527	21	1833	9

CUBIC YARDS OF DIGGING IN DRAINS.

Depth 6 feet.

Length.	Width 9 inches.		Width 10 inches.		Width 11 inches.		Width 1 foot.		Width 1 foot 6 inches.		Width 2 feet.		Width 2 feet 6 inches.		Width 3 feet.	
Yards.	Yds.	ft.	Yds.	ft.	Yds.	ft.	Yds.	ft.	Yds.	ft.	Yds.	ft.	Yds.	ft.	Yds.	ft.
½		7		7		8		9		13		18		22	1	
1		13		15		16		18	1		1	9	1	18	2	
2	1		1	3	1	6	1	9	2		2	18	3	9	4	
3	1	13	1	18	1	22	2		3		4		5		6	
4	2		2	6	2	12	2	18	4		5	9	6	18	8	
5	2	13	2	21	3	1	3	9	5		6	18	8	9	10	
6	3		3	9	3	18	4		6		8		10		12	
7	3	13	3	24	4	7	4	18	7		9	9	11	18	14	
8	4		4	12	4	24	5	9	8		10	18	13	9	16	
9	4	13	5		5	13	6		9		12		15		18	
10	5		5	15	6	3	6	18	10		13	9	16	18	20	
11	5	13	6	3	6	19	7	9	11		14	18	18	9	22	
12	6		6	18	7	9	8		12		16		20		24	
13	6	13	7	6	7	25	8	18	13		17	9	21	18	26	
14	7		7	21	8	15	9	9	14		18	18	23	0	28	
15	7	13	8	9	9	4	10		15		20		25		30	
25	12	13	13	24	15	7	16	18	25		33	9	41	18	50	
40	20		22	6	24	12	26	18	40		53	9	66	18	80	
55	27	13	30	15	33	16	36	18	55		73	9	91	18	110	
70	35		38	24	42	21	46	18	70		93	9	116	18	140	
85	42	13	47	6	51	25	56	18	85		113	9	141	18	170	
100	50		55	15	61	3	66	18	100		133	9	166	18	200	
200	100		111	3	122	6	133	9	200		266	18	333	9	400	
500	250		277	21	305	15	333	9	500		666	18	833	9	1000	
1000	500		555	15	611	3	666	18	1000		1333	9	1666	18	2000	

Depth 6 feet 6 inches.

Length.	Width 9 inches.		Width 10 inches.		Width 11 inches.		Width 1 foot.		Width 1 foot 6 inches.		Width 2 feet.		Width 2 feet 6 inches.		Width 3 feet.	
Yards.	Yds.	ft.	Yds.	ft.	Yds.	ft.	Yds.	ft.	Yds.	ft.	Yds.	ft.	Yds.	ft.	Yds.	ft.
½		7		8		9		10		15		19		24	1	2
1		15		16		18		19	1	2	1	12	1	22	2	4
2	1	2	1	5	1	9	1	12	2	4	2	24	3	16	4	9
3	1	17	1	22	2		2	4	3	7	4	9	5	11	6	13
4	2	4	2	11	2	17	2	24	4	9	5	21	7	6	8	18
5	2	19	3		3	8	3	16	5	11	7	6	9	1	10	22
6	3	7	3	16	3	26	4	9	6	13	8	18	10	22	13	
7	3	21	4	6	4	17	5	1	7	16	10	3	12	17	15	4
8	4	9	4	22	5	18	5	21	8	18	11	15	14	12	17	9
9	4	24	5	11	5	26	6	13	9	20	13		16	7	19	13
10	5	11	6		6	17	7	6	10	22	14	12	18	1	21	18
11	5	20	6	17	7	8	7	25	11	25	15	24	19	23	23	22
12	6	13	7	6	7	25	8	18	13		17	9	21	18	26	
13	7	1	7	22	8	16	9	10	14	2	18	21	23	13	28	4
14	7	16	8	11	9	7	10	3	15	4	20	6	25	7	30	9
15	8	3	9	1	9	25	10	22	16	7	21	18	27	2	32	13
25	13	15	15	1	16	15	18	1	27	2	36	3	45	4	54	4
40	21	18	24	2	26	13	28	24	43	9	57	21	72	6	86	18
55	29	21	33	3	36	11	39	19	59	16	79	12	99	8	119	4
70	37	25	42	3	46	9	50	15	75	22	101	3	126	10	151	18
85	40	1	51	4	56	7	61	10	92	2	122	21	153	13	184	4
100	54	4	60	5	66	5	72	6	108	9	144	12	180	15	216	18
200	108	9	120	10	132	11	144	12	216	18	288	24	361	3	433	9
500	270	22	300	25	331		361	3	541	18	722	6	902	21	1083	9
1000	541	18	601	23	662	1	722	6	1083	9	1444	12	1805	15	2166	18

CUBIC YARDS OF DIGGING IN DRAINS.

Depth 7 feet.

Length.	Width 9 inches.		Width 10 inches.		Width 11 inches.		Width 1 foot.		Width 1 foot 6 inches.		Width 2 feet.		Width 2 feet 6 inches.		Width 3 feet.	
Yards.	Yds.	ft.	Yds.	ft.	Yds.	ft.	Yds.	ft.	Yds.	ft.	Yds.	ft.	Yds.	ft.	Yds.	ft.
½		8		9		10		10		16		21		26	1	4
1		16		17		19		21	1	4	1	15	1	25	2	9
2	1	4	1	8	1	11	1	15	2	9	3	3	3	24	4	18
3	1	20	1	25	2	4	2	9	3	13	4	18	5	22	7	
4	2	9	2	16	2	23	3	3	4	18	6	6	7	21	9	9
5	2	25	3	6	3	15	3	24	5	22	7	21	9	19	11	18
6	3	13	3	24	4	7	4	18	7		9	9	11	18	14	
7	4	2	4	14	5		5	12	8	4	10	24	13	16	16	9
8	4	18	5	5	5	19	6	6	9	9	12	12	15	15	18	18
9	5	7	5	22	6	11	7		10	13	14		17	13	21	
10	5	22	6	13	7	3	7	21	11	18	15	15	19	12	23	9
11	6	11	7	3	7	23	8	15	12	22	17	3	21	10	25	18
12	7		7	21	8	15	9	9	14		18	18	23	9	28	
13	7	16	8	11	9	7	10	3	15	4	20	6	25	7	30	9
14	8	4	9	2	9	26	10	24	16	9	21	21	27	6	32	18
15	8	20	9	19	10	19	11	18	17	13	23	9	29	4	35	
25	14	16	16	5	17	22	19	12	29	4	38	24	48	16	58	9
40	23	9	25	25	28	14	31	3	46	18	62	6	77	21	93	9
55	32	2	35	17	39	6	42	21	64	4	85	15	106	25	128	9
70	40	22	45	10	49	24	54	12	81	18	108	24	136	3	163	9
85	49	16	55	2	60	16	66	3	99	4	132	6	165	7	198	9
100	58	9	64	22	71	8	77	21	116	18	155	15	194	12	233	9
200	116	18	129	17	142	16	155	15	233	9	311	3	388	24	466	18
500	291	18	324	2	356	13	388	24	583	9	777	21	972	6	1166	18
1000	583	9	648	4	712	26	777	21	1166	18	1155	15	1944	12	2333	9

For example, it is required to know how many cubic yards of earth are to be removed in making 100 yards of drain, 3 feet deep; the top width being 18 inches and the bottom 4 inches.

$$\begin{array}{r} 18 \\ 4 \\ \hline 2\,)\,22 \\ \hline 11 \end{array}$$

11 inches, mean width.

Find, under the head of 3 feet deep, in the column *length*, the number of yards; then under the column of 11 inches wide, opposite the number of yards, will be found the number of cubic yards in the proposed drain.

Many machines have been invented for the purpose of digging ditches, thus not only making them in a much shorter period of time, but much cheaper. We believe ditching, or drain plows (not *mole* plows) have been exhibited at every fair held by the Ohio State Board of Agriculture since 1856; but we do not remember of having seen a single one, among the entire lot, which we could recommend to the farmers as being an implement with which they would not be disappointed.

Messrs. Pratt, of Canandaigua, N. Y., have invented a ditching machine, which is highly recommended by the "*Rural New Yorker;*" but parties who have witnessed its operation are not so favorably impressed with it. Mr. B. B. Briggs, of Sharon, Medina county, Ohio, in 1859, invented a machine, which looks not very unlike a mole plow, to lay tile without digging a ditch. The following is Mr. Briggs' own account of the working capability of the machine. While we have no disposition to deny that this machine will do precisely what Mr. Briggs claims for it, we must, at the same time, be permitted to state that we would not recommend any one to undertake to underdrain any considerable quantity of land in this manner; because it is a matter of *impossibility* to have the tile as firmly and as correctly laid as if done by hand.

"My mode of taking round tile (which are considered the best by those most experienced) is this:—Make an excavation at the heel of the mole, with a gentle inclination backward; then fasten the first section of rope to the heel of the mole; then string said section with tile to within about four feet of the mole, and secure it by means of a set of clutches, and the hook of the succeeding section; each section to be about twenty-five feet long, though I have taken in upward of fifty feet to each set of clutches. The machine can then be moving forward, as the next section is being strung and secured as before, and so continue to do, until so much is taken in as the strength of the rope will justify, say four hundred or more feet. Next, dig down again at the heel of the mole (making the excavation as before), and detach the rope; then draw it out by hand from the place of entrance, and again proceed as before. The spaces left at the excavation can be filled in by hand, and the joints of tile set close together, by prying from either end, as one can move four hundred or more feet with an iron bar.

"This machine can be used in all places where the mole plow can, and for laying almost any kind of tile; for the arch and sole, or horseshoe, the clutches are made fast to one continuous rope, at such intervals as may be thought necessary. The main, or one into which the smaller ditches empty, should of course be larger, and put in by hand, as they could not be joined by the machine."

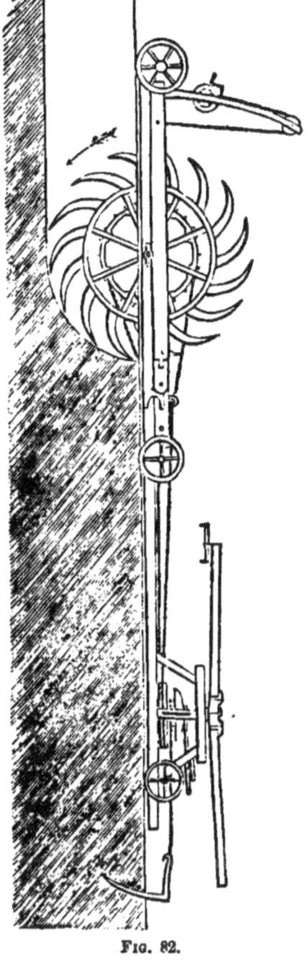

Fig. 82.

Mr. Paul, of Thorpe Abbots, near Scole, Norfolk, England, has lately invented an ingenious machine for cutting drains, of which we give an elevation, Fig. 82. It is drawn, as will be seen, by means of chain and capstan, worked by horses; and at the same time that it moves forward, it acts as a slotting machine on the land, the tools on the circumference of the acting wheel taking successive bites of the soil, each lifting a portion from the full depth to which it is desired that the trench should be cut, and laying the earth, thus removed, on the surface, at either side. There is a lifting apparatus at the end of the machine, by which the cutting wheel may be raised or lowered, according to the unevenness of the surface, in order to insure a perfectly uniform "fall" in the bottom of the drain. The whole process is carried on at the rate of about four feet per minute; and it results, on suitable soils, in cutting a drain from three to five feet deep, leaving it in a finished state, with a level bottom for the tiles to rest upon. It seems to present the right idea of a draining plow, and whether successfully developed in the present instance or not, we think it probable that a machine constructed on these principles, will yet be found cheaply effectual for the purpose which now involves such an enormous cost of manual labor. Mr. J. J. Thomas, one of the editors of the *Country Gentleman*, says:

"The writer has made many experiments with various ditching machines, with a hope of greatly reducing this heavy expense, and has at last attained the desired object in a considerable degree—so that ditches, costing at three feet in depth not less than 30 cents a rod in the hard clayey, tenacious soil operated on, have been cut for about 12 cents a rod; and it is believed that, with the practical knowledge now attained, 3-feet drains may be cut for 10 cents a rod, or at one third the cost when done wholly by hand.

"The process is a very simple one. A subsoil plow of peculiar construction, is so made that the draught-beam and handles may be successively elevated, as the ditch becomes deeper; with this plow and a pair of horses, the hard earth in the bottom of the drain, which is only loosened by the pick, in the common process, is broken up, and all the hand labor required is throwing out this loose earth. This labor is performed with the common long-handled, pointed shovels, and when the ditch has been cut to about one half of its intended depth, a similar shovel, with the sides bent up at a blacksmith's, to fit the narrow channel, is then made use of. A very hard or stony hard-pan requires considerable dressing off with the pick, to prepare the bottom for laying the tile; but where the soil is more favorable, such dressing is scarcely necessary. One two-horse team will commonly plow fast enough to keep from six to twelve men constantly shoveling, varying with the hardness of the soil.

"In an experiment performed the present autumn in cutting drains a mile and a fourth in aggregate length, a small portion was much intercepted with rocks and some quarry stone, with great numbers of smaller stones. Through these portions, the subsoil loosening plow could be used but imperfectly, and it was necessary to occupy eight days' work in quarrying, etc., and ten days more in dressing off these stony and hard bottoms with pick and crowbar.

"The following is the actual cost of 400 rods:—

4 days with two-horse team, - - - - -	$ 8,00
35 " shoveling, 87½ cents, - - - - - -	30,75
10 " dressing bottom, etc., - - - - -	8,75
8 " quarrying rocks, etc., - - - - - -	7,00
5 " laying tile and covering it, - - -	4,37
5500 tile, 95 cents, - - - - - - - - -	52,25
Drawing half mile, - - - - - - -	2,00
Plowing in ditches, . - - - - - -	1,50
	$114,62

or, 28½ cents a rod completed.

"Omitting the four last items, connected with the tile and laying it, the cost of merely cutting the drains is $54.50, or 13½ cents a rod; or, omitting the cost of quarrying the stone, and two thirds of dressing the bottom (this being confined to a very small portion), the expense would be 10 1-5 cents a rod.

"A part of the work was done during a severe drought, when the subsoil was very hard, and the loosening was consequently slower and more laborious. Earlier in the season, when the earth is softer, the loosening plow would do its work in less than half the time here required. This would be especially important where a fractious hard-pan exists. From one to six inches of earth are loosened at each passage of the plow. An "evener," or central whipple-tree, from five to seven feet long, is required, the horses walking on opposite sides of the ditch.

"It is also very obvious that no complex machine can ever succeed as a ditcher, especially among stones, which constantly tend to jar and break it, but that the very simplest form of excavators must be adopted, which are easy to handle, light in striking stones, not liable to breakage, and easily and cheaply repaired."

CHAPTER XII.

TIME TO CUT DRAINS, AND LAY TILE.

No one should ever undertake to make drains in wet weather, or in severe frost. When the land is unoccupied, and the weather dry, draining may be carried on with success. It can be managed to the best advantage when the land is in stubble or pasture, and is afterward to be plowed for a crop. After the removal of a crop in the fall, is a very favorable time, but it must be finished before the fall rains set in, or before hard frosts come. In any case, it should be done in the spring before a crop is put in, or before the land is fallowed in the fall, or rather one of the operations should follow it in a short time, that the superior condition into which the soil may then be brought may be realized at the earliest possible moment.

Tile lying may commence at either end of the drain. When the soil is sandy, and liable to fall in, it is usual to begin at the lower end, and fill in as the work progresses, taking care to have the last tile well stuffed with straw to prevent the mud from entering, while another piece is being dug. Where there is little danger of the sides falling in, it is decidedly better to have the whole drain dug out before a single tile is laid, and to have the tile laying commence at the upper end of the drain. In this way the tiles are kept clear of mud, and there is an opportunity to correct any defect in the digging, or to equalize the fall more perfectly than could otherwise be done. Tiles are usually laid with the instrument heretofore described. (See Tile Layer, Fig. 74, page 391.) They are laid as closely together, and the joints made to fit as perfectly as possible. When the warping of the tiles in burning makes it impossible to lay them absolutely straight, all deviations must be lateral, so as not to interfere with the true level of the bottom of the drain. Straw is sometimes put thinly upon the tiles before the earth is thrown in; and it is an excellent practice. Sometimes brush is laid upon the straw, so as to fill up the drain in part; some benefit is derived from this in deep drains, in very tenacious clay. Others lay in the turf next to the tiles, the grassy side downward; this is some trouble, but it answers an excellent purpose. Small stones are frequently laid upon the tiles; this brings the pressure so unequally upon the surface of the tiles, as to result in their fracture. When drains are dug on stony land, they are necessarily made wider on the bottom to allow of the use of the pick and shovel; it is then important to pack small stones by the side of the tiles, so as too keep them firmly in place.

If the drains are wider at the bottom than is required for the tiles, care must be taken in returning the earth not to disarrange them, or admit loose earth into them. Some pack earth or clay between the tiles and the sides of the ditch, or place a part of the turf or top spading upon them. In filling the drains great care should be taken not to leave a body of earth, and especially clay on the surface. In grass land the turf may be laid in its original position, so that no portion of the land will be made unproductive for a single season.

In some instances, it may be difficult to known how to dispose of the clay, when it can not be used in filling the drains. Mr. Donald[1] says: "If the surface soil is not too stiff, the clay may be spread over it, and after being fully broken down by the influence of the atmosphere,

[1] James McDonald (England) on Land Draining.

and separated by harrowing, it may be mixed with the old earth."

The filling of drains is often done too carelessly, as though this were of no consequence. The earth thrown directly upon the tiles, or upon their covering, ought to be put on with care, so as not to displace the tiles. It is also well to tread the earth a little as it is thrown in, otherwise it will be so loose as to admit the descent of the water too rapidly, carrying with it much sand into the drain. After the drain is full, the remainder should be carefully laid in a ridge on top, to sink down as that below settles. There is no great difficulty in rigging a plow so as to fill in most of the earth removed from drains.

Minor Drains.—These should *always* be cut in the direction or up and down the line of greatest descent, and should, when practicable, be cut parallel, or at most, having a slight angle only to the sub-main drain, E, J, Fig. 53, page 377. When the minor drains are led into a collecting drain, as G, G, same Fig. and page just referred to, the drain, G, should not be dug at right angles with the outlet of the minors, nor should it be dug at right angles with the sub-main, but should make a slight angle with both, as represented in the cut, so as to cause the least possible impediment in the flow of water from one into the other.

The point of intersection between the minor and collecting drain should be made at a considerably greater angle than the general direction of the drains respectively, as represented at Fig. 83. The collecting drains should be several inches lower than the minor drains, and the last joint of the minor should be lowered at the connecting end, so as to be on a level with the collecting

Fig. 83.

drain. With pipe tile the connections are best made at the joints, by breaking off a portion of the side of each piece of tile which is to receive the incoming drain. Where horseshoe tile are used the connection is made at the center of a piece of the receiving tile, as shown in Fig. 84, where *a* is the tile of receiving drain, and *b* the tile of the incoming drain. When tile are laid by commencing at the upper end of the drain, the upper end of the first tile laid should rest firmly with its entire end against a brickbat, or other close fitting surface, so as to prevent the ingress of sand or mud at a point where it is not likely to have a sufficient current of water to carry it off. It is always best to surround the points of junction between the drains by small stones, and these covered with straw, or turf, so as to prevent the introduction of sand, silt or mud.

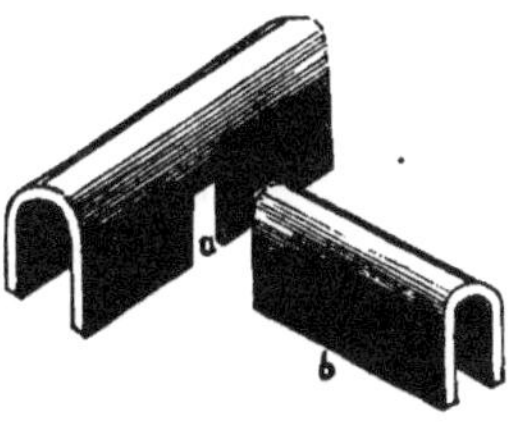

FIG. 84.

CHAPTER XIII.

OBSTRUCTIONS IN DRAINS.

The Central Society of Agriculture, at Paris, having investigated the causes of the obstructions in pipe drains, Mr. Barral said that there are three causes, viz: deposits of carbonate of lime, sediments of hydrate of peroxyd of iron, and intrusion of roots.

It was remarked that, in general, those obstructions had a primordial cause in the defective laying of the drains: some of them lacked sufficient declivity; others had pipes with imperfect joints; but the most frequent occurrence was the intervention of roots. However, a fall of 1-500 is sufficient to carry away the roots, and these are found in balls at the exit of the main drain.

Obstructions are easily detected by an extraordinary moisture which is manifested in the soil at the place where the pipe is obstructed.

Mr. Hervé Mangon, Drainage Engineer of the French government, says:

"I have found obstructions caused by sediments of carbonate of lime and oxyd of iron. I will present the result of my studies, on these two classes of deposits, and indicate the means by which I prevent them.

"*Calcareous Obstructions.*—Spring water sometimes contains carbonate of lime in sufficient quantity to produce incrustations; that is, it deposits calcareous salt; the same phenomenon takes place within drain pipes, the section of which rapidly decreases, until it does not allow any passage for water, and soon the profitable, wholesome effect of drainage, which is established at great expense, is entirely lost.

"Water, thus impregnated with carbonate of lime, does not dissolve it, unless it is acted upon by carbonic acid gas, which it also contains; water remains limpid as long as that gas is not disengaged. The calcareous deposit is produced only when the quantity of this gas is no longer in proportion with the calcareous salt present in water.

"In order to prevent the formation of the calcareous obstructions in drain pipes, all that is required is, to prevent the separation of the carbonic acid gas from the water which flows through the pipes. This may be easily accomplished by protecting the water in the pipes from communication with external air.

"The atmosphere which is confined within the subterranean ducts, soon becomes impregnated with a full proportion of carbonic acid gas, as compared with the volume that is dissolved in water; this latter gas does not then any more tend to disengage itself; water charged with its calcareous salt, preserves its limpidity; and it may flow forever without impediment.

"This theory is very readily put into practice. A *pneumatic* pipe, set upright a few yards above the exit, and others, if necessary, at the point of junction of the most important main pipe drains, will be sufficient. These pneumatic pipes are made of two or three large pipes, well joined together, laid over a flat stone, and covered with another. Some mason work ought to be laid around and beneath the upright, and the horizontal pipes connecting with it; those *flowing in*, must be placed a shade lower than those *flowing out;* water will thus intercept the air, and the object is attained; that is, carbonic acid gas will be retained.

"*Ferruginous Obstructions*, are formed by sediments more or less impregnated with oxyd of iron, and may be of a red, dark brown, or pale yellow color. When precipitation takes place in quiet water, there appear on its surface rainbow-like cuticles, which are sunk at the bottom by the slightest motion of the liquid. That sediment soon obstructs the pipes and completely stops the drain..

"Waters containing such deposits are met with, especially, in soils strongly impregnated with either oxyd or sulphuret of iron, in marshes, turf, and lands which are exposed to filtrations from woods situated on a higher level. The acids named crenic, and apocrenic also perform an important part in the formation of the above deposits. The elements of the soil have, of course, a great influence in the case; most of the deposits contain large quantities of clay, sand, or detritus of vegetables; so that all the analyses presented widely different results."

Without following the author in his minute chemical demonstrations, we proceed with a practical and very interesting experiment. He says:

"Having collected a fresh deposit, with the very water in which it was formed, I put it on a filter and obtained a perfectly clear liquid; which, being placed into flagons entirely full, and well corked, or within an atmosphere deprived of oxygen, remains transparent. Having exposed one flagon to the action of pure oxygen and another to the open air, both became dim after a short time, and allowed the ocher-like substance, which is the basis of the aforesaid obstructions, to settle or precipitate.

"This substance, which is the same that settles in the drains, was easily separated from the liquid; being exposed to the air, it became more and more reddish, until, after a few hours, no further change took place; being then inclosed within an air-tight flagon, it soon resumed a dark brown and almost black color. After a few weeks, the same sediment being placed again on the filter, the result was the same, that is, a clear liquid that became dim by contact with air, and deposited the identical yellow substance. On the other hand, the matter left on the filter resumed the reddish tint which it possessed when placed in the flagon."

The same operation may be repeated any number of times on the same sample, with the same result.

It is then evident that this body presents the double quality of becoming *insoluble* by its *oxydation*, and of *reducing* itself, when left alone, so as to become partly *soluble*.

The above may be summed up in the following two propositions: First, the water which causes ferruginous obstructions within drain pipes, preserves its limpidity, and gives no sediment when not in contact with the oxygen of the air; second, the deposit recently formed may exercise a reducing action upon itself, which causes it to resume in a great part its soluble property.

From these two facts, it is easy to conclude that pneumatic upright pipes, as described above, will prevent the formation of ferruginous obstructions by *excluding* the oxygen of the air, as well as calcareous sediments, by *including* carbonic acid gas.

Brandt had observed that the water impregnated with ferruginous matter collected from the bottom of a meadow, kept in open bottles, began to thicken at the end of three days, and to deposit flakes after five days. The occurrence in some experimental holes made in the meadow, produced similar result. As the drain water, even under the most unfavorable circumstances, does not admit of so long a stay in the pipes, Brandt, for the sake of further observation, made an experiment in which he employed three tubes of 120 feet in length, 3 feet in depth, and 6 inches fall (on the whole length), to be laid in the meadow in question.

The tube A was provided with a wooden discharge pipe two feet long, which was perforated in an oblique line, and placed so that the discharged water was compelled to fill the opening of the tube.

The tube B had a free discharge pipe.

The tube C had likewise a wooden discharge pipe, which for a length of 5 feet was stamped around with clay, in order to produce a damming of the water in the tube.

The works were undertaken in December, 1852; eight days, however, after the three tubes had been laid, all the drain water was turbid, the openings assumed an orange color, and a short time after, when it rained, the tubes discharged—owing to the more violent intrusion of bottom water—a large amount of oxyd of iron. After a minute investigation, the cause of this occurrence was found in the fact that the single tubes were not placed in the ground in a mathematical straight line; but that they, deviating from the latter more or less, had here and there some points of stoppage in which the water remained stationary, and the formation of oxyd of iron took place slowly but uninterruptedly. Stronger water currents in the tubes overcame these stopping points, and carried away the sedimentary matter.

The tubes A and B were obstructed in May, 1853; the third, C, was constantly kept clear by the frequent damming of its own water, effectuated by closing the discharge pipe with a tenon. In order to see how high the water was dammed in the tubes, the tenon was perforated, and a small glass tube placed in the perforation. Two or three days were generally sufficient to press the water to the margin of the small pipe. After the removal of the tenon the water, filling the entire space of the pipe, flowed off with the deposed substances of iron, and it did so, finally, in general very pure; which result justified the opinion that in this manner an arrangement had been found for protecting against obstructions from oxyd of iron. The draining of the meadows undertaken in the fall of 1853 and spring of 1854, was then executed by tubes or pipes 20 perches long, laid at the upper end 2½ feet, on the lower 3 feet deep. The tubes were laid with great care, and clay slightly stamped around; the discharge pipes were of wood, and led into a ditch, which latter could, by means of a dam, in two days be dammed up 1 foot above the highest point of the drain pipes. By alternate damming and discharging, repeated every fortnight, the drain tubes had up to the middle of 1855, remained free from any obstruction.

Tischendorff tries to remove the obstructions occurring in the drain pipes by pressing water into the tubes at the upper end of the obstructed pipes by means of a simple pump-work. (*Zeitschr. f. d. Landw.*, 1855, 64.)

There are no definite reports on the success of the funnel pipes recommended for the prevention of intrusion of *quicksand.* (*cf. Jahresb.*, 1854, *I*, 69.)

Dr. Motherby-Areusberg (East Prussia), reports that, in draining in quicksand he had left none of the means recommended untried, but found none always reliable, and that he now gives preference to the following plain method, the principle of which consists in as speedy a performance of the successive operations as possible, in order to prevent the movement of the quicksand. The contemplated ditch is first thrown out deep enough to allow only one more cut to the stratum of quicksand; into the walls of the yet shallow ditch leveling pegs are driven sideways, and to them is fastened a cord, by which the depth can at any instant be correctly ascertained—this being the most important item in the rapid succession of operation. The workmen now begin one after the other, and so close to each other that the necessary free movement only is allowed to each. The second workman commences only after the first one has made his first cuts; the rest proceed in the same way, so that they stand in their work entirely by steps, and the last must constantly be prepared with his hook, ready to receive the tiles and place them accurately and quickly, so that they may be immediately covered by a workman stepping over the ditch, with one foot of earth. In order to be perfectly sure as to the work being everywhere done right, stoppages are made from time to time, which, if arrested, furnish the best proof whether the work has been perfectly made, or where the mistake is which as yet can easily be remedied. In order to make these stoppages, the drain ditch is closed from distance to distance by a small loam dam; the pipe itself projecting from this dam is closed by a cork; the water is then permitted to gather in order to observe whether, after removing the cork, a complete discharge of water takes place.

As to the intrusion of roots, Mr. B. de Latour states that a pipe drain, four feet below the surface, being choked up, he ordered it to be repaired; that a great number of thread-like beet roots, ten to twelve feet long, had penetrated and filled the largest pipes; that in another field carrots had caused the same accident; that potatoes had not done it, and he feared nothing from the roots of fruit trees and vineyards.

Mr. L. Giraud and Mr. Th. Galos, from the neighborhood of Bordeaux, state that pipe drains, in the vineyards of that district, are protected against the intrusion of roots, by surrounding the pipes with straw, after having covered the joints with short pipes or collars.

CONCLUSION.

We have now discussed all the prominent principles involved in underdraining, and have given such practical directions for determining the construction of the drains, that, with a little experience, no one guided by them will be liable to commit serious errors.

It may be objected that we have not advocated any special system of underdraining—that we have not adopted Elkington's, Smith's of Deanston, Josiah Parkes', Pusey's, Wharncliffe's, Keythorpe's, Barrall's, Wauer's, Shoenermark's, Gropp's, Mollenkopf's, or any other special system; or that we have not introduced whole page engravings, exhibiting entire fields of underdrains, or introduced engravings representing Johnston's, Yeoman's, or some other farms as models. We have deemed it best to discuss simply the principles involved, and then let the reader apply the principles in practice as best suits his location and circumstances. We doubt very much whether twenty farms are drained precisely alike in any other respect than upon the general principles—the details necessarily differ in each according to soil, situation, finances, etc. We were induced to adopt this method when we learned the fact that, so far as crops are concerned, underdrains with the mole plows, where the nature of the soil would permit, produced the same *effects* that the system of frequent or thorough drains advocated by Gisborne and Parkes did. The advantage of tile drains over the mole plow consists in this, viz: tile drains can be made in *all soils;* are made with greater regard to precision; are *permanent;* while the mole plow drains can be made in clay soil only; are, from their manner of construction, unavoidably subject to irregularities; and what is more than all, are merely temporary expedients. But the physical conditions of the soil are rendered the same; and the increased productiveness is the same, whether made by the mole plow or laid with tile.

With systems differing so greatly in their details as pipe tile and the mole plow, and yet producing the same results, and involving the same *general* principles, it appeared to us like unmitigated prejudice to be partial to the details of one system and exclude all others, especially when we are fully aware that innovations, changes, and differences of detail are introduced by almost every one who undertakes to drain any considerable amount.

We would address ourselves particularly to the young men of the West, and suggest to them that it would not only be *well*, but honorable and profitable, for them to qualify themselves to take charge of drainage works on farms; that is, to examine the grounds, determine the proper depth and position of drains, and advise as to the best method of making them. Judging from the tenor of many letters addressed to the writer in his official capacity, making inquiries respecting "drainage engineers," he is convinced that in a few years those who qualify themselves for the position will have much better cause for congratulation than those who enter the ranks of professional life. Drainage will soon become a new field of industry, which will demand more engineers than the railways have done—more "surveyors" than the western wildernesses. It is a field in which thousands and tens of thousands will find employment, and will go on increasing until the greater portion of the whole North American continent will be underdrained.

Let young men of the present and "rising generation" turn aside from the overcrowded ranks of professional life—from the fascinations of the mercantile avocation, or the dazzling speculations of commercial enterprises—and become promoters of the productiveness of the soil.

APPENDIX.

LAWS OF OHIO RELATING TO DRAINAGE.

An Act to provide for locating, establishing and constructing ditches, drains and watercourses.

[*Passed and took effect March* 24, 1859. 56 *vol. Stat.* 58.]

SECTION I. *Be it enacted by the General Assembly of the State of Ohio,* That the county commissioners of any county shall have power, at any regular session, whenever, in their opinion, the same is demanded by, or will be conducive to the public health, convenience or welfare, to cause to be established, located and constructed, as hereinafter provided, any ditch, drain or watercourse, within such county.

SEC. II. That before the county commissioners of any county shall take any steps toward locating or establishing any ditch, drain or watercourse, there shall be filed with the county auditor a petition from one or more persons owning lands adjacent to the line of such proposed ditch, drain or watercourse, setting forth the necessity of the same, with a description of its proposed starting point, route and terminus, and shall, at the same time, file a bond with good and sufficient sureties, to the acceptance of the county auditor, conditioned to pay all expenses incurred, in case the commissioners shall refuse to grant the prayer of the petition, and it shall be the duty of the county auditor immediately thereafter, to place a correct copy of said petition in the hands of the county surveyor or a competent engineer, who shall thereupon, taking with him the necessary assistance, proceed to make an accurate survey of the route of such proposed ditch, drain or watercourse, and on the completion thereof, shall return a plat, or plat and profile of the same to said county auditor, and shall also set forth in his return a description of the proposed route, its availability and necessity, with a description of each separate tract of land through which the same is proposed to be located, how it will be affected thereby, and its situation and level as compared with that of adjoining lands, together with such other facts as he may deem material. It shall be the duty of the county auditor, immediately on said report being filed, to cause notice in writing to be given to the owner or one of the owners of each tract of land along the route of such proposed ditch, drain or watercourse, of the pendency and prayer of said petition, and of the time of the session of the county commissioners at which the same will be heard, which notice shall be served at least ten days prior to said session, and an affidavit of said service filed with the county auditor; and in case any such owner is not a resident of the county, or should any party or parties in interest, die during the pendency of said proceeding, such death shall not work an abatement of such proceeding, but the commissioners, on being notified thereof, shall make such order as they may deem proper, for giving notice to the person or persons succeeding to the right of such deceased party or parties, and notice of the pendency and prayer of said petition, and the time of hearing the same shall be given to such owner or persons, by publication for two consecutive weeks in some newspaper published or of general circulation in said county.

Sec. III. That any person or persons claiming compensation for lands appropriated for the purpose of constructing any ditch, drain or watercourse under the provisions of this act, shall make his, her or their application in writing therefor to the county commissioners, on or before the third day of the session, at which the petition has been set for hearing, and on failure to make such application, shall be deemed and held to have waived his, her or their right to such compensation.

Sec. IV. That said county commissioners, at the session set for the hearing of said petition, shall, if they find the requirements of the second section of this act to have been complied with, proceed to hear and determine said petition; and if they deem it necessary, shall view the premises, and if they find such ditch, drain or watercourse to be necessary, and that the same is demanded by or will be conducive to the public health, convenience or welfare, and no application shall have been made for compensation as provided in the third section of this act, they shall proceed to locate and establish such ditch, drain or watercouse on the route specified in the plat and return of said county surveyor or engineer. But if any application or applications for compensation as aforesaid, shall have been made, further proceedings by the county commissioners shall be adjourned till their next regular session; and the county auditor shall forthwith certify to the probate judge of said county a copy or copies of said application or applications, together with a description or descriptions of the property sought to be taken and appropriated, as contained in the plat or report of the county surveyor or engineers;

which shall be forthwith docketed by said probate judge, styling the applicant or applicants plaintiff or plaintiffs, and the county commissioners defendants; and such proceeding shall thereupon be had to assess and determine the compensation of such claimant or claimants, as are authorized and required by the act entitled "an act to provide for compensation to the owners of private property appropriated to the use of corporations," passed April 30, 1852, and the acts amendatory thereof and supplementary thereto, so far as the same may be applicable; and the compensation so found and assessed in favor of said claimant or claimants shall be certified by the probate judge to the county auditor and paid out of the county treasury, from the general fund, or remain deposited therein for the use of such claimant or claimants; and said county commissioners shall, at the next regular session after such compensation shall have been assessed and paid or deposited as aforesaid, proceed to locate and establish such ditch, drain or watercourse as herein before provided.

Sec. V. That said county commissioners, whenever they shall have established any such ditch, drain or watercourse, shall divide the same into suitable sections, not less in number than the numbers of owners of land through which the same may be located, and shall also prescribe the time within which the work upon such sections shall be completed.

Sec. VI. That the county auditor shall cause notice to be given of the time and place of letting, and of the kind and amount of work to be done upon said sections, and the time fixed by the commissioners for its completion, by publication for thirty days, in some newspaper printed, or of general circulation in said county, and shall let the work upon said sections respectively to the lowest bidder therefor; and the person or persons taking such work at such letting, shall, on the completion thereof to the satisfaction of the county commissioners, be paid for such work out of he county treasury upon the order of the county auditor; provided, that if any person or persons to whom any portion of said work shall be let as aforesaid, shall fail to perform said work, the same shall be re-let by the county auditor, in the manner hereinbefore provided.

Sec. VII. That the county auditor shall keep a full and complete record of all proceedings had in each case under this act.

Sec. VIII. That the auditor and surveyor or engineers shall be allowed such fees for services under this act, as the county commissioners shall, in each case, deem reasonable and allow; and all other fees and costs accruing under this act shall be the same as provided by law for like services in other cases, and all costs, expenses, costs of construction, fees and compensation for property appropriated, which shall accrue and be assessed and be determined under this act shall be paid out of the county treasury, out of the general fund, on the order of the county auditor, provided that no part of the same, except the compensation for property appropriated, shall be paid out of the county treasury till the sum shall have been levied and collected as provided in the next section of this act.

Sec. IX. That the county commissioners shall make an equitable apportionment of the costs, expenses, cost of construction, fees and compensation for property appropriated, which shall accrue and be assessed and determined under this act, among the owners of the land benefited by the location and construction of such ditch, drain or watercourse, in proportion to the benefit to each of them through, along the line, or in the vicinity of whose lands the same may be located and constructed respectively; and the same may be levied upon the lands of the owners so benefited, in said proportions, and collected in the same manner that other taxes are levied and collected for county purposes.

Sec. X. The act entitled "an act authorizing the trustees of townships to establish watercourses and locate ditches in certain cases," passed May 1, 1854, and the act amendatory thereto, passed April 14, 1857, and the original act, passed February 24, 1853, on the same subject, are hereby repealed: Provided, that no proceedings had or commenced under any law repealed by this act shall be affected by such repeal.

Sec. XI. This act to take effect from and after its passage.

An Act to authorize the making roads and drains in certain cases.
[*Passed February* 8, 1847. 45 *vol. Stat.* 50.]

Section I. *Be it enacted by the General Assembly of the State of Ohio*, That any person, persons, or company, having the ownership or possession of low lands, lakes, swamps, quarries, mines, or mineral beds that, by means of adjacent lands belonging to other persons or public highway, can not be approached, worked, drained, or used in the ordinary manner, without crossing said lands and highways, may be authorized to establish roads, drains, ditches, railways, or tunnels to said places, in the manner herein provided.

Sec. II. The party desiring to make such improvements shall file a petition therefor with the commissioners of the county where the premises are situated, setting forth, in detail, the proposed work, and the situation of the adjoining lands, accompanied by a bond, to the satisfaction of the county auditor, and made payable to him, conditioned to pay the expenses of the committee of view or review, as hereinafter provided.

SEC. III. The commissioners of the county, on the filing of said petition and bond, and at their first meeting thereafter, shall appoint a committee of view, and fix their compensation per day, to be composed of not less than three, nor more than five judicious, disinterested persons, to meet on the premises on a day named, within one month from the date of their appointment, and by examination and inspection, determine whether the proposed improvement is necessary to the ordinary working, occupation and beneficial use of said grounds, swamps, ponds, low lands, mines, or mineral beds; and if so, said committee shall proceed to lay out and establish the same, of a width not exceeding sixty feet, and in such a manner as to do as little injury as practicable, and shall, furthermore, fix and assess the amount of damages which any proprietor of adjacent lands will be likely to sustain, and report and return the same, with all their proceedings, to the county auditor, within ten days from the time when said appointment shall be completed; but before said committee shall proceed to said examinations, they shall be satisfied that three weeks' notice, setting forth the time and place thereof, has been published in some newspaper in general circulation in the proper county, prior to the day fixed upon by the commissioners.

SEC. IV. At the next meeting of the county commissioners, after the return of the committee is received, said commissioners shall proceed to consider the subject, and if they shall be of opinion, taking into view the public as well as private interests, that said improvements would be advantageous and desirable, they shall fix the same in the manner described in the petition and report, and cause a copy of said description and record to be made out for the benefit of the party praying therefor, unless either party shall, ten days before said meeting of the commissioners, file a petition for a committee of review and reassessment.

SEC. V. In case a petition for review is filed, as aforesaid, the party filing the same shall file a bond, as aforesaid, for the payment of the expenses of said committee, and the same shall be appointed and act, in all respects, in the manner pointed out for the committee of view, and on return and report of their proceedings of review, the commissioners shall take the same action as in the case of the committee of view.

SEC. VI. The party praying for said improvement, shall cause the final report of the commissioners to be recorded in the record of deeds, and shall pay or tender to each of the parties reported to be injured as aforesaid, the full amount of money assessed by said committee of view or review, before entering upon the premises in order to complete said works; and if the same shall be received, it shall be in full of said damages, but if it shall not be received, it shall be deposited with the county treasurer, for the use of the party injured.

SEC. VII. The party refusing said award and tender, shall not be debarred his action at law for damages, in the proper courts, but unless a larger amount is recovered than the tender aforesaid, or otherwise, the plaintiff shall pay his own costs.

SEC. VIII. Works constructed under the provisions of this act, shall be entitled to the benefit of all laws for the protection of railways and canals in this state.

An Act to amend the act entitled "an act to authorize the making of roads and drains in certain cases," passed February 8, 1847.

[*Passed March* 8, 1850. 48 *vol. Stat.* 48.]

SECTION I. *Be it enacted by the General Assembly of the State of Ohio*, That every petition filed with the county commissioners, under the law to which this is an amendment, shall set forth the names of all persons interested (if known to the petitioner), as well those whom it is supposed will be benefited as those who will be injured by the proposed improvement, and the notice required by the third section of said act shall also set forth the names of all the persons interested, as fully as the same are stated in said petition.

SEC. II. Whenever any committee appointed by the commissioners, either of view or review, shall determine that the proposed improvement is necessary, and shall lay out and establish the same, and shall find that damages will be sustained by any proprietor or occupant of any adjacent lands, and the amount which they will respectively sustain, said committee, either of view or review, shall then determine the proportion of said damages which shall be paid by each of the proprietors of the adjacent lands, having strict regard to the benefits which they will receive, and the award so made shall be held as conclusive upon each of the parties charged with such payment.

SEC. III. When any petitioners shall have paid over or deposited the full amount of all the damages so assessed, and after the improvement is finished in conformity with the details of the work as set forth in the petition, and in the manner contemplated by the viewer or reviewers, such petitioner may bring suit in any court of competent jurisdiction, and recover from each party the amount with which he stands charged by said award: Provided he has, before the commencement of such suit, made demand of such sum upon the party so charged by said award.

SEC. IV. Whenever it may be necessary to repair such work, any one of the persons benefited by it may cause such repairs to be made, and may compel contributions from each person benefited, on the basis of the award, the just and fair price of such repairs.

INDEX.

38

ROBERT CLARKE. R. D. BARNEY. J. W. DALE.

ROBERT CLARKE & CO.,

Publishers, Booksellers and Importers,

55 WEST FOURTH-STREET,

CINCINNATI, OHIO.

Have constantly on hand a large and varied stock of Standard American and British Works, in every department of Literature and Science. The following is a selection from those pertaining more particularly to Agriculture, Horticulture, Domestic Architecture, and Domestic Economy.

All orders will be promptly filled, and letters of inquiry will meet with careful attention.

Abell (Mrs. L. G.) Skillful Housewife........................$ 50
Agriculturists' Calculator, Land Measuring, Drainage, etc., *Glasgow*..2 25
Allen (J. Fisk) On the Culture of the Grape....................1 00
Allen (L. F.) Rural Architecture................................1 25
Allen (R. L.) American Farm Book...............................1 00
Allen (R. L.) Diseases of Domestic Animals.................... 75
American Farmers' Encyclopædia................................4 00
American Florists' Guide.. 75
Archer (T. C.) Economic Botany, colored plates, *London*.........1 75
Armstrong (John) A Treatise on Agriculture.................... 50
Balfour (J. H.) Botanists' Companion, *London*.................. 75
Balfour (J. H.) Manual of Botany, *London*.......................3 25
Barry (P.) Fruit Garden..1 25
Baucher (T.) Breaking and Training of Horses.............. ...1 25
Beck (Lewis C.) Botany of the United States......................1 25
Beecher (H. W.) Fruits, Flowers and Farming 1 25
Bement (C. N.) The American Poulterer's Companion.............1 25
Bement (C. N.) Rabbit Fancier................................... 50
Blake (John L.) Farmer at Home..................................1 25
Blake (J. S.) Every Day Book, or Life in the Country............2 25
Boussingault (J. B.) Rural Economy.............................1 25
Breck (Joseph) Book of Flowers...................................1 00
Bridgeman (Thos.) Young Gardener's Assistant..................1 50
Bridgeman (Thos.) Kitchen Gardener's Instructor................ 60
Bridgeman (Thos.) Florist's Guide............................... 60
Bridgeman (Thos.) Fruit Cultivator's Manual..................... 60
Browne. American Bird Fancier.................................. 50
Browne. American Poultry Yard...................................1 00
Brown (D. J.) Trees of America...................................4 50

BROWN (Thos.) The Taxidermist Manual, *Edinburgh*..............$1 00
BUCHANAN (Robert) On Grape Culture, and LONGWORTH (N.) On the Strawberry.. 63
BUIST (Robert) American Flower Garden Directory...... ....1 25
BUIST (Robert) Family Kitchen Gardener......................... 75
BUIST (Robert) The Rose Manual.................................. 75
BUEL (Judge) The Farmer's Instructor, 2 vols.,..................1 00
BUEL (Judge) The Farmer's Companion............................ 75
CAMPBELL (J. L.) Practical Agriculture...........................1 00
CARSON (J. C. L.) The Form of the Horse, *Dublin*.................. 75
CATLOW (A.) Greenhouse Botany, colored plates, *London*..........1 75
CATLOW (A.) Field Botany, colored plates, *London*.................1 75
CATLOW (A.) Garden Botany, colored plates, *London*...............1 75
CECIL and YOUATT on the Horse, *London*.......................... 88
CHAPTAL's Agricultural Chemistry................................. 50
CHORLTON (Wm.) Grape Growers' Guide........................... 60
CLATTER's Farriery, edited by Spooner, *London*......................1 25
CLEAVELAND and BACKUS. Village and Farm Cottages.............2 00
COBBETT's American Gardener.. 50
COCK (M. J.) The American Poultry Book........... 35
COLE (S. W.) American Fruit Book................................ 50
COLE (S. W.) American Veterinarian.............................. 50
COLEMAN (W. S.) Our Woodlands, Heaths and Hedges, colored plates, *London*...1 00
COLEMAN's Agricultural and Rural Economy.......................4 50
Comprehensive Farm Record..3 00
COPELAND (R. M.) Country Life..................................2 50
Cottage and Farm Bee-keeper....................................... 50
COULTAS (Herman) What We may Learn from a Tree............1 00
CURTIS (John) Farm Insects, *Glasgow*........................ ...8 50
DADD (Geo. H.) Modern Horse Doctor............................1 00
DADD (Geo. H.) American Cattle Doctor...........................1 00
DADD (Geo. H.) Anatomy and Physiology of the Horse, plain plates.2 00
The same work, colored plates..................................4 00
DANA (Samuel H.) Muck Manual.................................1 00
DARLINGTON's American Weeds and Useful Plants..................1 50
DAWBENY (Dr.) Geography of Plants, colored plates, *London*.......1 75
DELAMER (E. S.) The Kitchen and Flower Garden, *London*......... 65
DIXON (E. S.) Domestic and Ornamental Poultry, plain plates.......1 00
The same work, colored plates..................................2 00
DOWNING (A. J.) Cottage Residences..............................2 00
DOWNING (A. J.) The Fruits and Fruit Trees of America...........1 50
DOWNING (A. J.) Landscape Gardening............................3 50
DOWNING (A. J.) Lindley's Horticulture..........................1 25

Downing (A. J.) Loudon's Gardening for Ladies................$1 25
Downing (A. J.) Rural Architecture............................4 00
Downing (A. J.) Theory and Practice of Landscape Gardening.....3 50
Downing (A. J.) Rural Essays—Horticulture, etc................3 00
Doyle (Martin) Book of Domestic Poultry, colored plates, *London*.1 25
Eastwood (Benj.) On the Cultivation of the Cranberry............ 50
Edgeworth (Mrs.) Southern Gardener and Receipt Book..........1 25
Elliott (F. R.) Western Fruit Book...........................1 25
Emerson (Geo. B.) Trees and Shrubs of Massachusetts............2 00
English Forests and Forest Trees, *London*.........................1 75
Farmers' Practical Horse Farrier................................ 60
Fessenden (T. G.) Complete Farmer and Gardener...............1 25
Fessenden (T. G.) American Kitchen Gardener.................. 50
Field (Thos. W.) Pear Culture...................................1 00
Flint (Chas. L.) On Grasses and Forage Plants.................1 25
Flint (Chas. L.) Milch Cows and Dairy Farming.................1 25
French (Henry F.) Farm Drainage...........................1 00
Fry (W. H.) Artificial Fish Breeding.......................... 75
Gardeners' and Farmers' Reason Why, *London*.................... 75
Gardner. Farmer's Dictionary....................................1 50
Garlick (Theo.) Fish Culture.....................................1 00
Gaylord (W.) and Tucker (L.) American Husbandry, 2 vols......1 00
Glenny (Geo.) The Culture of Fruits and Vegetables, *London*......1 50
Glenny (Geo.) The Culture of Flowers and Plants, *London*........1 50
Glenny (Geo.) Manual of Practical Gardening, *London*............1 50
Glenny (Geo.) Gardener's Every Day Book, *London*..............1 50
Glenny (Geo.) The Properties of Flowers and Plants, *London* 30
Gray (Asa) Manual of Botany.....................................1 50
Gray (Asa) Manual of Botany with Mosses, illustrated...........2 50
Gray (Asa) Structural and Systematic Botany....................2 00
Gray (Asa) Lessons in Botany....................................1 00
Gray (Asa) "How Plants Grow"............................ 75
Green (F. H.) Class Book of Botany.............................1 50
Green (F. H.) Primary Botany.................................. 75
Guenon (Francis) On Milch Cows.............................. 60
Hale (Mrs.) New Cook Book.......................................1 00
Hall (Miss E. M.) American Cookery and Domestic Economy.....1 00
Harrison (W. C.) Bees and Bee-keeping.........................1 00
Haskell (Mrs. E. F.) Housekeepers' Encyclopædia................1 25
Herbert (H. W.) The Dog, by Dinks, Mayhew and Hutchinson... 2 00
Herbert (H. W.) Field Sports in the United States..............4 50
Herbert (H. W.) Hints to Horse-keepers..........................1 25
Herbert (H. W.) The Horse and Horsemanship of the United States, 2 vols...10 00

Hind's Farrier and Stud Book, by Skinner........................ 75
Horses and Hounds, by Scrutator, with Rarey on Horse-taming, *Lon.* 1 25
Hovey (C. M.) The Fruits of America, colored plates............12 00
Hooper (E. J.) Western Fruit Book............................1 00
Jaeger (Prof.) Life of North American Insects...................1 25
Jennings (Robert) The Horse and his Diseases...................1 25
Johnson (Louisa) Every Lady her own Flower Gardener.......... 50
Johnston (Jas F. W.) The Chemistry of Common Life, 2 vols.....2 00
Johnston (Jas. F. W.) Elements of Agricultural Chemistry and Geology..1 00
Johnston (Jas. F. W.) Lectures of Agricultural Chemistry........1 25
Kemp (Edward) On Landscape Gardening.........................1 50
Klippart (John H.) Principles and Practice of Land Drainage......1 25
Klippart (John H.) The Wheat Plant............................1 50
Langstroth (L. L.) On the Hive and Honey Bee..................1 25
Leslie (Miss) New Cookery Book..............................1 25
Leuchar (P. B.) How to Build and Ventilate Hot-houses..........1 25
Liebig (J.) Agricultural Chemistry.............................1 00
Liebig (J.) Familiar Lectures on Chemistry...................... 50
Liebig (J.) Letters on Modern Agriculture.................... .. 75
Lindsay (D. C.) Morgan Horses................................1 00
Lindsey (W. L.) British Lichens, colored plates, *London*...........1 75
Loudon (J. C.) Arboretum et Fruticetum Britanicum, 4 vols. text, 4 vols. plates, 8 vols., *London*...............................25 00
Loudon (J. C.) Encyclopædia of Agriculture, *London*.............7 50
Loudon (J. C.) Encyclopædia of Cottage, Farm and Villa Architecture, *London*..10 00
Loudon (J. C.) Encyclopædia of Gardening, *London*..............7 50
Loudon (J. C.) Encyclopædia of Plants, *London*.................13 50
Loudon (J. C.) The Villa Gardener, *London*........... ....3 00
McCulloch. Land Measurer's Ready Reckoner, *Glasgow*.......... 50
Mackintosh (Charles) Book of the Garden, 2 vols., *London*......17 00
McMahon. American Gardener.................................2 00
Magne (J. H.) How to Choose a good Milk Cow, *Glasgow*......... 63
Mahew (E.) On Dogs, their Diseases and Treatment, *London*...... 65
Mahew (E.) Illustrated Horse Doctor...........................2 50
Marsh (Geo. P.) The Camel, its application to the United States, etc. 63
Mason. Farrier and Stud Book, by Skinner......................1 00
Mechi (Alderman J. J.) How to Farm Profitably, *London*......... 75
Milburn (M. M.) On the Cow and Dairy Husbandry............. 50
Miles. On the Horse's Foot, and How to Keep it Sound........... 50
Miner (T. B.) Bee-keeper's Manual.............................1 00
Moore (T.) British Ferns, colored plates, *London*.................1 75
Morfit (C.) On Manures...................................... 38

Morrell (L. A.) The American Shepherd.....................$ 90
Morton (J. C.) Cyclopædia of Agriculture, 2 vols., *Glasgow*.....15 00
Morton (W. J. T.) Veterinary Pharmacy, *London*................3 15
Munn (B.) Practical Land Drainer.............................. 50
Nash (J. A.) Progressive Farmer................................ 60
Neill (Patrick) Fruit, Flower and Kitchen Gardener's Companion.1 00
Norton (John P.) Elements of Scientific Agriculture............. 60
Olcott (Henry S.) Sorgho and Imphee...........................1 00
Our Farm of Four Acres, and the Money we made by it............ 50
Pardee (R. C.) On Strawberry Culture........................... 60
Parsons (S. B.) The Rose, its Culture, etc.......................1 00
Paxton (J. P.) Botanical Dictionary, *London*....................5 00
Pedder (James) Farmer's Land Measurer......................... 50
Quinby (M.) Mysteries of Bee-keeping Explained.................1 00
Randall (Henry S.) Sheep Husbandry...........................1 25
Reemelin (Chas.) Vine Dresser's Manual......................... 50
Repton (H.) Landscape Gardening, edited by Loudon, *London*.....4 00
Rham. Dictionary of the Farm, *London*..........................1 25
Rhind (Wm.) Vegetable Kingdom, colored plates, *Glasgow*........6 00
Richardson (H. D.) On Dogs, their Origin and Varieties.......... 50
Ritch (John W.) American Architect.............................6 00
Ritchie (R.) The Farm Engineer, A Treatise on Barn Machinery, etc., *Glasgow*..3 00
Rivers (Thos.) The Orchard House.............................. 40
Robbins (R.) Produce and Ready Reckoner....................... 60
Roessle (T.) How to Cultivate and Preserve Celery... ....1 00
Rundell (Mrs.) Domestic Cookery, *London*.... 50
Saxton. Rural Handbooks, 4 vols., each.........................1 25
Schenck. Gardener's Text Book................................. 50
Seeman (Dr. B.) Natural History of Palms, colored plates, *London*..1 75
Smith (C. H. J.) Landscape Gardening, Parks and Pleasure Grounds..1 25
Springer (J. S.) Forest Life and Forest Trees................... 75
Spooner (W. C.) Veterinary Art, *London*........................ 75
Stephens (Henry) Book of the Farm, 2 vols.....................4 00
Stewart (John) Stable Book....................................1 00
Stewart (John) Stable Economy, *Edinburgh*.....................1 75
Stonehenge. On the Dog, in Health and Disease..................4 50
Talpa, the Chronicles of a Clay Farm........................... 75
Thaer (Albert D.) The Principles of Agriculture... ....2 00
Thomas (John J.) Farm Implements.............................1 00
Thomas (John J.) American Fruit Culturist......................1 25
Thompson (R. D.) On the Food of Animals....................... 75
Thomson (Robert) Gardener's Assistant, *Glasgow*................8 50
Thomson (S.) Wild Flowers, colored plates, *London*.............1 60

Todd (S. E.) Young Farmer's Manual and Workshop............$1 25
Turner (J. A.) Cotton Planter's Manual...... ....1 00
Vaux (Calvert) Villas and Cottages.............................2 00
Vegetable Substances used for the Food of Man.................. 45
Walden (J. H.) Soil Culture....................................1 00
Walsh (Dr.) English Cookery Book, *London*................ ... 75
Walsh (Dr.) Economical Housekeeper, *London*...................1 00
Walsh (Dr.) Manual of Domestic Economy, *London*...............2 00
Warder (J. A.) Hedges and Evergreens...........................1 00
Waring (Geo. E. jr.) Elements of Agriculture.................. 75
Watson (Alex.) The American Home Garden.......................1 50
Webb (James) The Farmer's Guide, *Glasgow*..................... 75
Weeks (John M.) Manual on Bees................................ 50
Wheeler (G.) Homes for the People..............................1 50
White (John) Rural Architecture...................... ....13 00
White (W. N.) Gardening for the South..........................1 25
Widdifield. New Cook Book......................................1 00
Wilson (John) Our Farm Crops, *London*.........................2 00
Wilson (John M.) The Farmer's Dictionary, 2 vols., *Edinburgh*...12 00
Wood (J. G.) Common Objects of the Country, colored plates, *Lon* 1 00
Yale Agricultural Lectures.................................... 50
Youatt (W.) On the Horse, enlarged by E. N. Gabriel, *London*....3 25
Youatt (W.) On the Pig, edited by Sidney, *London*.............1 25
Youatt (W.) On Sheep.. 75
Youatt (W.) and Martin. On Cattle..............................1 25
Youatt (W.) and Martin. On the Hog............................ 75
Youatt (W.) and Randall. Shepherd's Own Book...................2 00
Youatt (W.) and Spooner. On the Horse..........................1 25

BRITISH PERIODICALS.

List of the most important British Periodicals, relating to Agriculture, Horticulture, etc., and the subscription price per annum at which they are supplied:

Cottage Gardener (weekly)....................................$5 00
Curtis' Botanical Magazine (monthly)....................................13 50
Edinburgh Veterinary Review (quarterly)....................................4 00
Farmers' Herald (monthly)....................................1 50
Farmers' Magazine (monthly)....................................8 00
Floral Magazine (monthly)....................................9 00
Floral World (monthly)....................................1 50
Florist, Fruitest, etc. (Turner's), (monthly)....................................4 00
Gardeners' Chronicle (weekly)....................................9 00
Gardeners' and Farmers' Journal (weekly)....................................10 00
Gardeners' Weekly Magazine....................................2 50
Glenny's Gardeners' Gazette (monthly)....................................1 25
Gossip for the Garden (monthly)....................................2 00
Quarterly Journal of Agriculture....................................4 00
Scottish Gardener (monthly)....................................2 00
Veterinarian (monthly)....................................6 00

www.ingramcontent.com/pod-product-compliance
Lightning Source LLC
Chambersburg PA
CBHW020904210326
41598CB00018B/1766

* 9 7 8 3 7 4 4 6 7 8 5 6 8 *